T0353452

Uncertain Inference

Coping with uncertainty is a necessary part of ordinary life and is crucial to an understanding of how the mind works. For example, it is a vital element in developing artificial intelligence that will not be undermined by its own rigidities. There have been many approaches to the problem of uncertain inference, ranging from probability to inductive logic to nonmonotonic logic. This book seeks to provide a clear exposition of these approaches within a unified framework.

The principal market for the book will be students and professionals in philosophy, computer science, and artificial intelligence. Among the special features of the book are a chapter on evidential probability, an interpretation of probability specifically developed with an eye to inductive and uncertain inference, which has not received a basic exposition before; chapters on nonmonotonic reasoning and theory replacement that concern matters rarely addressed in standard philosophical texts; and chapters on Mill's methods and statistical inference that cover material sorely lacking in the usual treatments of artificial intelligence and computer science.

Henry E. Kyburg, Jr. is the Burbank Professor of Philosophy and Computer Science, at the University of Rochester and a research scientist at the Institute for Human and Machine Cognition at the University of West Florida. He is the author or editor of a number of books on induction, probability, and statistical inference, as well as an editor (with Ron Loui and Greg Carlson) of *Knowledge Representation and Defeasible Reasoning*. He is a Fellow of the American Academy of Arts and Sciences and a Fellow of the American Association for the Advancement of Science.

Choh Man Teng is a research scientist at the Institute for Human and Machine Cognition at the University of West Florida. She is interested in uncertain reasoning and in machine learning.

Uncertain Inference

HENRY E. KYBURG, Jr.

University of Rochester and
Institute for Human and Machine Cognition,
University of West Florida

and

CHOH MAN TENG

Institute for Human and Machine Cognition,
University of West Florida

CAMBRIDGE
UNIVERSITY PRESS

CAMBRIDGE
UNIVERSITY PRESS

32 Avenue of the Americas, New York NY 10013-2473, USA

Cambridge University Press is part of the University of Cambridge.

It furthers the University's mission by disseminating knowledge in the pursuit of education, learning and research at the highest international levels of excellence.

www.cambridge.org
Information on this title: www.cambridge.org/9780521800648

First published 2001

A catalogue record for this publication is available from the British Library

Library of Congress Cataloguing in Publication data
Kyburg, Jr., Henry Ely, 1928–
Uncertain inference / Henry E. Kyburg, Jr. & Choh Man Teng.
p.   cm.
Includes bibliographical references and index.
ISBN 0-521-80064-1
1. Uncertainty (Information theory).   2. Probabilities.   3. Logic, Symbolic and mathematical.   I. Teng, Choh Man.   II. Title.
Q375 .K93   2001
003'.54 – dc21   00-052954

ISBN 978-0-521-80064-8 Hardback
ISBN 978-0-521-00101-4 Paperback

Contents

Preface

This book is the outgrowth of an effort to provide a course covering the general topic of uncertain inference. Philosophy students have long lacked a treatment of inductive logic that was acceptable; in fact, many professional philosophers would deny that there was any such thing and would replace it with a study of probability. Yet, there seems to many to be something more traditional than the shifting sands of subjective probabilities that is worth studying. Students of computer science may encounter a wide variety of ways of treating uncertainty and uncertain inference, ranging from nonmonotonic logic to probability to belief functions to fuzzy logic. All of these approaches are discussed in their own terms, but it is rare for their relations and interconnections to be explored. Cognitive science students learn early that the processes by which people make inferences are not quite like the formal logic processes that they study in philosophy, but they often have little exposure to the variety of ideas developed in philosophy and computer science. Much of the uncertain inference of science is statistical inference, but statistics rarely enter directly into the treatment of uncertainty to which any of these three groups of students are exposed.

At what level should such a course be taught? Because a broad and interdisciplinary understanding of uncertainty seemed to be just as lacking among graduate students as among undergraduates, and because without assuming some formal background all that could be accomplished would be rather superficial, the course was developed for upper-level undergraduates and beginning graduate students in these three disciplines.

The original goal was to develop a course that would serve all of these groups. It could make significant demands on ability and perseverance, and yet it would have to demand relatively little in the way of background—in part precisely because relatively little could be assumed about the common elements of the backgrounds of these diverse students. In this event, the only formal prerequisite for the course was a course in first order logic. At the University of Rochester, this is the kind of course designed to produce a certain amount of "mathematical maturity."

The course has been taught for two years from earlier versions of this book. It is a difficult course, and students work hard at it. There are weekly exercises and a final research paper. All the material is covered, but some students find the pace very demanding. It might be that a more leisurely approach that concentrated, say, on probability and nonmonotonic acceptance, would be better for some groups of students.

The most common suggestion is that probability theory, as well as logic, should be required as a prerequisite. At Rochester, this would make it difficult for many students to fit the course into their schedules. Furthermore, although the axiomatic foundations of probability and a few elementary theorems seem crucial to understanding uncertainty, the need for an extensive background in probability is questionable. On the other hand, because the whole topic, as we view it, is a kind of logic, a strong background in logic seems crucial.

Support for the research that has allowed us to produce this volume has come from various sources. The University of Rochester, the University of New South Wales, and the Institute for Human and Machine Cognition have each contributed time and facilities. Direct financial support has come from the National Science Foundation through award ISI-941267.

We also wish to acknowledge a debt of a different kind. Mark Wheeler was a graduate student in Philosophy at the University of Rochester many years ago when this volume was first contemplated. At that time he drafted the first chapter. Although the book has been through many revisions and even major overhauls, and has even changed authors more than once, Mark's chapter has remained an eminently sensible introduction. He has allowed us to include it in this, our final version. We are very grateful, and herewith express our thanks.

Henry E. Kyburg, Jr.
University of Rochester and the Institute for
Human and Machine Cognition,
University of West Florida

Choh Man Teng
Institute for Human and Machine Cognition,
University of West Florida

1

Historical Background

MARK WHEELER
San Diego State University

The theory of Induction is the despair of philosophy—and yet all our activities are based upon it.

Alfred North Whitehead: *Science and the Modern World*, p. 35.

1.1 Introduction

Ever since Adam and Eve ate from the tree of knowledge, and thereby earned exile from Paradise, human beings have had to rely on their knowledge of the world to survive and prosper. And whether or not ignorance was bliss in Paradise, it is rarely the case that ignorance promotes happiness in the more familiar world of our experience—a world of grumbling bellies, persistent tax collectors, and successful funeral homes. It is no cause for wonder, then, that we prize knowledge so highly, especially knowledge about the world. Nor should it be cause for surprise that philosophers have despaired and do despair over the theory of induction: For it is through inductive inferences, inferences that are uncertain, that we come to possess knowledge about the world we experience, and the lamentable fact is that we are far from consensus concerning the nature of induction.

But despair is hardly a fruitful state of mind, and, fortunately, the efforts over the past five hundred years or so of distinguished people working on the problems of induction have come to far more than nought (albeit far less than the success for which they strove). In this century, the debate concerning induction has clarified the central issues and resulted in the refinement of various approaches to treating the issues. To echo Brian Skyrms, a writer on the subject [Skyrms, 1966], contemporary inductive logicians are by no means wallowing in a sea of total ignorance and continued work promises to move us further forward.

1.2 Inference

In common parlance, an inference occurs when we make a judgment based on some evidence. We make inferences all the time: If we know that Adam had ten cents and later learn that he found another thirty cents, then we infer that he has a total of forty cents; if we know that all farmers depend on the weather for their livelihood, and we know that Salvatore is a farmer, then we infer that Salvatore depends on the weather for his livelihood; if we have won at the track when we have brought our lucky rabbit's foot in the past, we infer that we will win today, since we have our rabbit's foot; if we see Virginia drop her cup of coffee, we infer that it will fall to the floor and spill.

Some of the inferences we make are good ones, some bad. Logic is the science of good inferences, and for logicians, a correct inference occurs when we derive a statement, called a *conclusion*, from some set of statements, called *premises*, in accordance with some accepted rule of inference. In a given instance of inference, the set of statements constituting the premises and the conclusion, perhaps together with some intermediate statements, constitute an *argument*, and good arguments, like correct inferences, are those in which the conclusion is derived from the premises according to accepted rules of inference.

Traditionally, deductive logic is the branch of logic that studies inferences that are both sound (all the premises are true) and valid. If it is impossible for both the premises of an argument to be true and the conclusion of that argument to be false, then the inference from those premises to that conclusion is considered deductively valid. Valid arguments have the following three important features:

(1) All of the information contained in the conclusion of a valid argument must already be implicitly "contained in" the premises of that argument (such arguments are thereby termed *nonampliative*).

(2) The truth of the premises of a valid argument guarantees the truth of its conclusion (thereby making such arguments *truth preserving*).

(3) No additional information, in the form of premises, can undermine the validity of a valid argument (thus making such arguments *monotonic*).

For example, suppose that we have the following three claims:

(a) all arguments are tidy things;
(b) all tidy things are understandable; and
(c) all arguments are understandable.

The inference from claims (a) and (b) to the claim (c) is a valid inference: If (a) and (b) are true, then (c) must be, too. Notice that the information expressed in (c) concerning arguments and understandable things is in some sense already to be found in the conjunction of (a) and (b); we just made the connection explicit in (c). Also note that no further information concerning anything whatever will render (c) false if (a) and (b) are true.

Sound arguments are valid arguments with the following additional property: All of the premises of sound arguments are true. Valid arguments may have either true or false premises; validity is not indicative of the truth of the premises of an argument; nor does validity ensure the truth of the conclusion of an argument. Validity, rather, concerns the relationship that is obtained between the premises and the conclusion of an argument, regardless of their actual truth values, with one exception: If it is impossible for the conclusion of an argument to be false on the supposition that the premises are true, then the argument is valid; otherwise it is invalid. Soundness does concern the truth value of the premises and the conclusion. A sound argument is a valid argument with true premises, and since it is a valid argument, a sound argument also has a true conclusion.

The main branches of deductive logic are well established, and the rules for deductive inference well known.[1] Not so for inductive logic. *Induction* is traditionally understood as the process of inferring from some collection of evidence, whether it be of a particular or a general nature, a conclusion that goes beyond the information contained in the evidence and that is either equal to or more general in scope than the evidence. For example, suppose an ethologist is interested in whether or not owls of the species *Bubo virginianus* will hunt lizards of the species *Sceloperous undulatus*. The scientist exposes some of the lizards to ten different owls of the species, and all of the owls respond by capturing and hungrily eating the lizards.

In setting up her eleventh experiment, the researcher is asked by her assistant whether or not she believes that the next owl will hunt and eat the released lizards. The animal behaviorist tells the assistant that, based on the prior tests, she believes the next owl will eat the lizards. Here the ethologist has used evidence about the particular owls that were tested to infer a conclusion which is equally general in scope—the conclusion is about a particular owl, the next one to be tested—but which goes beyond the information implicit in the premises. After having completed the eleventh test, the ethologist returns to her laboratory and opens her laboratory notebook.

She writes, "After having completed the eleven field studies and having a positive response in each case, I conclude that all owls of the species *Bubo virginianus* will eat lizards of the species *Sceloperous undulatus.*" In this case, the scientist has for her evidence that the eleven owls hunted and ate the lizards. She concludes that all owls of that species will eat that species of lizard. Her evidence is about particular owls; her conclusion is a generalization about all owls of a certain species. The generality and the informational content of the conclusion are greater than those of the premises.

Again, suppose a market researcher does a demographic study to determine whether or not most people in America shop for food at night. He collects data from ten major American cities, ten suburban American towns of average size, and ten average size rural American towns. In each case, the data shows that most people shop for their food at night. The market researcher concludes that most Americans shop for their food at night. Here, the researcher's evidence consists of thirty generalizations concerning the shopping habits of most people in specific locations in America. The researcher's conclusion generalizes from this evidence to a claim about the shopping habits of most Americans in all locations of America.

Inductive logic is the branch of logic that inquires into the nature of inductive inferences. Many writers distinguish two general types of simple inductive inferences—those by enumeration and those by analogy—and we can divide each of these into two further kinds: particular enumerative inductions and general enumerative inductions; and particular analogical inductions and general analogical inductions.

A particular enumerative induction occurs when we assert that a particular individual *A* has the property of being a *B* on the basis of having observed a large number of other *A*s also being *B*s. For example, if each of the twenty dogs we have encountered

---

[1]Whereas this is true generally, there are extensions of deductive logic, for example, modal logic, concerning which there is less agreement. There are also arguments over the foundations of logic. See, for example, [Haack, 1996].

has barked, and we infer from this that the next dog we encounter will also bark, we have performed a particular enumerative induction.

If we conclude from a number of observed As being Bs the more general claim that most As are Bs or that all As are Bs, then we have performed a general enumerative induction. Continuing our dog example, we might have inferred from our having observed twenty barking dogs that most or all dogs bark.

In analogical induction, we use the fact that a given A (which may be a particular thing or a class of things) possesses properties $P_1, \ldots, P_n$ in common with some C (another particular thing or class of things) to support the conclusion that A possesses some other property P_{n+1} that C also possesses. If our conclusion is about a particular A, then we have performed a particular analogical induction; if the conclusion is general, then we have performed a general analogical induction. Here the warrant for our conclusion about A possessing the property P_{n+1} is derived not from the number of As observed, but rather from the similarities found between A and something else.

For example, we know of geese that they are birds, that they are aquatic, and that they migrate. We also know that they mate for life. Ducks are birds, are aquatic, and migrate. We might infer, by general analogical induction, that they also mate for life.

Inductive inferences are notoriously uncertain, because in an inductive inference the conclusion we infer from our premises could be false even if the evidence is perfectly good. We can put this in terms of arguments by saying that inductive arguments are not truth preserving—the truth of the premises in an inductive argument does not guarantee the truth of the conclusion.

For example, suppose we know that a very small percentage of the U.S. quarter dollars in circulation are silver quarter dollars; it would be reasonable to infer from this that the next U.S. quarter dollar we receive will not be made of silver. Nevertheless, the next one *might* be made of silver, since there are still some silver quarter dollars in circulation. Our inference concerning the next U.S. quarter dollar we receive, while reasonable, is uncertain; there is room for error. Our inferred conclusion could be false, even though our premises are true. Such inferences are not truth preserving. Because the conclusion of an inductive argument could be false even if the premises are true, all inductive arguments are invalid and therefore unsound.

The reason that some, indeed most, of our inferences are uncertain is that often the information contained in our conclusions goes beyond the information contained in our premises. For instance, take the following as exhausting our premises: Most crows are black, and the bird in the box is a crow. These premises make it reasonable to conclude that the bird in the box is black. But, of course, the bird in the box might not be black. Our premises don't tell us enough about the particularities of the bird in the box to make it certain that it will be a black crow. Our premises present us with incomplete information relative to our conclusion, and as a result the truth of our conclusion is uncertain. Inductive arguments have been called *ampliative* to describe the fact that their conclusions contain more information than their premises.

Scientific inferences are ampliative in nature. For example, when a scientist collects data about some phenomenon, analyzes it, and then infers from this data some generalization about every instance of that phenomenon, the scientist is making an ampliative inference. Thus, suppose a scientist tests for the resistance to an antibiotic of some gonococcus he has prepared in a number of Petri dishes, analyzes the results of the experiment, and then (perhaps after a number of similar trials and controlling for error) infers from his results that all gonococci of the type under investigation are resistant to the antibiotic. His inferred conclusion concerns all instances of that type of gonococcus, but his evidence is only about a limited number of such instances.

In cases of uncertain inference, true premises make the truth of the conclusion more or less plausible. Whereas deductive arguments are monotonic, inductive arguments are *nonmonotonic* in character. If we gather more information and add it to our set of premises, our conclusion may become either more tenable or less tenable.

For example, suppose Gumshoe Flic, an able detective, is investigating a reported shooting. The victim was a successful, single, middle-aged woman named Sophia Logos who was a prosecuting attorney in the community. Sophia was found by her housecleaner dead in the living room some hours ago. She had been shot in the chest repeatedly. No gun was found near the body. Gumshoe, upon inspecting the body, comes to the conclusion that Sophia was murdered by someone who shot her. Later, while searching the house, Gumshoe finds an empty bottle of barbiturates on the counter in the bathroom and a small caliber handgun hidden in the basement. He has these sent to the forensic laboratory for fingerprints and ballistic testing.

Back at the office Gumshoe hears that, the day before the shooting, a criminal, Hunde Hubris, had escaped from the nearby prison. Hunde, Gumshoe knew, had been sent to prison for life six years before in a sensational murder case brought to court by prosecutor Sophia Logos. Gumshoe remembers that Hunde had vowed, as he left the courtroom, to kill Sophia if it was the last thing he did. Gumshoe tells his colleague over a coffee and doughnut that he thinks Hunde was most likely the murderer.

After leaving the office and upon questioning one of Sophia's closest friends, Gumshoe learns that, recently, Sophia had been suffering from severe depression and had mentioned to several friends her despair and her occasional thoughts of suicide. Gumshoe puts this information together with that of the empty bottle of barbiturates and pauses. He asks the friend if Sophia had been taking any prescription drugs. The friend replies that Sophia was notorious for refusing medication of any sort. "She practiced Eastern medicine. You know, acupuncture, Taoism, that kind of thing. She often said, usually as a joke of course, that she would only take drugs in order to kill herself." Gumshoe begins to wonder. On calling the forensic lab and inquiring about the blood analysis done by the coroner, it turns out that Sophia had died from an overdose of barbiturates two hours before she had been shot. Well, Gumshoe thinks to himself, that blows the theory that she died from the shooting, and it's beginning to look like Sophia committed suicide by taking barbiturates. But the gunshot wounds. ... It still seems likely that she was murdered.

As he ponders, he gets a call on his cellular phone from a fellow detective at the station. Hunde has been apprehended in a nearby hotel. He had called the authorities and turned himself in. Under questioning, he said he was innocent of the shooting. Instead, Hunde claimed that he and Sophia had been in correspondence while he was in prison, that she had made him see the error in his ways, and that he had escaped from prison in order to help her after having received her last letter in which she mentioned her depression and growing fear of her housecleaner. Gumshoe takes this in and cynically responds, "And I'm Santa Claus." The other detective replies by explaining how he had checked with forensics and found out that the handgun found in the basement matched the weapon used in the killing and that, most surprisingly, the handgun failed to have Hunde's prints on it; but, also surprisingly, both the handgun and the bottle of barbiturates had the fingerprints of the housecleaner all over them. "And get this, Hunde has the last letter he got from Sophia, and I'm telling you, Gum, it looks authentic."

After hanging up, Gumshoe revises his previous conclusions on the basis of his new evidence. It now seems less plausible that Hunde killed Sophia, though given the wacky story Hunde is telling, Hunde is still a prime suspect. The housecleaner now seems to be a likely suspect, and it looks like this is indeed a murder, not a suicide. If the housecleaner skips town, then there's more reason to think she did it. Maybe she heard that Hunde had escaped and was trying to make it look like Hunde did it by shooting Sophia. But what is the motive?

Note that as Gumshoe's evidence changes, so does the plausibility of the conclusions he previously inferred. Sometimes the new evidence undermines the plausibility that a previous conclusion is correct, sometimes it strengthens the plausibility of the conclusion. If a deductive conclusion is entailed by its premises, nothing can weaken that relation. The kind of inference we are looking at is thus nonmonotonic, and therefore not deductive. Attempting to provide canons for nonmonotonic reasoning is one of the major challenges for an inductive logic.

1.3   Roots in the Past

The first system of logic was presented to the Western world by Aristotle in the fourth century B.C. in his collected works on logic entitled *Organon* (meaning, roughly, "a system of investigation"). Aristotle's was a syllogistic deductive logic, taking the terms of a language to be the fundamental logical units. Whereas Aristotle emphasized deductive inferences, he did allow for syllogisms that had for premises generalizations that asserted what happened for the most part (i.e., generalizations of the form "Most *A*s are *B*s"), and Aristotle was clearly aware of both enumerative and analogical inductions. Nevertheless, Aristotle did not present a systematic study of inductive inferences.

During the Hellenistic period of Greek philosophy, the Stoics extended Aristotle's work and were the first to develop a propositional logic, according to which the propositions of a language are taken as the basic logic units. As Aristotle had before them, these later thinkers devoted their attention to deductive inference and did not develop a logic of uncertain inference. Likewise, during the Middle Ages such figures as Peter Abelard (1079–1142) and William of Ockham (ca. 1285–1349) made contributions to the work done by the ancients on deductive logic but did little work on the theory of uncertain inference.

From Aristotle to Newton and into the present, philosophers and scientists have modeled science upon mathematics. This way of analyzing science in terms of deductive demonstration may be called the *axiomatic approach* to science, and one of its primary characteristics will help to motivate our discussion of the historical development of inductive logic. The axiomatic approach to science can be characterized by its emphasis on the need to model science and scientific language upon the exact methods and languages of mathematics. Historically, this has meant presenting scientific theories in terms of axioms which express the fundamental laws and concepts of the science. The substantive content of the science is then embodied in theorems derived from these axioms.

Euclidean geometry provides a prime example of this approach. Euclidean geometry is a theory expressed in a limited vocabulary (point, line, plane, lies on, between, congruent, etc.) and is based on a small number of statements accepted without proof. In many presentations, a distinction is made between *axioms* and *postulates*. For example, it is regarded as an axiom that things that are equal to the same thing are equal to each other, while it is a postulate peculiar to geometry that a straight line is determined by two points. Non-Euclidean geometry is based on postulates that are different from the standard Euclidean postulates. Euclidean geometry accepts the postulate that there is one and only one parallel to a given line through a given point. Lobachevskian geometry allows for more than one line parallel to a given line through a given point. Riemannian geometry denies the existence of parallel lines. All these geometries are internally perfectly consistent. But which is true?

An important component of any axiomatic account of science is how we come to know the axioms or first principles of a science. For example, which set of geometric postulates captures the truth about lines and planes? Here we discern a source of one of the great schisms in Western philosophy—that between the *rationalist* approach to scientific knowledge and the *empiricist* approach.

The rationalists, for example, Plato, Augustine, Descartes, and Leibniz, typically assume that our knowledge of the axioms of science is independent of any empirical evidence we may have for them. Different rationalist thinkers have proposed different views concerning how we come to know the axioms. Plato, in the *Phaedo*, asserted that we gained knowledge of the axioms before birth, forgot them at birth, and then rediscovered them during our lifetime. Augustine believed that it was through the presence of God's light in the mind that the axioms were rendered certain and known. Descartes held that the knowledge of the axioms of science was derived from self-evident truths, and Leibniz maintained that we possess knowledge of the axioms of science in virtue of certain conceptual truths about God. Reliance on some supersensible ground is characteristic of rationalist accounts of our knowledge of the axioms of science.

Empiricists, on the contrary, typically argue that all scientific knowledge is based, ultimately, on our sensible experience of the world—experience which is always of the particular and immediate. Given their assumption that all knowledge is ultimately based on experience, empiricists need to explain how we can achieve the generality of our scientific knowledge of phenomena given that our experience is always of particular instances of phenomena. Typically, but not always, induction will play an important part in any such empiricist explanation. To the extent that an empiricist who relies on induction as the source of our scientific knowledge cannot provide an account of the nature of induction, his view of science, and of our knowledge of the world, is incomplete. In other words, the empiricist is under a strong compulsion to provide an account of inductive inference.

1.4 Francis Bacon

The English statesman, philosopher, and moralist Francis Bacon (1561–1626) was well aware of the need for an inductive logic that would serve as a canon for scientific inference. To meet this need, Bacon presented the first explicit account of the methods of induction in his *Novum Organum* [Bacon, 1620], and with Bacon the story of inductive logic begins.

Long before Bacon, ancient and medieval natural scientists had investigated and amassed data concerning the phenomena of nature, but Bacon for the first time attempted to codify the methods of gathering empirical data and making inferences based on it. Bacon believed that previous scientific and philosophical efforts had been marred by unswerving devotion to the Aristotelian approach to science laid out in the *Organon*. This approach stressed deduction from first principles as the main mode of scientific inquiry. Bacon believed that, while such deduction was appropriate at the stage where science was nearly, if not already, complete, scientists needed to free themselves from the Aristotelian paradigm during the developing stages of science and adopt a different and inductive approach to science. He thus proposed a new system of investigation in his *Novum Organum*.

Bacon sets the stage for his investigation by first explaining the available options and the current tendencies:

There are and can be only two ways of searching into and discovering truth. The one flies from the sense and particulars to the most general axioms, and from these principles, the truth of which it takes for settled and immovable, proceeds to judgment and to the discovery and middle axioms. And this is the way now in fashion. The other derives axioms from the senses and particulars, rising by a gradual and unbroken ascent, so that it arrives at the most general axioms last of all. This is the true way, but as yet untried. [Bacon, 1620, Book I, Aphorism xix or Section xix.]

Bacon identifies the "true way" of searching into and discovering truth with induction, and by "induction" he means a method by which science "shall analyze experience and take it to pieces, and by a due process of exclusion and rejection lead to an inevitable conclusion" [Bacon, 1620, preface.] For Bacon, it is knowledge of causes that is true knowledge, and Bacon's is an Aristotelian conception of causes, including the material, efficient, formal, and final causes; but most important for science are the efficient causes. Thus, every individual in the world has, in Bacon's view, a *material cause* (the stuff of which it is constituted), an *efficient cause* (that which caused, in our sense of the term, the thing to occur), a *formal cause* (the essential properties of the thing), and a *final cause* (the purpose of the thing). Metaphysics has for its task the determination of the formal causes of things, and science the determination of the material and efficient causes.

Bacon's scientific method (not his inductive method) consists of three stages:

(1) the amassing of experimental data (termed "a natural and experimental history"),
(2) an ordered arrangement of the experimental data (termed "tables and arrangements of instances"), and
(3) the principled inference from the ordered data to more and more general axioms (termed "induction").

Let us follow Bacon, and suppose we are investigating the nature of heat. So, first, we gather a lot of data about heat. Once we complete this first stage of our investigation, Bacon tells us we are to order the experimental data we have collected into three tables:

(a) the *table of essence and presence* (wherein we list all those instances of things that possess the property of heat),

(b) the *table of deviation, or of absence in proximity* (wherein we list all those instances of things that fail to possess the property of heat), and

(c) the *table of degrees or comparison* (wherein we list the observed changes in the property of heat as we vary another property).

Having completed the second stage and ordered our information, we can then apply Bacon's method of induction.

Bacon's method of induction involves three parts—a negative part, an affirmative part, and a corrective part. The negative part, termed "rejection or process of exclusion," is applied first and involves rejecting as distinct from the nature of heat anything that fails to always be correlated with the property of heat. So, for example, if there are cases in which stones are not hot, then we can exclude the property of being a stone from being a part of the nature of heat. If there are cases in which heat attends water, then we can discount the property of solidity as being a part of the nature of heat. Again, if there are cases of change in brightness without a change in heat, then we can reject brightness as being a part of the nature of heat.

Following this negative enterprise, or concomitant with it, we are to engage in an effort to determine which among the properties correlated with heat has a nature "of which Heat is a particular case" [Bacon, 1620, Book 2, Aphorism xx or Section xx]. That is to say, of which of these latter properties is heat a species? Bacon is optimistic that there will be a few instances of heat which are "much more conspicuous and evident" [Bacon, 1620, Book 2, Aphorism xx or Section xx] than others and that we can therefore, by choosing among these, eventually determine which is the correct one. Bacon is aware that this affirmative part of induction is prone to error, since we might initially pick the wrong general property, and he aptly terms it "indulgence of the understanding." (Bacon, in fact, proposed the property of motion as that property of which heat is a species.) But Bacon believes that in putting forward the hypothesis that heat is a species of some more general property and, as a consequence, focusing our attention on our data concerning heat and this other property, we shall be more efficient in determining whether or not heat is in fact a species of it.

Once we have completed the processes of exclusion and the indulgence of the understanding, we shall have arrived at a provisional definition of the phenomenon under investigation (Bacon terms this the *first vintage*), which will serve as the raw material for the next stage of the inductive process, that of *correction*. What we have termed the "corrective part" of Bacon's method was to involve ten distinct steps. Of these Bacon presents an exposition of only the the first, *prerogative instances*, having been unable to complete his entire project. This exposition amounts to twenty-seven different and special kinds of instances purported by Bacon to facilitate the refinement of the first vintage.

Bacon believed that if his methods were followed scientists would discover the laws that govern the phenomena of nature. Bacon clearly maintained the major tenets of the Aristotelian method with regard to completed science, but he recognized that completed science was a long way off and that, during the interim between incipient science and completed science, the mainstay of scientific activity would be inductive methods.

After the publication of Bacon's work in 1620, it was roughly two hundred years before further work on inductive logic was pursued again in a systematic fashion by John Herschell, William Whewell, and John Stuart Mill. During this interim there were developments in the mathematical treatment of chance and statistical approaches to observational evidence that play an important part in the history of uncertain inference.

1.5   The Development of Probability

Attempts to mathematize decisionmaking under uncertainty were not originally undertaken primarily with an eye to clarifying problems within natural science, although recent scholarship suggests that such problems were part of the motivating force. Rather, interest in a systematic understanding of probabilities is traditionally considered to have arisen in the seventeenth century in connection with gambling.

As history preserves the story of the development of the mathematical theory of probability, Antoine Gombaud (1607–1684), who is known as the Chevalier de Mere and who was an author, councillor, and prominent figure in the court of Louis XIV, proposed some questions to Blaise Pascal (1623–1662), the noted and brilliant French mathematician, concerning some problems surrounding games of chance. Pascal, in working out the solutions to these problems, corresponded with Pierre de Fermat (1601–1665), the preeminent mathematician of the time, and the work done by the two of them is usually viewed as the start of the mathematical theory of probability.

The two problems posed to Pascal by the Chevalier were concerned, respectively, with dice and with the division of stakes. These are of sufficient interest to repeat. The dice problem goes as follows: When one throws two dice, how many throws must one be allowed in order to have a better than even chance of getting two sixes at least once? The division problem (also known as the *problem of points*) involves the following question: How shall one divide equitably the prize money in a tournament in case the series, for some reason, is interrupted before it is completed? This problem reduces to the question: What are probabilities for each player to win the prize money, given that each player has an equal chance to win each point?

Dice problems such as the one above were well known by Pascal's time; Geralmo Cardano (1501–1576), around 1525, had discovered and presented in his *De Ludo Aleae* rules for solving the dice problem for one die. Problems similar to the *problem of points* had also been around for some time. An early example of the division problem can be found in Fra Luca Paciuoli's *Summa* (1494), and has been found in Italian manuscripts as early as 1380. Notwithstanding the fact that these problems were current before Pascal and Fermat attended to them, the mathematical techniques, such as the *arithmetical triangle*, marshaled by the two in solving them were novel. However, Pascal's life was a short one (he lived but 39 years), and neither Pascal nor Fermat felt the need to publish their mathematical results. All we possess of Pascal's and Fermat's efforts is found in their few surviving letters to each other.

The importance of Pascal's and Fermat's investigations far outstrips the apparently trivial concerns of dicing, for in their attempts to solve the difficult combinatorial problems posed to them, they developed techniques for analyzing situations in which a number of alternative outcomes are possible and in which knowledge of the probabilities of the various alternatives is important for practical decisions. After the work of Pascal and Fermat little progress was made on the mathematical theory of probability for about 50 years.

At the time of Pascal and before, the term "probable" was connected closely with the notion of authoritative opinion and meant something like "worthy of approbation". "Probabilism", during the sixteenth century, was a principle of deciding arguments advanced by the Jesuit order. At the time, there were an increasing number of conflicts among the opinions of the Church authorities. According to probabilism, any opinion of a Church authority is probable to some degree in virtue of its source, and, in the case of conflicting opinions, it was sufficient to adopt some one of these probable opinions, not necessarily the one which has the most authority (i.e., probability). Around the middle of the seventeenth century, the term "probability" took on the two meanings which we more readily connect with it—that of the likelihood of an occurrence and that of the degree of reasonable belief. When the *Port Royal Logic or the Art of Thinking* was published in 1662 by Arnauld and Nicole, it included mention of probabilities in connection with both numerical measurement of chance and rational belief in the form of a summary of Pascal's famous wager.

Pascal's wager is an example of early mathematical decision theory, the theory of what to do when the outcome of your action is uncertain. In simple outline, decision theory tells us that if we take the probabilities of the various possible outcomes of our actions together with the utilities of each of those outcomes, we can determine through the comparison of all possible acts the most rational course of action.

For example, if the probability of rain is 0.3, and your problem is to decide whether or not to go to a football game, you may reason as follows. If I go to the game and it rains, I'll be miserable—measured by a utility of −10 units. If I go to the game and it is sunny, I'll have a great time: +70 units. If I don't go to the game, and curl up with a good book, I'll enjoy myself somewhat whether it rains or not: 35 units. So if I go to the game, my *expected* utility is $0.30(-10) + 0.70(+70) = 46$; while if I stay home my expected utility is $1.0(35) = 35$. I should therefore take the chance and go to the football game.

In Pascal's wager, the decision at issue is whether or not to believe in God. Pascal argued in the following way: Either God exists, or God doesn't exist, and the chance that God exists is equal to the chance that God doesn't exist. If God exists and you decide to believe in God, then you can expect infinite happiness. If God exists and you decide not to believe in God, you can expect to exhaust your happiness upon dying. If God doesn't exist and you decide to either believe in God or not believe in God, then you can expect to exhaust your happiness upon dying. Given these alternatives, Pascal argued, deciding to believe in God is the most rational decision. For, if your belief is true and God does exist, then you get infinite happiness; if your belief is false and God doesn't exist, then you are no worse off than you would have been had you not believed in God. Whereas one certainly can question the assumptions Pascal makes in setting up his decision problem, the method he displays for making decisions under conditions of uncertainty is marvelous in its economy and force and has been subsequently developed into a powerful decisionmaking tool.

Another important development in the history of uncertain inference during the latter half of the seventeenth century was the birth of modern statistical theory. The work of John Graunt (1620–1674), William Petty (1623–1687), and John de Witt (1625–1672) on annuity data is the first recorded example of using statistical data as a basis for inferences. Their work made an important contribution to the problem of taking care of the social needs of large groups of people, as well as to the problem of setting competitive insurance rates, important for commercial development. Whereas these thinkers did not attempt to analyze the nature of such inferences, their work set the stage for an increased emphasis on the nature of statistical inference.

With the work of Gottfried Wilhelm Leibniz (1646–1716), we see a shift toward the modern mathematical approach to logic. Leibniz both envisioned a universal language for science based on an analysis of the functions of the various parts of speech (thus foreshadowing contemporary formal treatments of grammar) and worked toward mathematizing deductive logic by experimenting with logical algebras (a goal achieved by modern logicians). Although Leibniz was interested in the mathematical theory of probability, he did not contribute to its development; but his work on the calculus, along with Newton's, as well as his conception of an algebraic inductive logic, was to have lasting impact on the theory of probability.

Although he did not contribute to the development of the mathematical approaches to probability, Leibniz was familiar with the notion of "equipossibility" in connection with probability and is thought by some, e.g., [Hacking, 1975, p. 122ff.], to have employed the principle of indifference (also known as the principle of insufficient reason), whereby we can treat as equally probable the occurrence of events among which we are unable to choose one as more likely than another. As we shall see in Chapter 4, this conception of probability has played a major role in the development of uncertain inference.

Most importantly, Leibniz was familiar with the concept of evidence as used in jurisprudence, where information is presented as partially supporting a proposition. This notion of a hypothesis being partially supported by evidence can be traced back to the Roman scales of laws, and Leibniz's theory of partial implication (which he conceived of as an inductive logic) is an attempt to mathematize this relation of partial support. At least superficially, Leibniz's notion of inductive logic is different from Bacon's conception, focusing on the relation between a hypothesis and a body of evidence as opposed to focusing on causal conditions. Work done in this century by Carnap and others on inductive logic (see Chapter 4) has followed Leibniz's approach.

It was the work of Jacob Bernoulli (1654–1705) in his posthumously published *Ars Conjectandi* [Bernoulli, 1713] that brought the mathematical theory of probability beyond its incipient stages. The first part of this work is a reprinting of Huygens's *De Ratiociniis in Ludo Aleae* with commentary by Bernoulli; the second is devoted to the theory of combinations and permutations; the third is devoted to certain problems concerning games of chance; and the fourth attempts to apply the theory of probability to larger social issues. The most remarkable contribution made in Bernoulli's text is now known as *Bernoulli's theorem*.[2] The theorem is discussed both in its direct form and most importantly, in its inverse forms. The early work on the mathematical theory of probability had focused on problems of combination and permutation involving a priori probabilities (probabilities known independently of experience) of the occurrences under consideration. Bernoulli's theorem in its inverse form is an important contribution to the attempt to determine the a posteriori

---

[2]This theorem will be discussed in Chapter 3.

probability of an occurrence (the probability of an occurrence given some evidence from experience).[3]

Jacob Bernoulli's work on inverse probabilities was the first attempt to apply the mathematical theory of probability to the question of how we learn from experience. His investigation was continued by Abraham De Moivre (1667–1754) through his work on the binomial approximation, and by Thomas Simpson (1710–1761) through his work on the distribution of error. It was Pierre Simon Laplace (1749–1827), however, who first successfully applied the mathematical theory of probability to statistical data of practical import (viz., astronomical data), developed in an analytic fashion the theory of observational error, and presented a convincing account of inverse probability, which improved upon that of Thomas Bayes (1701–1761). With Laplace, the mathematical theory of probability becomes firmly connected with statistical theory [Laplace, 1951].

1.6   John Stuart Mill

Bacon had believed that his inductive methods would lead to axioms of science which were known with certainty. He was not alone in this belief. The notion that inductive arguments would provide certain conclusions was prevalent until the end of the eighteenth century. In 1739, the Scottish philosopher David Hume anonymously published *A Treatise of Human Nature*, in which he presented a now classic criticism of inductive argument. Hume's criticism concerns the justification for accepting inductive conclusions. Why is it reasonable to accept the conclusion of an inductive argument? What is the justification for believing that a rule for inductive inference will lead to acceptable conclusions? We accept the conclusions of valid deductive arguments on the basis that such arguments are truth preserving. What similar ground do we have for accepting the conclusions of inductive arguments? Hume was unable to find an answer to his question that upheld induction, and he advocated skepticism with regard to the truth of inductive inferences.

Much of the force of Hume's analysis of induction stemmed from his analysis of causation. Hume had argued that causal connections are merely the chimera of human imagination based on the psychological feeling of expectation and not real connections among phenomena in the world. Every time we see lightning, we hear thunder; we come to expect thunder when we see lightning; and according to Hume, through this habituation we come to believe that lightning causes thunder.

As we have seen, Bacon's inductive methods were intended to result in the general knowledge of the causes of phenomena. Thus, our inductive investigation into instances of heat would ultimately give rise to generalizations concerning the causes of heat. Hume's objection to Bacon's method would thus be twofold. Your inductive methods, Hume might say to Bacon, are never sufficient to confer justification upon such causal generalizations. On the one hand, we fail to observe the necessary connection requisite for causal connections, and on the other hand, the possibility is

---

[3]The Bernoulli family is one of the most remarkable families in the history of mathematics; some scholars estimate that as many as twelve members of it made contributions to some branch or other of mathematics and physics [Stigler, 1986, p. 63]. Besides the contributions to the mathematical theory of probability made by Jacob Bernoulli, other contributions were made by Nicolas, John, and Daniel Bernoulli.

always open that these inductive generalizations may be defeated by further evidence. Today there is still no generally accepted answer to Hume's question, just as there is no definitive view of inductive inference, and there have been a variety of different responses ranging from the rejection of induction to the rejection of Hume's question as illegitimate.

Bacon's project of presenting an inductive logic was not continued until the middle of the nineteenth century, when a number of authors published works on the topic. Of these, John Stuart Mill's *A System of Logic* [Mill, 1843] has dominated the discussion. This book was until recently the standard text on inductive logic. Mill was obviously familiar with Bacon's *Novum Organum*, as well as the writings of his contemporaries, among them John Herschell's *A Preliminary Discourse on the Study of Natural Philosophy* (1831) and William Whewell's *History of the Inductive Sciences, from the Earliest to the Present Times* (1837). Mill assumed, *pace* Hume and with Bacon, that causal regularities obtained in the world and that inductive methods would lead to the discovery of these regularities. Mill also assumed that every phenomenon in the world is caused by some state of affairs which immediately precedes the given phenomena in time; Mill referred to these antecedent states of affairs as "circumstances."

The first step in the inductive process for Mill is to decide which of the many circumstances we will take as possible causes of the phenomena under investigation. We certainly can't take every circumstance as a possible cause of the phenomena, or we could never get started. Let's suppose we want to investigate the cause of cancer. Among the circumstances in a given instance of cancer, we're certain that the distant stars are causally irrelevant; we're almost certain that the local ambient sound waves don't cause cancer; we're fairly confident that the person's heart rhythm doesn't, but we're getting closer to real possibilities; something related to genetic makeup is a possibility; another possibility might be the exposure of cells to certain toxins; etc. This set of assumptions concerning which circumstances are and which are not possible causes constitutes our *background knowledge* in a given investigation. Given our set of possible causes, our next step is to gather information concerning these possible causes of the phenomena under investigation through experiment and observation.

Mill presented four fundamental methods of induction, which he termed "the methods of experimental inquiry," of which he combined two to make a fifth method. Mill's methods are termed "eliminative methods of induction" because they lead to their conclusion through the elimination of all irrelevant circumstances. In a given investigation, we will be looking to establish a conclusion of the form X is a cause of events or phenomena of kind Y; that is, we will be searching for some circumstance or set of circumstances X which, when present, is either sufficient for the occurrence of a Y, necessary for the occurrence of Y, or necessary and sufficient for the occurrence of a Y.

Necessary and sufficient conditions play a role in both modern inductive logic and philosophy in general. We may characterize them in first order logic as follows:

- S is a *sufficient condition* of E: If S occurs, or characterizes an event, then E occurs or characterizes the event. Formally,

$$(\forall x)(S(x) \rightarrow E(x)).$$

- N is a *necessary condition* of E: N is necessary for E; or, if N doesn't occur (or characterize an event) then E can't occur; or, if not-N ($\neg N$) does occur, then $\neg E$ must occur. Formally,

$$(\forall x)(\neg N(x) \rightarrow \neg E(x))$$

or what is the same thing,

$$(\forall x)(E(x) \rightarrow N(x)).$$

- C is a *necessary and sufficient condition* of E: You can have E if and only if you have C. Formally,

$$(\forall x)(C(x) \leftrightarrow E(x)).$$

Notice that if S_1 is a sufficient condition of E, then $S_1 \wedge S_2$ is also a sufficient condition of E. What interests us most is a *maximal* sufficient condition in the sense that it is not the conjunction of a sufficient condition with something else.

Similarly, if N_1 is a necessary condition of E, then $N_1 \vee N_2$ is a necessary condition of E. We may focus on *minimal* necessary conditions: Those not equivalent to a disjunction of a necessary condition with something else.

For example, suppose we know that, whenever someone is exposed to high doses of gamma radiation, he suddenly loses his hair. Then, we also know that exposure to high levels of gamma radiation is sufficient to cause sudden hair loss. Suppose we also find out that there is never an instance of sudden hair loss unless there is among the circumstances exposure to high doses of gamma radiation. We can conclude from this that exposure to high levels of gamma radiation is not only sufficient but is necessary for there to be sudden hair loss.

Mill's five methods of eliminative induction are:

(1) the method of agreement,
(2) the method of difference,
(3) the joint method of agreement and difference,
(4) the method of residues, and
(5) the method of concomitant variations.

Let us assume we're investigating the phenomenon of heat again. The *method of agreement* involves comparing different instances in which heat occurs and has for its evaluative principle that we eliminate as causally irrelevant any circumstance which is not found in all compared instances of heat. The method of agreement gives rise to the *first canon* of Mill's inductive logic:

If two or more instances of the phenomenon under investigation have only one circumstance in common, that common circumstance is the cause of the given phenomenon.

Mill's method of agreement is similar to employing Bacon's table of essence and presence in conjunction with the process of exclusion.

The *method of difference* requires that we compare instances in which heat occurs with instances in which heat does not occur and has for its evaluative principle that we eliminate as causally irrelevant any circumstance which both attends and fails to attend to occurrence of heat. The method of difference gives rise to the *second canon* of Mill's inductive logic:

If an instance in which the phenomenon under investigation is present possesses every circumstance possessed by an instance in which the phenomenon is absent and, in addition, possesses an additional circumstance, then that additional circumstance is the cause of the phenomenon.

Mill's method of difference is similar to using Bacon's table of deviation in conjunction with the process of exclusion.

Mill was skeptical that our background knowledge would be sufficiently fine grained to avoid erroneously ignoring actual causes in deciding what shall count as the possible causes. This was particularly true, he thought, in those cases where we didn't exert experimental control over the instances but, rather, used as instances those from ordinary observation. Mill thought that the exactness demanded by the method of difference was suited to experimental method, which he believed implied the introduction into one of two exactly similar circumstances the phenomenon to be tested. Thus, Mill believed the method of difference to be accurate in its conclusions. The method of agreement, on the other hand, did not require the exactness of the method of difference. Any two instances of the phenomenon, be they natural occurrences or experimentally produced, were available for comparison. As a result, he did not believe that the method of agreement would suffice to demonstrate a causal connection between a circumstance and a phenomenon—we might hit upon an accidental correlation, and not a true causal connection.

The *joint method of agreement and difference* is used when, as a result of the uniqueness of the phenomenon under investigation, the first two methods by themselves are insufficient for determining the cause of the phenomenon. This situation arises when we can find no instances in which the phenomenon is absent that are similar to those instances in which it is present, thus making the method of difference inapplicable by itself. In such cases, the joint method of agreement and difference directs us to take a number of instances in which the method of agreement has shown a given circumstance to be uniquely associated with the phenomenon and to remove that circumstance from those instances. If we are successful in creating this situation, we are then able to use the method of difference to determine if the phenomenon is still present in the altered instances. If it is not, then by the method of difference, we will have located the cause of the phenomenon. The joint method of agreement and difference gives rise to the *third canon* of Mill's inductive logic:

If two or more instances in which the phenomenon occurs have only one circumstance in common, while two or more instances in which it does not occur have nothing in common save the absence of that circumstance, that circumstance is the cause of the phenomenon.

The *method of residues* is a special case of the method of difference. According to this method, we are to determine which of the circumstances in a given instance are known to cause which part of the phenomenon; we then eliminate these circumstances and their effects from the phenomenon and focus on the remaining circumstances and parts of the phenomenon. What remains of the phenomenon is known to be caused by what remains of the circumstances. This is the *fourth canon* of Mill's inductive logic:

Subduct from any phenomenon that which is known by previous inductions to be the effect of certain circumstances, and the residue of the phenomenon is the effect of the remaining circumstances.

The last of Mill's methods, the *method of concomitant variation*, is essentially the same as employing Bacon's table of degrees or comparison in conjunction with the process of exclusion and can be understood through reference to the *fifth canon* of Mill's inductive logic, to which it gives rise:

Whatever phenomenon varies in any manner whenever another phenomenon varies in some particular manner, is either a cause or an effect of that phenomenon, or is connected with it through some fact of causation.

Each of Mill's methods admits of a number of variations, depending on the assumptions we make concerning our background conditions. For example, we can sometimes assume that one and only one of our possible causes is the actual cause of X, or we could assume that a conjunction of possible causes is the actual cause of X, or a negation, etc., and which of these assumptions we make will determine the strength of our conclusion, the strongest assumption being that there is one and only one actual cause among the possible causes, and the weakest being that there is some combination of the possible causes which is the actual cause.

1.7 G. H. von Wright

These possibilities of arriving at general laws by exploiting the fact that a singular statement can refute a general statement have been systematically explored by G. H. von Wright [von Wright, 1951]. It turns out that in the absence of substantive assumptions (for example, that there is exactly one simple cause of E among our candidates), the methods that depend on elimination, as do all of Mill's methods, lead to very uninteresting conclusions.

Suppose that C has a sufficient condition among $S_1, \ldots, S_n$. Consider the ith possibility:

$$(\forall x)(S_i(x) \rightarrow E(x)).$$

For the moment, let us assume that these properties are all observable: We can tell whether or not S is present. Thus, we can *observe* something like

$$S_i(a) \wedge \neg E(a).$$

This entails

$$(\exists x)(S_i(x) \wedge \neg E(x)),$$

which is equivalent to

$$\neg(\forall x)(S_i(x) \rightarrow E(x)).$$

Under what conditions will empirical evidence reduce the set of possible conditioning properties—the set of possible sufficient conditions—to a single one? It is shown by von Wright that if we consider all the logical combinations of the conditioning properties $S_1, \ldots, S_n$, it is possible to eliminate all the possible sufficient conditions but one. The one that remains at the end, however, is the logical conjunction of all the conditioning properties, since if S_1 is a sufficient condition, so is $S_1 \wedge S_i$. If we allow negations of properties to be causally efficacious, the single remaining candidate among possible sufficient conditions is a conjunction of n properties or their negations. That is, it will be a property of the form $\pm S_1 \wedge \cdots \wedge \pm S_n$, where "$\pm$" is replaced by a negation sign or nothing in each clause. Similarly, if we seek a necessary condition, we are assured of finding one: It will be an n-fold disjunction of the negated or unnegated initial properties $N_1, \ldots, N_n$.

Of course, having a list of properties from which we are sure we can construct a sufficient condition for the event E is just to be sure that one of a finite number of possible general statements whose antecedent is constructed from the list of properties $S_1, \ldots, S_n$ and their negations, and whose consequent is E, is true. But it can be argued that in any realistic circumstance, we do not need to consider that many cases: Of the enormous number of possible laws that can be constructed, only a much smaller number will be in the least plausible.

On the other hand, to be sure that one of a finite number of possible general statements is true is already to know something that goes beyond what experience can tell us without presupposing induction. Thus, while Mill's methods and their eliminative procedures can certainly illuminate practical scientific inference, they hold no solution to "the problem of induction." To have the complete story of inductive inference calls for something more. This is particularly true if we take account, as we must, of inferences to laws that themselves embody some kind of uncertainty.

There is also no natural way, on the basis of Mill's canons alone, to evaluate the *degree* to which evidence supports a generalization. As we saw in the case of Gumshoe the sleuth, there are degrees of cogency of inference that we should take account of. In the remainder of this book, we shall look at various more formal ways to treat the logic of uncertain inference, as well as at ways of assigning measures to the inferences or the results of inference. Note that these are two different things. To conclude with certainty that the housekeeper is probably the murderess is quite different from concluding with some intermediate degree of cogency that she *is* the murderess. If five people admit that they drew lots to see who would commit the murder, we can be quite certain that the probability is 0.20 that individual A is the murderer. If there is a long and weak chain of argument leading from the evidence to the unqualified conclusion that A is the murderer, then the argument may be looked on with suspicion, though the conclusion is unqualified by anything like "probably." Both kinds of uncertainty will be examined in the sequel.

In the chapters that follow, some of the major contemporary approaches to inductive logic will be presented and investigated. The tools provided by modern mathematical logic have allowed recent theorists to present their positions rather clearly and exactly, and we will look at these theories with an eye to their motivations, strengths, and weaknesses. A variety of interpretations of the meaning of the term "probability" and the inductive logics that arise from those interpretations will be investigated. Various attempts to explain the nature of evidential support will be considered, as will attempts to explain uncertain inference in terms of statistical knowledge. We shall also look at a number of different nonmonotonic logics.

1.8 Bibliographical Notes

A general and quite complete text on logic before 1940 is [Kneale & Kneale, 1962]. A review of some of the philosophical literature concerning induction, covering mainly work since 1950, will be found in [Kyburg, 1983].

1.9 Exercises

(1) Many examples of the use of "inductive" arguments involve troubleshooting of machines. Why doesn't the car start? Well, it might be out of gas, or the spark plugs might be fouled, or ..., or some combination of these. Imagine, or report, a fairly complex example of this sort of reasoning. Identify the necessary and sufficient conditions. What are the instances that did, or could, lead to a conclusion? Which of Mill's methods would you say is being used?

(2) Over the past few days you have no doubt been exposed to many cases of uncertain reasoning: Your friends have wanted to persuade you of some conclusion, or you have wanted to persuade them of some conclusion. The reasons offered would rarely suffice to yield a deductive argument. Sometimes plausible implicit premises could turn the argument into a deductive one, and sometimes not; sometimes the argument is inherently uncertain or inductive. Cite four to six of these instances of uncertain argument, and discuss them with respect to these two possibilities.

(3) Give a possible example of the use of one of Mill's methods to establish an empirical conclusion. Try to make as good a case as you can for it. State the assumptions carefully, give the evidence, and trace the argument. Does the example really work? What can you say about the assumptions? Can they also be established by Mill's methods?

(4) Would it make sense for inductive argument to be "self-supporting" in the sense that argument A might legitimately make use of premise P to establish conclusion C, while argument B might make use of premise C to establish conclusion P? Thus, both C and P could be defensible by inductive argument. Can you think of a case, say in physics, where this is what *appears* to be what is going on? (Well-known philosophers have argued on both sides of this question; either view can be defended.)

(5) Show that if A is a necessary condition of B, then A is a necessary condition of $B \wedge C$.

(6) Show that the following inference is *valid*, i.e., not inductive in the sense that the conclusion goes beyond the premises:
Either C_1 or C_2 or C_3 is a necessary condition of E.
$\neg C_3(a) \wedge E(a)$.
$\neg C_1(b) \wedge E(b)$.
Therefore C_2 is a necessary condition of E.

(7) Describe an incident from your own experience when a piece of reasoning or argument was used to establish the "sufficiency of a condition."

(8) Show that if A is a necessary condition for E and B is a necessary condition for E, then A and B are a necessary condition for E.

(9)　　Show that if $\neg C$ is a sufficient condition for $\neg D$, then C is a necessary condition for D.

(10)　　Being out of gas is a sufficient condition for a car's not starting. Name another sufficient condition. Derive the corresponding *necessary* condition for the car's starting.

(11)　　Show that if C_1 is a sufficient condition for E, and C_2 is a sufficient condition for E, then the disjunction $C_1 \vee C_2$ is a sufficient condition for E.

(12)　　Suppose that both N_1 and N_2 are necessary conditions for the effect E (for example, both heat and oxygen are necessary for combustion). Show that bringing about either the failure of N_1, i.e., $\neg N_1$, or the failure of N_2, i.e., $\neg N_2$, will prevent E.

(13)　　Why is it hard to come up with a complete list of the minimal (shortest) sufficient conditions for a kind of event E, but not so hard, sometimes, to find a necessary condition for E? (Think of fire, or a disease for D.)

Bibliography

[Arnauld & Nicole, 1662] Antoine Arnauld and Pierre Nicole. *Logic or the Art of Thinking*. Cambridge University Press, 1996 (1662).

[Arnauld & Lancelot, 1660] Antoine Arnauld and Claude Lancelot. *The Port–Royal Grammar*. Mouton, The Hague, 1975 (1660).

[Bacon, 1620] Francis Bacon. *The Great Instauration/Novum Organum*. Modern Library, 1955 (1620).

[Bernoulli, 1713] Jacob Bernoulli. *Ars Conjectandi*. 1713.

[Haack, 1996] Susan Haack. *Deviant Logic, Fuzzy Logic*. University of Chicago Press, Chicago, 1996.

[Hacking, 1975] Ian Hacking. *The Emergence of Probability*. Cambridge University Press, 1975.

[Kneale & Kneale, 1962] William Kneale and Martha Kneale. *The Development of Logic*. Oxford University Press, Oxford, 1962.

[Kyburg, 1983] Henry E. Kyburg, Jr. Recent work in inductive logic. In Kenneth G. Lucey and Machan, Tibor R., editors, *Recent Work in Philosophy*, pp. 87–152. Rowman and Allenheld, Totowa, NJ, 1983.

[Laplace, 1951] Pierre Simon Marquis de Laplace. *A Philosophical Essay on Probabilities*. Dover, New York, 1951.

[Mill, 1843] John Stuart Mill. *A System of Logic*. Longmans Green, London, 1949 (1843).

[Skyrms, 1966] Brian Skyrms. *Choice and Chance: An Introduction to Inductive Logic*. Dickenson, Belmont, CA, 1966.

[Stigler, 1986] Stephen M. Stigler. *The History of Statistics*. Harvard University Press, 1986.

[von Wright, 1951] G. H. von Wright. *A Treatise on Induction and Probability*. Chapman and Hall, London, 1951.

2

First Order Logic

2.1 Introduction

Traditionally, logic has been regarded as the science of correct thinking or of making valid inferences. The former characterization of logic has strong psychological overtones—*thinking* is a psychological phenomenon—and few writers today think that logic can be a discipline that can successfully teach its students how to think, let alone how to think correctly. Furthermore, it is not obvious what "correct" thinking is. One can think "politically correct" thoughts without engaging in logic at all. We shall, at least for the moment, be well advised to leave psychology to one side, and focus on the latter characterization of logic: the science of making valid inferences.

To make an inference is to perform an act: It is to *do* something. But logic is not a compendium of exhortations: From "All men are mortal" and "Socrates is a man" do thou infer that Socrates is mortal! To see that this cannot be the case, note that "All men are mortal" has the implication that if Charles is a man, he is mortal, if John is a man, he is mortal, and so on, through the whole list of men, past and present, if not future. Furthermore, it is an implication of "All men are mortal" that if Fido (my dog) is a man, Fido is mortal; if Tabby is a man, Tabby is mortal, etc. And how about inferring "If Jane is a man, Jane is mortal"? As we ordinarily construe the premise, this, too is a valid inference. We cannot follow the exhortation to perform all valid inferences: There are too many, they are too boring, and that, surely, is not what logic is about.

A better way to construe logic is to think of it as a standard to which inferences *ought* to conform. Again, however, an inference is an act, and it is difficult to tell when an act conforms to the canons of logic. If I infer that Jane is mortal from "All men are mortal," have I erred? I can't add the premise that Jane is a man, but I could add the premise that Jane is a woman. Now is the argument valid? It is if I construe "men" in the general premise as the collective "humans"—which no doubt would be intended—and took "humans" as comprising both men and women.

"All men are male chauvinist pigs" does imply that Tom is a male chauvinist pig. But now we would want to reject the broad reading of "men," since Jane is not required to be a male chauvinist pig in virtue of her humanity.

It is easy to see that in ordinary argument premises are left out that are taken to be obvious to all, words like "men" are construed in a way that makes the argument valid, according to a natural principle of generosity, and in general there is a distance between the words that people use and the objects (terms, predicates, functions) with which formal logic is concerned. In favorable cases, the translation from the language of the argument—ordinary English, for example—to the language of our formal logic is not difficult. In unfavorable cases, there can be as much difference of opinion about the translation as there is about the question of validity.

Nevertheless, there are some disciplines—mathematics, philosophy, statistics, computer science, some parts of physics—in which the problem of translation is not very difficult (it is easy to agree that "+" is a function and ">" is a relation, and easy to agree on their meanings in a given mathematical or quantitative context). In these disciplines, the role of logic seems relatively clear. (Indeed, it was in the context of mathematical argument that modern logic was developed.) One characteristic of these disciplines that makes their arguments easily open to evaluation from the point of view of formal logic is their emphatic concern with making their terms clear. The problem of translation into a canonical logical symbolism is not a great one in certain parts of these disciplines. The problem in these disciplines centers on the arguments themselves. What is an argument, and how does it differ from an inference?

An inference is an act. We infer a conclusion C from a set of premises $P_1, \ldots, P_n$. An argument is a collection of sentences that is intended to trace and legitimize an inference. An argument is basically social, rather than psychological: It is the means by which I attempt to persuade you that the inference of the conclusion from the premises is legitimate. ("You," of course, may be my other or more skeptical self.) More or less detail may be given, as appropriate to the context. Certain steps in an argument are common enough to have been given their own names: *modus ponens*, for example, is an argument of the form "If S, then T; S; therefore T." From here it is a natural step to want to characterize legitimate arguments in general; and thus we develop formal logic.

Everything we have said so far, except for the examples and the existence of a formal logic, could as well apply to uncertain inference or inductive inference as well as to deductive inference. Deductive arguments, of the kinds found in mathematics, are the best understood, and are the focus of most courses in formal logic. Inductive arguments, arguments that confer partial support on their conclusions, have received less formal attention, though their importance has been widely recognized, both in philosophy for centuries and in artificial intelligence for decades.

Formal logic is a systematization and codification of all the legitimate arguments of a certain sort. Thus, the system of formal logic you are assumed to have studied, first order logic, provides a complete characterization of valid arguments involving predicates and relations, functions and names, making use of quantification over a single class of entities. Left out are modal operators ("It is necessary that …"), epistemic operators ("John knows that …"), and the like.

The logic is specified in the abstract—a vocabulary and formation rules are given that characterize the sensible (or meaningful) sentences of the logic; axioms and rules of inference are given that characterize the set of theorems of the logic. A central notion in first order logic is that of *proof*. A proof is a mechanized form of argument. It is a sequence of formulas of the logic, each of which is either (a) an axiom, or (b) inferrable from previous formulas in the sequence by one of the rules of inference of the logic. A theorem is a sentence that may appear as the last line of a proof.

This idea can be extended to that of proof from premises: A proof of a conclusion C from premises $P_1, \ldots, P_n$ is a sequence of formulas of the logic, each of which is either (a) a theorem of the logic, or (b) one of the premises, or (c) inferrable from previous formulas in the sequence by one of the rules of inference of the logic.

This is a long way from the psychological process of inferring. What is the relation between inference and logic? We can make a connection through *argument*. Even psychologically, an inference may not take place in a single leap of intuition. One may think through the consequences of a set of premises or assumptions in a number of steps before arriving at a conclusion. If one is persuading someone else of the truth of the conclusion, given the premises, one generally must provide intermediate steps. Arguments admit of many degrees of detail, according to the size of the intervening steps. The ultimately most detailed form of argument, in a formal language, is exactly a *proof*, in the sense just spelled out: a sequence of formulas, each of which is a premise, an accepted logical truth, or the result of applying one of the rules of inference of the underlying logic.

No one, except perhaps for illustrative purposes in a logic course, presents arguments as proofs. Rather one seeks to persuade the reader or the listener that a proof exists. Formal logic provides a general framework within which we can systematically refine our arguments toward the ultimate standard: *proof* in our formal logic. Given a set of premises in a formal language—arriving at such a set of premises, as we noted, can be problematic in the sense that formalization is not a mechanical process—the claim that a sequence of formulas is a proof of a conclusion from those premises is mechanically decidable. All that is needed is the ability to decide when two patterns are the same and when they are different. If three lines have the forms S, $S \to T$, and T, then the third is a consequence of the first two, whatever S and T may be. In the ordinary course of events, we achieve agreement about whether or not there is a proof long before we actually produce one; at that point it is generally no longer of interest to exhibit a proof. But the possibility of exhibiting a proof is what renders our arguments constructive and useful.

Other claims regarding the virtues of logic have been made, often with some legitimacy. It is perhaps true that a person trained in logic will be able to see more consequences of a set of premises more quickly than a person lacking that training. It is surely true that training in logic will enhance one's ability to see flaws in the arguments of others. It has also been argued that logic is a discipline that "trains the mind" somewhat in the manner of weight lifting or running: a discipline that is probably unpleasant in itself, but that has beneficial effects in the long run. This is a view of logic that is hard to either attack or defend. Another defense of logic rests on its inherent interest: it is interesting in itself, and provides the student with objects of significant beauty and elegance. But not all students find the aesthetic qualities of logical systems sufficient motivation for their study.

It is assumed that the users of this text will be familiar with first order logic. This will usually mean having had a course in first order logic, though some students may have studied logic on their own. Even those who have had a course, however, may have had quite different courses, for logic is taught in very different ways at different institutions, and even at the same institution by different instructors. This chapter therefore contains a review of some of the useful material of a first order logic course. Although the construction of formal proofs is an appropriate emphasis for a first order logic course, our basic concerns in this chapter are more general, and we shall have little concern with the construction of formal proofs.

Our approach to uncertain inference will be to explore the possibilities of a logic of uncertain inference in the sense characterized earlier: a canon of standards against which the cogency of arguments may be measured, whether or not those arguments purport to lend "certainty" to their conclusions. Because classical first order logic is our standard and our inspiration, and since we are proposing to take it for granted, the present chapter will provide a review of some of the substance of that logic. We will at the same time make use of this chapter to remind the student of some useful facts *about* logic.

2.2 Syntax

Informally, we may say that a good or valid inference is one that leads to true conclusions from true premises. It is a form of inference that preserves truth, if it is there to start with. We leave open the question of what it leads to when the premises on which it is based are not true. We leave open, also, the question of what are to count as premises. And, for the time being, we leave open the question of whether—and if so, how—to weaken this standard of validity.

It is surely a good thing if one's inferences preserve truth. For example, from the premises "If Jane's net income exceeds $32,000, and she is a single parent, then Jane must pay a tax of 25% on the difference between her income and $32,000," and "Jane's net income exceeds $32,000, and Jane is a single parent," we (that is, anyone) may infer "Jane must pay a tax of 25% on the difference between her income and $32,000."

Actually, this is a pretty boring inference. To find something interesting, we must consider premises that are quite complicated. Thus, from "Every dog has fleas," "Some fleas have no dog," and "No flea has more than one dog," we may infer that there are more fleas than dogs. More importantly, if we take the axioms of a mathematical system (abstract algebra, geometry, number theory—even probability theory or set theory) as premises, and the theorems of that theory as conclusions, the theorems may be inferred from the premises. Correct inference in mathematics is valid inference. It is legitimized by the existence of formal *proofs* and often presented by *informal* arguments designed to convince us of the existence of the formal proofs.

But now we must face the question of what valid inference is, and how to tell a valid from an invalid inference. In the examples of valid inferences we just looked at, the premises and conclusions were given in ordinary English. We needed no arguments, since the inferences were so direct. In mathematics, arguments are needed to buttress the inferences being made. Mathematical arguments are generally given in a more formal and structured language than English; they make use of such predicates as "$=$", "is prime", "$\geq$", "$\in$", and so on. Legal arguments fall somewhere in between, involving a certain amount of technical vocabulary as well as a certain amount of ordinary English. In each case, though, there is a core of "logic" that is the same, and the concept of validity is the same.

We begin by characterizing a formal language within which arguments and proofs may be presented. We will be talking *about* this language; the language we talk about is called the *object language*. For example, in a formal language for representing number theory, we would have the mathematical predicates mentioned in the last paragraph; in a formal language for discussing contracts, we would need "... is a contract," "... is bound by ...," and so on.

The language we use for talking about the object language is called a *metalanguage*. This language needs (among other things) terminology for talking about the expressions—the terms, the predicates, the sentences, the formulas, etc.—of the object language. As we shall see almost immediately, the metalanguage is also often taken to include the object language as a sublanguage, as well as to have the ability to talk about the object language. The metalanguage gives us the means for talking about proofs in the object language: the characterization of proof that we gave in the last section was in our informal metalanguage, English. For example, one might say:

A proof of a sentence is a sequence of formulas, to which the sentence belongs, each of which is an axiom or inferrable from earlier formulas in the sequence by one of the official rules of inference.

There is a canonical procedure for characterizing the language of a formal system. The language consists of a vocabulary, formation rules, axioms, and rules of inference.

The *vocabulary* consists of predicates (corresponding to the phrases "... is red" and "... is even" in English, for example), relation expressions (corresponding to "... is brother of ..." and "... lies between ... and ..."), operations (such as "the sum of ... and ..." and "the mother of ..."), names (such as "John" and "25"), and usually an infinite set of variables (for example, $x_1, x_2, \ldots x_n, \ldots$). The infinity of the set of variables is no problem, since we can give a recursive characterization of them; for example, we can say "x is a variable, and x followed by any number of primes (such as x''') is a variable." Note that this gives us a mechanical procedure for discovering if an expression is a variable. The answer is "yes" if it is "x" or is a variable followed by a prime. Now we have the shorter problem of determining whether what is followed by the prime is a variable.

We may introduce an infinite number of constants (for example, the natural numbers, or the real numbers) as well as an infinite number of predicates ("the length of ... is r," where r is a rational constant), provided we can do it in a finitary way.

It should be observed that though we have allowed for very complex languages, for many purposes quite simple languages will serve. Thus, if we want to characterize the family relationships of a primitive society, we need a collection of names (code numbers will do); one predicate, "female"; and one relation, "sibling of"; and one function, "parent of." For convenience, we may then introduce relations like "maternal uncle of," but these terms can be defined away.

In some cases we will have several *sorts* of variables in the language: thus we might use Greek letters for real number variables, and Latin letters for variables that range over physical objects. If we do so, then it may be that there are correspondingly only some predicates and relations that make sense when applied to individuals and variables of a given sort. Thus, "x is heavy" makes no sense, in ordinary talk, applied to "25", and "α is even" doesn't make sense if α is taken to vary over ponderable bodies. Note that it is quite possible to have sensible expressions that combine both sorts: Thus "the height of John in inches ≥ 72" could correspond to a mixed sentence that makes perfectly good sense, because the function "the height of ... in inches" could be a function having a domain consisting of people, and a range in the reals.

The formation rules tell us what a well-formed formula or well-formed term of the calculus is. These rules must be (in ordinary cases) recursive, because we want to allow for an unlimited number of sentences. Furthermore, since we want to introduce statements by *kinds* rather than one at a time, we use the device (due to Quine) of quasiquotation. Thus, to form a general descriptive name for conjunctions we write "⌜ $\phi \wedge \psi$ ⌝", by which we mean the set of formulas we could (at length) describe as: the formula ϕ, followed by the symbol "$\wedge$", followed by the formula ψ. Note that "$\phi \wedge \psi$" denotes something quite different: the Greek letter "ϕ" followed by "$\wedge$" followed by the Greek letter "ψ". But we want to say something about formulas of a certain sort, not something about Greek letters. Traditionally, the formation rules run as follows:

(1) An n-place predicate followed by n terms is a well-formed formula.
(2) If ϕ and ψ are well-formed formulas and α is a variable, then ⌜$\neg\phi$⌝ (negation),
     ⌜$\phi \wedge \psi$⌝ (conjunction), ⌜$\phi \vee \psi$⌝ (disjunction), ⌜$\forall\alpha\phi$⌝ (universal quantification),
     and ⌜$\exists\alpha\phi$⌝ (existential quantification) are well-formed formulas. The conditional
     and the biconditional are, similarly, sentential connectives. If τ and σ are terms,
     ⌜$\tau = \sigma$⌝ is a well-formed formula. (Later we will both impose new restrictions
     and allow new forms; these are the traditional ones.)
(3) A variable is a term.
(4) A 0-place function is a term (specifically, a name).
(5) If σ is a k-place function, and $\tau_1, \ldots, \tau_k$ are terms, then ⌜$\sigma(\tau_1, \ldots, \tau_k)$⌝ is a term.

Notice that we have to know what a term is before we understand what a formula is. It is also the case that we may need to know what a formula is in order to know what a term is. This is not as awkward as it sounds. One case in which we need to understand formulas before we understand terms is that of definite descriptions. A definite description is an expression of the form ⌜$(\imath x)\phi$⌝, for example:

$(\imath x)(\text{Husband}(x, \text{Jane}))$

The inverted Greek iota, $\imath$, is a variable binding operator: it binds the two occurrences of x in the expression displayed. The expression thus created is a term denoting the unique object in the domain of quantification—the set of objects that are the *values* of the variables—who is the husband of Jane, provided there is exactly one. If there is none (if Jane is unmarried), or if there is more than one (if Jane is polyandrous), then we shall take the displayed expression to denote the empty set. Different treatments of nondenoting definite descriptions are possible, but this treatment, due to Bertrand Russell, is relatively straightforward. It might be thought that taking a definite description to be meaningless when it applies to less than or more than a single object would be more natural, but this approach carries with it the difficulty that we must know a *fact* about the world before we can tell whether or not an expression is meaningful. It seems more desirable to separate questions of meaning and questions of truth. Note that what follows the variable binding operator $(\imath x)$ must be a well-formed formula for the expression to be a well-formed term.

There is no vicious circularity here, since a definite description must be longer than the well-formed formula that it contains. Thus, if ϕ is a well-formed formula, any term that it contains must be shorter than it; and if some term in ϕ is a definite description, the formula that it contains must be shorter than ϕ. The shortest formulas don't contain descriptions, and the shortest terms don't contain formulas.

Thus, in English the complex but meaningful sentence, "The shortest man in town is married to the tallest woman in town" is meaningful (on the Russellian construction) precisely because "the shortest man in town" corresponds to a definite description: the unique x such that x is a man in town, and for every y, if y is a man in town other than x, then x is shorter than y; and "the tallest woman in town" can be parsed similarly. Thus, the sentence is *meaningful*. Is it true? That depends, of course. If there are two or more men in town who are shorter than all the other men, the sentence is false because there *is* no shortest man, similarly for the tallest woman. But even if there is a shortest man and a tallest woman, they might not be married—they might not even know one another.

In addition to definite descriptions, we will assume that our languages contain the machinery to refer to set-theoretical objects. Thus, "$\{x : \text{Integer}(x) \land 3 < x < 8\}$" will refer to the set of integers between 3 and 8, or $\{4, 5, 6, 7\}$; $\{\langle x, y \rangle : x$ and y are married$\}$ is the set of married couples. The expression "$\{x : \ldots\}$" is, like the description operator, a variable binding operator. The variable or variables that it binds are those preceding the colon; as in the case of the description operator, for the term to be well formed, the expression following the colon must be a well-formed formula.

Because we will not be concerned with sophisticated issues in set theory, we will simply suppose that the rudiments of manipulating sets are familiar.

The *axioms* of logic consist of a set of well-formed formulas that can be recognized mechanically. In some systems they are taken to be finite: thus a system for propositional logic might include variables $X, Y, Z, \ldots$ for propositions, and take the following four formulas as axioms:

(1) $X \lor X \to X$,
(2) $X \to X \lor Y$,
(3) $X \lor Y \to Y \lor X$,
(4) $(X \to Y) \to (Z \lor X \to Z \lor Y)$.

These four axioms, taken from Hilbert and Ackermann [Hilbert & Ackermann, 1950], will yield the calculus of propositional logic only if we have available a rule of substitution as well as *modus ponens* for making inferences.

An alternative approach, adopted by W. V. O. Quine in *Mathematical Logic* [Quine, 1951], is to introduce axioms in infinite groups. Thus, Quine takes the closure of any truth-functional tautology to be an axiom (axiom group *100), where the closure of a formula consists of the universal quantification of all its free variables. It is, in fact, nearly as intuitive and straightforward to verify that a formula is a truth-functional tautology as it is to verify that it has the form $X \to X \lor Y$.

One advantage of Quine's approach is that one need not have a rule of substitution among the inference rules. Quine's system Mathematical Logic, for example, has only one rule of inference: *modus ponens*. In this system, the single sentential connective (neither–nor) is taken as primitive, together with the universal quantifier. The usual connectives and the existential quantifier are introduced as abbreviations for

expressions that are officially to be thought of as spelled out in primitive notation. This allows our basic system to have a very simple (if impenetrable) structure, and simplifies the analysis of proof theory—that part of logic concerned with
provability.

2.3   Semantics

In addition to being concerned with what statements are provable in a system, we
may also be concerned with what statements are *true*. To speak of truth requires
more than the syntactical machinery we have introduced so far. It requires semantics, which in turn makes use of *models* of the language. The concept of a model
allows us to define truth, and derivatively to characterize *validity* and *invalidity*. Informally we have characterized validity in terms of the preservation of truth; model
theory gives us a way to do this formally, as well as being, as we shall see, useful in
other ways.

A model (in the simplest case) for a language with a given vocabulary consists of a
domain of objects $\mathcal{D}$ over which the variables range, and an interpretation function $\mathcal{I}$
that *interprets* the expressions of the language as entities in $\mathcal{D}$ or as entities constructed
from entities in $\mathcal{D}$.

(1) The interpretation function $\mathcal{I}$ assigns objects in the domain $\mathcal{D}$ to names in the
   language. To each proper name ("John", "2", "Tweety", "The Wabash
   Cannonball", "Venus", etc.) the function $\mathcal{I}$ must assign exactly one item in $\mathcal{D}$.
   (Sometimes we may be concerned with a special class of models that satisfy the
   unique names constraint. In such models, $\mathcal{I}$ must assign distinct objects to distinct
   names.)

(2) To each one-place predicate in the language ("... is red", "... is even", "... is
   mortal", etc.) $\mathcal{I}$ must assign a subset of $\mathcal{D}$. Allowable assignments include the
   empty set and $\mathcal{D}$ itself. Similarly, to each k-place predicate ("... loves ...", "... is
   longer than ...", "... lies between ... and ...", "... is the greatest common divisor
   of ... and ...", etc.), $\mathcal{I}$ assigns a subset of $\mathcal{D}^k = \mathcal{D} \times \cdots \times \mathcal{D} = \{\langle x_1, x_2, \ldots x_k \rangle :
   x_i \in \mathcal{D}\}$, including the possibility of the empty set $\emptyset$, and $\mathcal{D}^k$ itself.

(3) To each k-place function symbol or operator ("... + ...", "mother of ...", etc.) $\mathcal{I}$
   assigns a function from $\mathcal{D}^k$ to $\mathcal{D}$.

As an example, let us consider a language with the following simple vocabulary, in addition to the conventional logical machinery of first order logic. There
are two constants: "Sally" and "James". There are two one-place predicates: "has
red hair" and "is female." There is one function symbol: "mother of." There is
one relation, " ... likes ...". (Note that we use quotation to form the *names* in
the metalanguage of the expressions in the object language.) A natural empirical domain consists of a set of people, say six of them, $d_1, \ldots, d_6$. (Note that we
do not use quotation: d_1 is a person in our domain, not the name of the person.)
Here is an example of an interpretation function: Let the interpretation of "Sally"
be d_1 and of "James" d_2. Let the interpretation of the predicate "... has red hair"
be the set consisting of d_6 alone: $\{d_6\}$. Let the interpretation of "... is female" be
$\{d_1, d_4, d_5\}$. Now "mother of" must be a function from our domain to our domain.
That is, given any object in the domain, the interpretation of "mother of," applied

to that object, must yield an object in the domain. We can give this function by a table:

Object	Mother of object
d_1	d_4
d_2	d_4
d_3	d_4
d_4	d_5
d_5	d_1
d_6	d_1

Thus, d_4 is the mother of d_1, d_2, and d_3; d_5 is the mother of d_4; etc. Note that every object in our domain has a mother. Because there are only a finite number of objects in the domain, that entails that there is a closed cycle of motherhood: somebody is the mother of somebody who is the mother of ... who is the mother of the original person. In this case, d_1 is the mother of d_5, who is the mother of d_4, who is the mother of d_1. If we want it to be the case that "mother of" is a noncyclical function, and yet that our domain is finite, we can accomplish this by choosing an element of our domain to be the *universal mother*, including mother of herself. A natural candidate in this example is d_1: put $\langle d_1, d_1 \rangle$ in place of $\langle d_1, d_4 \rangle$.

Finally, we need an interpretation for "likes." This may be any set of pairs: Let's say $\{\langle d_1, d_3 \rangle, \langle d_1, d_6 \rangle\}$. We have described a world—a small one, it is true—in complete detail, or rather in as much detail as is permitted by our language.

We can now define truth for quantifier-free sentences of the language. A sentence is a statement with no free variables; so a quantifier-free sentence is a formula with neither free nor bound variables. In our example, we can see that "James has red hair" is false, because the only object that has red hair is d_6 and "James" is being interpreted as the object d_2.

We want truth to be defined for all sentences, and to accomplish this we must have truth defined also for sentences containing free and bound variables. We consider first free variables: e.g., the variable x in the formula "x has red hair."

The basic auxiliary notion we need is that of a variable assignment. A *variable assignment* is a function V that assigns to each of our infinite number of variables an object in the domain $\mathcal{D}$. Because in our example $\mathcal{D}$ only contains six objects, the function V must repeat itself a lot; each object may be the assignee of an infinite number of variables, and at least one object must be.

We define the central semantical notion of the *satisfaction* of an open atomic formula $P(t_1, \ldots, t_k)$ (a formula consisting of a k-place predicate followed by k terms) by employing the same procedure we employed in defining the syntactical notion of a well-formed formula. Given an interpretation function $\mathcal{I}$, and a variable assignment V, the k-tuple assigned to the terms $t_1, \ldots, t_k$ is a sequence of k objects in $\mathcal{D}$. If one of these terms is a name, it is the object assigned by $\mathcal{I}$ to that name; if it is a definite description, it is the corresponding object, if there is a unique one; otherwise it is the empty set. If it is a variable, the object is that assigned to that variable under V. We say that a formula is *satisfied* by the variable assignment function V, just in case the k-tuple assigned to the k terms $t_1, \ldots, t_k$ (under both $\mathcal{I}$ and V) belongs to the set of k-tuples picked out by the interpretation of P. In turn, of course, the object in $\mathcal{D}$ assigned to a particular term having the form of a function or description may depend on the formulas that occur in that function or description. But in view of the fact that the formulas appearing in the terms must be shorter than the formula whose satisfaction we are concerned with, this process comes to a halt. It will halt when the terms that require interpretation are either names (whose interpretation is given by $\mathcal{I}$) or variables (to which objects are assigned by $\mathcal{V}$).

To continue our example, the formula "likes$(x,(\imath y)(y$ has red hair))" will be true just in case the object that $\mathcal{V}$ assigns to x, paired with the object assigned to "$(\imath y)(y$ has red hair)", is in the interpretation of "likes". The object assigned to "$(\imath y)(y$ has red hair)" is d_6 according to the interpretation of "... has red hair." The initial sentence will be true just in case $\mathcal{V}$ assigns d_1 to the variable x.

Given a definition of satisfaction for atomic formulas, the rest of the process of defining truth is simple, except for quantification. An interpretation $\mathcal{I}$ and a variable assignment $\mathcal{V}$ satisfy a formula of the form $\ulcorner \neg\phi \urcorner$ if and only if $\mathcal{I}$ and $\mathcal{V}$ fail to satisfy ϕ; they satisfy $\ulcorner \phi \vee \psi \urcorner$ if and only if they satisfy one formula or the other; and so on. Of course, if there are no free variables in ϕ, an interpretation $\mathcal{I}$ and a variable assignment $\mathcal{V}$ satisfy ϕ if and only if ϕ is already true under $\mathcal{I}$.

The new case concerns quantification: if the variable α is free in the formula ϕ, then $\ulcorner \forall\alpha\phi \urcorner$ is satisfied by a variable assignment $\mathcal{V}$ and interpretation $\mathcal{I}$ if ϕ is satisfied by every variable assignment that is like $\mathcal{V}$ except possibly for the variable assignment made to the variable α.

We have thus generally defined truth in a model of our language, both for sentences containing free variables and for closed sentences, in terms of a domain ($\mathcal{D}$), an interpretation function ($\mathcal{I}$), and a variable assignment ($\mathcal{V}$). The ordered pair $\langle \mathcal{D}, \mathcal{I} \rangle$ is a model. Another structure that turns out to be of interest is the set of all models of our language when we fix the domain. One name for such a set of models is a "set of possible worlds." If we have two objects in our domain, $\mathcal{D} = \{d_1, d_2\}$, and one predicate, "... is red," in our language, then there are four possible worlds: that in which no objects are red, that in which only d_1 is red, that in which only d_2 is red, and that in which both objects are red.

If we disregard the internal structure of the sentences of the language, we can identify a possible world with a truth assignment to the atomic sentences of the language. Suppose our language consists of four atomic propositions, p_1, p_2, p_3, and p_4, and their truth-functional combinations. There are an infinite number of sentences in the language: $p_1 \wedge p_2$, $p_1 \wedge p_2 \wedge p_1$, etc. But there are only a finite number of possibilities, 16, that we can distinguish. These correspond to the truth value assignments we can make to the atomic propositions: these are the 16 "possible worlds" we can distinguish in this language.

2.4   W. V. O. Quine's Mathematical Logic

The formalization of first order logic we will look at is that given by Quine in Mathematical Logic. This is not an easily accessible system, and surely not one in which proofs are easy to construct. On the other hand, it has a very streamlined axiomatization, and a single simple rule of inference (*modus ponens*). It is thus a good system for talking *about* logic. It is this that is of primary importance to us, since we are going to be looking at proposals regarding uncertain inference as proposals for an extension of logic, as well as proposals for a formalization of uncertainty within logic.

The axioms for first order logic given by Quine make use of the notion of the *closure* of a formula. The closure of a formula is the sentence obtained by prefixing the formula with universal quantifiers, in alphabetical order (i.e., binding the alphabetically earliest free variable first), for each free variable in the formula. Thus, "has-red-hair(James)" is its own closure, "$\forall x$(likes(x, mother-of(x)))" is the closure of "likes(x, (mother-of(x)))," etc. Quine's four groups of axioms are:

*100 If ϕ is tautologous, its closure is a theorem.

*101 The closure of $\ulcorner(\forall\alpha)(\phi \to \psi) \to ((\forall\alpha)\phi \to (\forall\alpha)\psi)\urcorner$ is a theorem.

*102 If α is not free in ϕ, the closure of $\ulcorner\phi \to (\forall\alpha)\phi\urcorner$ is a theorem.

*103 If ϕ' is like ϕ except for containing free occurrences of α' wherever ϕ contains free occurrences of α, then the closure of $\ulcorner(\forall\alpha)\phi \to \phi'\urcorner$ is a theorem.

These four groups of axioms, together with a single rule of inference, *modus ponens*, suffice to yield all the logical truths of the first order predicate calculus. *Modus ponens* has the form:

*104 If ϕ is a theorem, and $\ulcorner\phi \to \psi\urcorner$ is a theorem, then ψ is a theorem.

Three further groups of axioms, involving membership, provide for the rest of set theory (and thus mathematics), the logic of identity, definite descriptions, and the like. We shall not display these axioms, since in Quine's system abstraction and identity can be handled independently of the axioms of membership, although they do involve the membership relation "$\in$". We shall, as the occasion demands, assume that we have available whatever mathematics and set theory we require.

No formulas of the system are displayed in the axioms. What are displayed are schemas. The closure of any formula fitting one of the schemas (or patterns) displayed as axioms is a theorem. Thus, the closure of "likes(x, y) → likes(x, y)" is a theorem under *100; the closure of "$\forall x$(likes(James, x)) → likes(James, y)" is a theorem under *103. The development in *Mathematical Logic* is given mainly in terms of metatheorems, that is, it is shown that the closure of any formula of a certain form is a theorem. This is done by establishing the existence of a pattern that, in the case of any formula of the form in question, would constitute a formal proof.

For example (the translation of) "If all animals have hearts, then every animal in this cage has a heart" is a theorem of our language. A proof of it is provided by the following three sentences (with the obvious interpretations of the letters):

(1) $(\forall x)((Ax \to Hx) \to (Ax \wedge Cx \to Hx)) \to$
     $\quad ((\forall x)(Ax \to Hx) \to (\forall x)(Ax \wedge Cx \to Hx)).$ \hfill *101

(2) $(\forall x)((Ax \to Hx) \to (Ax \wedge Cx \to Hx)).$ \hfill *100

(3) $(\forall x)(Ax \to Hx) \to (\forall x)(Ax \wedge Cx \to Hx).$ \hfill *modus ponens*

The system just outlined is sound in the sense that every theorem of the system is true in every model of the language: i.e., for every interpretation in every nonempty domain.

We can show that every sentence in a given language that is the closure of a formula fitting the schemata of the axioms is true in every model. For example, "$\forall x$(has-red-hair(x) $\vee \neg$ has-red-hair(x))" is true in a model, under a variable assignment $\mathcal{V}$, just in case "(has-red-hair(x) $\vee \neg$ has-red-hair(x))" is satisfied by every variable assignment that is like V except possibly for the assignment made to x. But whatever member of D is assigned to x, either it will be in the interpretation of "has-red-hair," making the left hand side of the disjunction true, or it will not be in the interpretation of "has-red-hair," making the right hand side of the disjunction true. Thus, the originally displayed sentence is true in every model—i.e., in every domain under every interpretation.

It is also clear that the property of being true in every model is inherited under *modus ponens*. If ϕ and $\ulcorner \phi \to \psi \urcorner$ are each true in every model of the language, then ψ must be true in every model.

Here is a more complete argument for soundness.

*100 The closure of a tautology must be true in every model, since its truth does not depend on whether a variable assignment satisfies any particular component of the tautology.

*101 The closure of $\ulcorner (\forall \alpha)(\phi \to \psi) \to ((\forall \alpha)\phi \to (\forall \alpha)\psi) \urcorner$ must be true in every model, since for it to be false the antecedent must be true and the consequent false; for the consequent to be false $\ulcorner (\forall \alpha)\phi \urcorner$ must be true and yet, for some variable assignment V, ψ must be false. This very variable assignment will fail to satisfy $\ulcorner \phi \to \psi \urcorner$.

*102 If α is not free in ϕ, the closure of $\ulcorner \phi \to (\forall \alpha)\phi \urcorner$ is true in every model, since just the same variable assignments will satisfy ϕ as satisfy $\ulcorner (\forall \alpha)\phi \urcorner$.

*103 If ϕ' is like ϕ except for containing free occurrences of α' wherever ϕ contains free occurrences of α, then the closure of $\ulcorner (\forall \alpha)\phi \to \phi' \urcorner$ is true in every model, because for the antecedent to be satisfied by a variable assignment V, ϕ must be satisfied by every variable assignment that assigns any object to the variable α, including what V assigns to α', and keeps all other assignments the same.

The axioms of each kind are therefore true in each model. We now show that this property of being true in every model is preserved under the only rule of inference of this system, *modus ponens*.

Specifically (since there is only a single rule of inference), we must show that if ϕ is true in every model, and $\ulcorner \phi \to \psi \urcorner$ is true in every model, then ψ is true in every model. Suppose that the premises are true in every model, but that there is some model in which the conclusion ψ is false. This is just to say that there is some variable assignment in some model that fails to satisfy ψ. Because every variable assignment in every model satisfies ϕ, that particular variable assignment, in that particular model, will fail to satisfy $\ulcorner \phi \to \psi \urcorner$, contrary to our supposition that $\ulcorner \phi \to \psi \urcorner$ is true in every model.

We conclude that the theorems of this system are true in every model—i.e., every nonempty domain D and every interpretation $\mathcal{I}$—of the language. This establishes the *soundness* of the system.

What is of even more interest is that quantification theory is *complete* as well as sound. Every sentence that is true in every (nonempty) model of the language can be proved as a theorem. We will not prove this here (though proofs can be found in many books on first order logic).

It is also worth observing that in general there is no mechanical procedure for deciding whether or not a sentence has this property of being true in every model. We must be fortunate enough to find a proof. On the other hand, we should not make too much of this undecidability in principle: very broad classes of sentences admit of a decision procedure.[1]

---

[1] A compendium of general constraints that yield decidable systems is provided by [Ackermann, 1962].

2.5 Arguments from Premises

From our point of view, arguments from premises are more interesting than proofs of theorems. The examples with which we started are arguments from premises: Given the premises $P_1, \ldots, P_n$, can we conclude C? We can if there is a proof of C from the premises. The connection between arguments from premises and theorems is provided by the *deduction theorem*. Let us call an argument to the conclusion C from premises $P_1, \ldots, P_n$ *valid* if the conclusion is true in every model in which the premises are true. (Note that calling an argument valid is different from calling a formula valid.) As a special case, of course, if we have in mind a collection of possible worlds, the conclusion of a valid argument will be true in every world in which the premises are true.

Theorem 2.1 (The Deduction Theorem). *An argument is valid if and only if the conditional whose antecedent is the conjunction of the premises and whose consequent is the conclusion of the argument is a theorem.*

> *Proof:* If the conditional $P_1 \wedge \cdots \wedge P_n \rightarrow C$ is true in all models, then clearly C must be true in every model in which the antecedent $P_1 \wedge \cdots \wedge P_n$ is true. The antecedent is true in every model in which each of the premises $P_1, \ldots, P_n$ is true. The conclusion is therefore true in every model in which each of the premises is true.
>
> Going the other way, to say that the conclusion of the argument is true in every model in which the premises are true is to say that every model is such that the conclusion of the argument is true if the conjunction of the premises is true, or, equivalently, that the conditional whose antecedent is the conjunction of the premises and whose consequent is the conclusion is true in every model; and this, in view of the completeness of quantification theory, is to say that the corresponding conditional is a theorem. ∎

We call a system *argument sound* if the conclusions of arguments it warrants are true in every model in which the premises are true. It is *argument complete* if, whenever a conclusion is true in every model in which the premises are true, it is also the case that the conclusion is derivable from the premises by means of the rules of inference of the system.

For example, in the present language, we might want to show that it is legitimate to infer that John's mother likes everyone whom he likes, in virtue of the fact that mothers all like their offsprings, and John's mother is a mother and likes anyone whom anyone she likes likes. In our official language, the premises are:

(1) $(\forall x)(M(x) \rightarrow (\forall y)(C(y, x) \rightarrow L(x, y)))$,

(2) $M(m(j)) \wedge C(j, m(j)) \wedge (\forall x)[(\exists y)(L(m(j), y) \wedge L(y, x)) \rightarrow L(m(j), x)]$,

and the conclusion is

(3) $(\forall x)(L(j, x) \rightarrow L(m(j), x))$.

One way to show that this inference is valid is to form the corresponding conditional and to construct a formal proof of the conditional—that is, a sequence of formulas in the language, each of which is an axiom or is obtained by *modus ponens* from earlier formulas in the sequence, and that culminates in the conditional in question.

Another way to show that this inference is valid is to provide an argument to the effect that a proof is possible. For this, we may take advantage of the deduction

theorem and exhibit a sequence of formulas, each of which is a premise or a theorem
of the logical system, and that culminates in statement 3. The following formulas can
clearly be expanded into such a sequence:

(1) $M(m(j)) \rightarrow (\forall y)(C(y, m(j)) \rightarrow L(m(j), y))$
First premise, instantiation.

(2) $M(m(j))$
Second premise, simplification.

(3) $(\forall y)(C(y, m(j)) \rightarrow L(m(j), y))$
Modus ponens, lines 1 and 2.

(4) $C(j, m(j)) \rightarrow L(m(j), j)$
Line 3, instantiation.

(5) $C(j, m(j))$
Second premise, simplification.

(6) $L(m(j), j)$
Lines 4 and 5, *modus ponens*.

(7) $(\forall x)[(\exists y)(L(m(j), y) \wedge L(y, x)) \rightarrow L(m(j), x)]$
Second premise, simplification.

(8) $(\exists y)(L(m(j), y) \wedge L(y, x)) \rightarrow L(m(j), x)$
Line 7, instantiation.

(9) $L(j, x) \rightarrow L(m(j), x)$
Lines 6 and 8, existential generalization, and *100 (tautologies).

(10) $(\forall x)(L(j, x) \rightarrow L(m(j), x))$
The deduction theorem says that the closure of the conditional whose antecedent is
the conjunction of the premises and whose consequent is line 9 is a theorem. The
only free variable in that conditional is x, which occurs free only in the
consequent. *101, *102, and an application of the deduction theorem yield line 10,
the conclusion.

This sequence of formulas is not a proof from premises. It should be thought of,
rather, as a recipe for producing such a proof. The annotations are intended to suggest
(a) which metatheorem(s) of *Mathematical Logic* would be relevant, and (b) how they
are related to the formulas in question. This general idea is assumed to be familiar to
readers of this text, whatever specific system of logic they have studied.

We will suppose that the logic of uncertain inference includes the standard first
order deductive logic we have outlined. There are, however, various ways that first
order logic can be expanded. One direction in which it can be expanded is to develop
a *modal logic* to take account of modal locutions, such as "S is possible" and "S is
necessary." Another way in which first order logic can be expanded is in the direction
of *epistemic logic*: a logic in which "it is known that S" can be expressed. It turns out,
as we shall see in our treatment of nonmonotonic logic, that some of the structure of
modal logic may be quite directly reflected in or made use of in nonmonotonic logic.
That will be a more useful place in which to present the rudiments of modal logic.

2.6 Limitations

What's the matter with logic? As a guide to "correct thinking," it is not very effective.
Nobody thinks in formal proofs, nor should they; it would be a dreadful waste of time.

Good habits of reasoning may be inculcated by the study of formal logic, or by any other formal subject, such as abstract algebra, but it is doubtful that any particular course can claim pride of place in having beneficial effects on habitual patterns of reasoning.

Logic is important to certain formal disciplines—mathematics and philosophy, to name two—and the study of logic certainly contributes to the understanding of those disciplines. Even here, although one cannot be effective in these disciplines without being able to construct valid inferences, or without being able to tell valid from invalid arguments, one need not, on that account, develop facility in any particular system of formal logic. There are many successful mathematicians and philosophers who have never had a formal course in logic.

The main importance of logic in these formal disciplines, and in other areas as well, is as a standard of last resort for valid inference. Valid inference is important because argument and proof are important. In order to get on with science, in order to agree on a course of action, we must agree on what our axioms and our premises entail. In a court of law, we need to know whether or not the evidence entails the guilt of the accused. In science, we need to know whether the outcome of the experiment could have been predicted on the basis of the accepted theory (that is, be derived from the axioms and the boundary conditions). In order to cooperate in practical life, we need to be able to agree that if proposition S is made true by an act, then, given what else we know, proposition T will also be true, since S, together with background knowledge, entails T. In each case, if we can come to agreement about whether or not an argument establishes the existence of a proof, we can be assured of agreeing on the entailment.

Argument is a social activity. Margaret asserts a proposition; Theodolphus doubts it; Margaret offers an argument from premises that Theodolphus agrees with to the conclusion in question. If Theodolphus accepts the premises, then exactly what is at issue is whether or not there is a valid argument from those premises to that conclusion.

Formal logic bears on this issue in an important way, but only indirectly and as a standard of last resort. Let us see how this happens. The results may help us to form an idea of what we might demand of inductive logic.

(1) In order to *apply* a formal first order logic, Margaret and Theodolphus must agree on a formal representation of the premises and conclusion. This involves translating the predicates, names, functions, etc. of the ordinary language of the premises and conclusion into a fragment of a formal language. This is a matter of degree. If what is at issue is an argument in the propositional calculus, the arguers need not be at all concerned with the internal structure of the sentences involved. If the argument involves the full predicate calculus—quantification theory, names, functions—some amount of internal structure must also be agreed on.

(2) Margaret and Theodolphus must agree not only on the *representation* of the premises, but on the *acceptability* of the premises. They must agree not only that the formula "$(\forall x)(Hx \rightarrow Mx)$" correctly represents the ordinary language sentence "All men are mortal," but that this sentence is true and acceptable as a premise.

(3) Given a relatively formal representation of the premises and conclusion, and agreement on the acceptability of the premises, the issue becomes that of whether

there is a formal proof leading from those premises to that conclusion. But no one, except perhaps in a class in formal logic, gives formal proofs. What one offers is a more or less complete sketch of a formal proof—an indication of how (one thinks) a formal proof *could* be constructed in a first order predicate logic.

(4) Formal logic represents a *standard* in an important way. Margaret and Theodolphus are assured that if they can agree on a representation of the conclusion and the premises, then they can come to agreement about whether Margaret's argument is compelling. If they cannot do this on the basis of her initial argument sketch, they can consider a more detailed sketch; ultimately Margaret may be required to provide an actual formal proof: a sequence of formulas of the language fragment, each of which is a theorem of the first order logic, or a premise, or the result of applying *modus ponens* to earlier formulas in the sequence. (For simplicity, we assume a logic like Quine's in which the only rule of inference is *modus ponens*.)

(5) All first order predicate logics[2] have the same theorems, and the inference from premises to conclusion is valid, according to the deduction theorem, if and only if the conditional whose antecedent is the conjunction of the premises and whose consequent is the conclusion is a theorem. Therefore, the particular version of the predicate logic employed is irrelevant.

Nevertheless, systems of logic like Quine's *Mathematical Logic* have serious short-comings, which are often reflected in the naive questions asked by beginning logic students. A good argument is one that is both *valid* (the conclusion is true whenever the premises are true) and *sound* (the premises are true, or at least acceptable in the sense that we take them to be true). This means that the conclusion of a good argument is true, or at least acceptable.

Most of the arguments we employ in communicating with our fellows are, strictly speaking, invalid. Margaret asserts that the check will not be honored. The argument is that there are insufficient funds in the bank. There are a lot of lacunae to be filled in here: that the occasion in question falls on a certain date and time (next week there might be sufficient funds); that it is the account the check is drawn on that lacks sufficient funds. We can spell out what amounts to sufficiency in terms of the amount for which the check is written and the amount in the corresponding account, etc.

But we will not achieve a valid argument until we incorporate the premise "The bank will not honor any check drawn against insufficient funds." If we include this premise, the argument can (perhaps) be made valid; but we have still not shown it to be sound, since now the premise may be untrue. It is true in general, of course. But banks, even with computers, make errors. Theodolphus could conceivably walk up to the teller's window and cash the check with no difficulty.

Alternatively, we can replace this premise with a true one: The bank *almost never* honors checks drawn against insufficient funds. Now the argument has acceptable premises, but it is no longer valid. What "almost never" happens does happen on (rare) occasions. It is possible for the premises to be true and the conclusion false.

---

[2]Strictly speaking, this is not true. Philosophers and mathematicians have raised objections to standard first order logic, for example on grounds that it is committed to nonempty universes, that it is committed to exactly two truth values, etc. Nevertheless, the resulting variants on first order logic are rare enough to be called deviant and will be ignored for the time being.

Here is another example: Theodolphus is a medical researcher, who has just isolated a bacterium that causes pimples on the tails of mice. Margaret is curious as to how he knows that he has done that. He offers the argument: The mice have been raised in a sterile environment for many generations. They are standard experimental mice. All and only those mice exposed to the bacterium in question develop pimples on their tails, regardless of how their environments are varied otherwise. (Shades of Mill's methods!)

Filling out this argument would require spelling out many details, most of which need not concern us. One detail that does concern us is this: We assume that a mouse does not develop pimples on its tail for no reason—there is a "cause" of tail pimples. More specifically, since this is a rather vague assertion, we take as a premise that the cause of tail pimples is included among the variables that Theodolphus varied in his experiments. (Mill's principle of universal causation.)

This is clearly a substantive assumption, and one that could be false. That it *could* be false does not mean that it is unacceptable. Before performing the experiments, we might have had every reason to believe it to be true; we might have taken it to be acceptable but uncertain. After the experiments, we might have had reason to think it false: for example, if the experiments had not yielded a positive result, it might have been natural to think that this substantive premise was false. Again, we have an argument that can be rendered valid by the addition of premises, but those added premises may render the argument unsound. What good is validity if, when an argument has had its set of premises expanded to render it valid, it turns out to be no longer sound?

The arguments that are of practical interest in real life are like these. If we insist on validity, we must expand the set of premises to include many that are ordinarily left implicit. When we do expand our set of premises enough to establish validity—to make it *impossible* for the premises to be true and the conclusion false—we often find that we have rendered soundness suspect. The premises that are needed to render the formal analogues of our informal arguments valid are simply not premises that we can in good conscience accept as true. If we demand premises that we can in good conscience jointly affirm as true, it is rare that we can construct a valid argument with an interesting conclusion. The world is too full of exceptions. Note that it is not a matter of winking at a few exceptions: the premises required for validity, if there are known to be exceptions to them, are simply, flatly false. That definitively undermines the application of deductive logic to many areas in which we might imagine that it could usefully be applied.

Consider a basic physical law—say that action and reaction are equal and opposite. This entails that if two bodies exert forces F and G on each other, they will be equal in magnitude and opposite in direction.

Let us measure the forces F and G. In general, the results of the measurements of these forces will not be identical. It is easy to say why: the difference is due to *errors of measurement*. But now we face a puzzle: The law of physics says nothing about measurement; it talks about forces. What we observe, however, is the result of measurement. There must be some connection, and of course there is. There is a (fairly) well-established theory of errors of measurement that connects the values of physical quantities, such as force, with the results of measuring them by specific procedures. According to the classical theory of error, first established by Gauss, errors of measurement may be treated as characterized by statistical distributions. Specifically, in many cases, including cases like that under discussion, the error is assumed to have a normal, or Gaussian, distribution—the familiar bell-shaped curve.

Suppose we are testing some hypothesis H, according to which the force F is double that of G. We measure F as 3.582, and G as 1.634. If we took our measurements literally, we would have a deductive refutation of H. F is not twice G. But if we take the possibility of errors of measurement seriously, we have at most an inductive or uncertain argument against H. To have a deductive argument refuting H we would need something like the following premise: If F is measured as 3.582, and G is measured as 1.634, then F is not twice G. But whether or not this premise is acceptable depends on the distribution of errors of measurement.

There are many arguments that we find perfectly persuasive that cannot be rendered deductively valid even by the addition of premises that are possibly true. "Coin A has just now yielded a hundred successive heads. Therefore, coin A is not a fair coin." The implied premise is that if a coin is fair, it doesn't yield a hundred heads in succession. But if fairness is interpreted in probabilistic terms, then that premise is flatly false: Probability theory, with the usual assumptions about coins, actually *entails* that a fair coin *will* yield a hundred successive heads on some (rare) occasions.

How much of ordinary argument is like this? That is, how much could only correspond to valid inference if we were to add premises that we know to be false? It is hard to give a quantitative answer, but it seems reasonable to suppose that it is a significant amount.

This fact has led a number of writers to investigate ways of representing the arguments and inferences that are not comfortably forced into the mold of classical first order validity. There are two main directions in which we can go to accommodate these inferences.

(1) One of them is to focus on *invalid* inference. That is, we could take for granted ordinary logic, as captured by *Mathematical Logic*, for example, and worry mainly about inferences in which the premises do not entail the alleged conclusion. This is exactly to say that there may be some models in which the premises are true and the conclusion false. One might look for plausible ways to exclude from consideration the models that are not relevant, and thus be able to demand that the conclusion hold in all the models that are relevant to the argument. We could bite the bullet, take certain forms of argument to be both invalid and rationally persuasive, and try to find principles characterizing those forms of argument. This will be the thrust of the developments of Chapters 6 through 12.

(2) Another way to go is to build on classical first order logic in one of several ways. We could adhere to the classical view of logic, but alter the objects of argument and inference. Thus, we might take the *objects* of ordinary argument and inference to be beliefs, or degrees of belief, and we might take their alteration in the light of evidence to be a matter of quantitative change. In the example involving the coin, we might say that the best representation of the argument would be in terms of probability, rather than in terms of categorical assertions: we should conclude, after the hundred heads, that the chance that the coin is fair is small, or even negligible. (On this view we might have difficulty specifying what "fair" comes to.) This development, associated with the mathematical treatment of probability,

which has mainly occurred in the last few hundred years (after more than two thousand years of the study of deductive logic), has become an important part of our treatment of uncertain inference. These quantitative approaches to uncertain inference, including probability and several other measures of uncertainty, will be explored in the next three chapters.

There are many important and standard results in the theory of probability, some of which will be developed in the next chapter. There have been several suggestions, which we will explore in subsequent chapters, as to how to use this machinery, and as to what it really means. In addition, there have been suggestions of alternative mechanisms to take account of quantitative changes in our epistemic attitudes toward statements. We shall also explore some of those.

2.7 Summary

First order logic, like most of its variants, is concerned with valid arguments—that is, arguments whose conclusions are true whenever their premises are true. We will take first order logic for granted, despite the fact that most arguments encountered in the wild are either unsound (have false or unacceptable premises) or invalid (cannot be represented by formally valid arguments). There are two directions in which to go from here: We may focus on arguments that are rationally compelling, even though they are invalid, and seek to explore their foundations. Alternatively, we may focus on valid and sound arguments that concern, not facts in the world, but our beliefs; the conclusions of such arguments are characterized in some quantitative way—for example, by probability, or by some other measure of uncertainty.

2.8 Bibliographical Notes

There are many fine texts on mathematical logic; among them are [Enderton, 1972], [Quine, 1951], [Mendelson, 1979], and [Eberle, 1996]. For getting some familiarity with the process of *doing* formal logic, one of the best books is [Kalish & Montague, 1964]. For a glimpse of a number of variations on classical logic, see [Haack, 1996]. Modal logic receives a thorough treatment in [Hughes & Cresswell, 1996]; modal logic will be of concern mainly in our discussion of nonmonotonic logics.

2.9 Exercises

(1)   For each of the following arguments, either show that the conclusion holds in every possible world (i.e., under every possible assignment of truth values) in which the premises hold, or exhibit a possible world in which the premises are true and the conclusion false:

(a)   If Jane studies (S), she receives good grades (G). If she does not study, she enjoys college (C). If she does not receive good grades, then she does not enjoy college. Therefore, Jane studies.

(b)   If Mary joins a sorority (J) and gives in to her inclinations (I), then her social life will flourish (F). If her social life flourishes, her academic life will suffer (A). Mary will give in to her inclinations, but her academic life will not suffer. Therefore, she will not join a sorority.

(c)   If Sam studies (S), then he receives good grades (G). If he does not study, then he enjoys college (C). If he does not receive good grades, then he does not enjoy college. Therefore, Sam enjoys college.

(d)   If prices are high (P), then wages are high (W). Prices are high or there are price controls (C). If there are price controls, there is not an inflation (I). There is an inflation. Therefore, wages are high.

(2)   Let Q represent weak preference: $Q(x, y)$ means that x is preferred to y or that x and y are equally preferable. Two natural axioms governing this relation are transitivity:

$$\forall x, y, z(Q(x, y) \wedge Q(y, z) \rightarrow Q(x, z))$$

and completeness:

$$\forall x, y(Q(x, y) \vee Q(y, x)).$$

Give a model of these axioms that shows that they are consistent, and a model that renders them both false. Show that $\forall x(Q(x, x))$ is true in every model of these axioms.

(3)   In the same system, define $P(x, y)$ by the additional axiom

$$\forall x, y(P(x, y) \leftrightarrow Q(x, y) \wedge \neg Q(y, x)).$$

Show that $\forall x, y(P(x, y) \rightarrow \neg P(y, x))$ holds in every model of the axioms.

(4)   Give a model of the preference axioms in which $\forall x, y\, \exists z(Q(x, z) \wedge Q(z, y))$ is true, and another in which the same sentence is false.

(5)   Show that from

$$\forall x, y(x \circ y = y \circ x)$$

and

$$\forall x, y, z(x \circ z = y \circ z \rightarrow x = y),$$

we can derive

$$\forall x, y, z(z \circ x = z \circ y \rightarrow x = y).$$

(6)   In the example of Section 2.3, it was observed that construing "mother of" as a function required that there be some closed cycle of motherhood. Suppose that there is a domain $\mathcal{D}$ in which no such cycles exist. What can you say about the cardinality of $\mathcal{D}$?

(7)   In the semantics of conventional first order logic, it is stipulated that a formula $\ulcorner \neg \phi \urcorner$ is satisfied in a model, under a variable assignment, just in case ϕ is not satisfied in that model under that assignment. This amounts to the assumption that a sentence is either true or false. This assumption has been objected to. Explore some possible grounds for this objection, and consider various responses that might be made to the objection.

(8)   Some people argue that standards of validity in logical argument are essentially social, and reflect prevailing social standards. Thus, it is claimed, for example, that standards of scientific evidence are determined by white, male practitioners of science, and thus reflect their biases and interests rather than any abstract conception

of rationality. Discuss this argument, and in particular discuss the standards being invoked by those who make the argument.

(9)   As a first step toward a logic that admits degrees of truth, suppose we assign the values 0, $\frac{1}{2}$, and 1 to statements, where 0 corresponds to "false", 1 corresponds to "true", and $\frac{1}{2}$ represents an intermediate truth value. Carry on with this idea: if these are the truth values, how should we assign truth to compound statements (negation, conjunction, disjunction, the conditional) in such a way as to preserve classical logic as far as statements having values of 0 and 1 are concerned? See what you can come up with, and give reasons for your choices.

Bibliography

[Ackermann, 1962] W. Ackermann. *Solvable Cases of the Decision Problem*. North Holland, Amsterdam, 1962.

[Eberle, 1996] Rolf Eberle. *Logic and Proof Techniques*. New Central Book Agency, Calcutta, 1996.

[Enderton, 1972] Herbert Enderton. *A Mathematical Introduction to Logic*. Academic Press, New York, 1972.

[Haack, 1996] Susan Haack. *Deviant Logic, Fuzzy Logic*. University of Chicago Press, Chicago, 1996.

[Hilbert & Ackermann, 1950] David Hilbert and Wilhelm Ackermann. *Mathematical Logic*. Chelsea, New York, 1950.

[Hughes & Cresswell, 1996] G. E. Hughes and M. J. Cresswell. *A New Introduction to Modal Logic*. Routledge, London, 1996.

[Kalish & Montague, 1964] Donald Kalish and Richard Montague. *Logic: Techniques of Formal Reasoning*. Harcourt Brace and World, New York, 1964.

[Mendelson, 1979] Elliott Mendelson. *Introduction to Mathematical Logic*. Van Nostrand, New York, 1979.

[Quine, 1951] W. V. O. Quine. *Mathematical Logic*. Harvard University Press, Cambridge, MA, 1951.

3

The Probability Calculus

3.1 Introduction

The opposing football captains watch as the coin arcs, glinting, through the air before landing on the turf at the referee's feet. Heads. Whatchamacallit U. kicks off.

For thousands of years, people have depended on devices that yield outcomes beyond the control of human agency; it has been a way of consulting the gods, or the fates. For us, the point of tossing a coin to determine who kicks off is that the outcome is a matter of chance. The probability of heads is one-half: The coin could, with equal conformity to the laws of the physical world, have landed tails.

Typically, for probability, matters are not as simple as they seem. In ancient times the outcome of chance events—the toss of a knucklebone or a coin—was often taken to reveal the will of the gods. Even today many people take the outcome of a chance event, at least if they have wagered on it, to be a matter of "luck," where luck plays the role of the old gods, and can be cajoled, sacrificed to, encouraged with crossed fingers and rabbit's feet. In most cases, however, chance events are understood to be outside of human control, and to yield the outcomes they yield in accord with the laws of probability.

The early probabilists (Pascal, Fermat, the Bernoullis, and Laplace) believed in a deterministic world in which chance events did not really exist. Our belief in any chance event (say the outcome of the toss of a six-sided die) is less than certain only as a consequence of the limitations of our knowledge. I do not know whether the next card to be dealt to me will be a heart—even though it is predetermined, given the present order of the deck, to be a heart or not to be a heart. Were we to play with the deck turned over, so that all of the cards could be read, we would have no problem in our decision procedure. The urge to gamble would also be frustrated.

This deterministic view can be seen explicitly in the work of Laplace [Laplace, 1951, p. 4]:

Given for one instant an intelligence which could comprehend all the forces by which nature is animated and the respective situation of the beings who compose it—an intelligence sufficiently vast to submit these data to analysis—it would embrace in the same formula the movements of the greatest bodies of the universe and those of the lightest atom; for it, nothing would be uncertain and the future, as the past, would be present to its eyes.

Unfortunately (or perhaps fortunately), in poker, as in life, we do not have the cards showing and are cast into ignorance and uncertainty.

The classicalists sought to develop a formal system for manipulating the uncertainties concerning the outcomes of chance events. Such a system was to provide a set of rules and criteria for making rational inferences under conditions of uncertainty.

One part of this program was the development of what we now call the probability calculus. The probability calculus is often held to begin with the correspondence of 1654 between Blaise Pascal and Pierre de Fermat, as mentioned in Chapter 1. Like the developers of other parts of mathematics at the time, the early probabilists developed the probability calculus in response to real world problems. The mathematics therefore required a real world interpretation—something to tell the user what could be put into the calculus, and to help the user to interpret the results provided by the calculus.

There are currently several interpretations of the probability calculus, and we will discuss some of them in Chapter 4. But the original interpretation of the probability calculus is what is now called the classical interpretation. The basic formulation of the classical interpretation (usually attributed to Laplace) is:

The probability of an outcome is the ratio of favorable cases to the total number of equally possible cases.

The rule is straightforward in some applications. There are six ways in which a die can land; these "ways" are equally possible; dice are manufactured to make this true; the probability of an odd number showing uppermost is therefore the ratio of the number of favorable cases (3) to the total number of cases (6), or $\frac{1}{2}$.

But in other cases, the definition is somewhat problematic. Whereas it states what the probability of an outcome is, given the equipossible cases, it does not tell us what the equipossible cases are. The classical probabilists were aware of this problem and attempted to address it by positing the following principle, which we will call the *principle of indifference*:

The elements of a set of outcomes are equally possible if we have no reason to prefer one of them to any other.

As we shall see, there are overwhelming difficulties with this principle if we take it generally and attempt to apply it broadly. Nevertheless, in the context of games of chance and other examples in which the basic cases are clearly defined, and are intuitively equipossible or equiprobable, the principle can have heuristic value. "Take a card, any card. Good. Now there are two possible cases: You got the ace of spades, or you got some other card. Two cases of which one is favorable: the probability that you got the ace of spades is a half, right?" Wrong. Those two cases are not equipossible: the *cases* in drawing a card are fifty-two. These are the equipossible cases, and only one of the fifty-two corresponds to having gotten the ace of spades. The probability is $\frac{1}{52}$.

An important feature of any attempt to calculate the probability of an event (on the basis of this definition) is the ability to calculate both the total number of outcomes and the number of favorable outcomes. One method of doing so is based on the theory of combinations and permutations—*combinatorics*. The theory of combinatorics was familiar to the classical probabilists. The second part of Jacob Bernoulli's

Ars Conjectandi [Bernoulli, 1713] provided a development and description of the (then) current state of the theory of combinatorics.

3.2   Elementary Probability

Although probability theory is a deep, rich, and beautiful part of our mathematics, and it would be quite unreasonable to expect to achieve a thorough grasp of it as a side effect of studying uncertain inference, there are a number of parts of the theory that are both essential to the discussion of uncertain inference, and quite straightforward and simple.

One part is the theory of combinations and permutations. Again, there is a rich mathematical literature on these topics; but we shall need only the most elementary considerations.

3.2.1   Combinations and Permutations

A *permutation* of a set is an ordering of that set (called a *sequence*). We may represent the set containing the first two integers by either {1, 2} or {2, 1}. We can use both representations interchangeably because they represent precisely the same set. When we are working with sets we are not concerned with the order of a set's elements, but when we are working with permutations the order of the elements is important. The order of the elements of a permutation is one of the characteristics, which is used to distinguish one permutation from another. The permutation $\langle 1, 2 \rangle$ differs from the permutation $\langle 2, 1 \rangle$ precisely because the order of the elements is different.

The number of permutations of a finite set depends only on the number of elements in the set. If there are n elements in the set, there are n ways of choosing the first element of the permutation. There are then $n - 1$ ways of choosing the second element, $n - 2$ ways of choosing the third, etc. When we have chosen the first $n - 1$ elements of our permutation, there is only one way of choosing the last element. The number of permutations of a set containing n elements is thus

$$n \times (n - 1) \times (n - 2) \times \cdots \times 2 \times 1 = n!$$

The expression "$n!$" is pronounced "n-factorial", and may be inductively defined as follows:

Definition 3.1.

$$0! = 1$$
$$1! = 1$$
$$n! = n \times (n - 1)!$$

Example 3.1. *How many ways are there to order the set* $\{a, b, c\}$?

There are three elements in this set. When we ask for the number of ways of ordering a given set we are asking for the number of permutations of that set. In this case, $n = 3$ and $3! = (3)(2)(1) = 6$. There are six ways to order this set.

The primary value of being able to calculate the number of permutations in a set is that it is useful in calculating combinations: the number of ways of *choosing* m out of n objects. How many ways are there of choosing four out of 10 objects? One way to figure this out is to imagine the 10 objects laid out in a line. There are 10! ways of doing this. If we take the four leftmost objects to be those chosen, and the six remaining ones to be those not chosen, each permutation gives rise to a selection of four objects from the ten. These are not all distinct, however. The permutation that begins $\langle a, b, c, d, e, \ldots \rangle$ gives us the same set of four chosen objects as the permutation that begins $\langle b, a, d, c, f, \ldots \rangle$. The number of ways of choosing four items out of ten is 10! divided by 4! (because the order of the items selected doesn't matter) divided by 6! (because the order of the items not selected doesn't matter either).

In general, the number of ways of choosing m objects out of n is

$$\frac{n!}{m!(n-m)!} = \binom{n}{m}.$$

The expression $\binom{n}{m}$ is pronounced "n choose m" and is sometimes written "^nC_m" or "C_m^n".

Example 3.2. *Let us compare the number of ways of getting five heads out of ten tosses of a coin with the number of ways of getting two heads out of ten tosses of a coin. The former is $\binom{10}{5} = 252$; the latter is $\binom{10}{2} = 45$. Note that there are many more ways of getting five heads than of getting two heads.*

Example 3.3. *A poker hand is a set of five playing cards drawn or dealt from a standard deck of 52 playing cards.*

(a) Find the number of distinct poker hands.

How many ways are there of choosing five from 52 cards?

$$\binom{52}{5} = \frac{52 \times 51 \times 50 \times 49 \times 48}{5!} = 2,598,960.$$

(b) What proportion of poker hands consist of five cards all of the same suit?

The number of poker hands consisting of five spades is $\binom{13}{5}$; the number consisting of five hearts is $\binom{13}{5}$; etc. Thus, the number of poker hands consisting of five cards all of the same suit is $4 \times \binom{13}{5}$, or 5148, and the proportion is

$$\frac{5148}{2598960} = 0.00198.$$

The reason combinatorics is so useful in the probability calculus is that it provides a rapid and easy way of counting cases. When we are unsure of how to use the method of combinatorics to calculate the probability of an event, we can use any method that provides accurate counts of the number of successful outcomes and the total number of outcomes. As an example, we provide Fermat's solution to the problem of points.[1]

Example 3.4. *Two players of equal ability are unable to continue a tournament. The first player needs two games to win and the second player needs three. How do we divide the prize money?*

Fermat's solution is as follows: There are at most four games to be played before one of the players is guaranteed to win. Count up the number of ways player 1 can win, and

---

[1]This was one of the problems posed by the Chevalier de Mére that sparked the interest of Blaise Pascal and

divide by the total number of outcomes of the four games needed to be played. Because there are four games, each of which has two possible outcomes, the total number of outcomes is $2^4 = 16$. The following table represents the possible outcomes. We denote a win for player 1 by a W, and a loss for player 1 (which corresponds to a win for player 2) by an L:

								Outcome								
Game	1	2	3	4	5	6	7	8	9	10	11	12	13	14	15	16
1	W	W	W	W	W	W	W	L	L	L	L	L	W	L	L	L
2	W	W	W	W	L	L	L	L	W	W	W	W	L	L	L	L
3	W	W	L	L	W	W	L	W	W	W	L	L	L	W	L	L
4	W	L	W	L	W	L	W	W	W	L	W	L	L	L	W	L

From this table it is easy to see that outcomes 1–11 are favorable to player 1. Thus, player 1 has an $\frac{11}{16}$ chance of winning. Fermat contends that we should divide the prize money according to each player's chance of winning. Therefore, player 1 gets $\frac{11}{16}$ of the prize money and player 2 gets $\frac{5}{16}$.

There is one basic difficulty with this solution. What do we count as a case? Some of these "cases" cannot occur. For example, since player 1 only needs two games to win, the first and second cases come to the same thing. If player 1 wins the first two games, the third and fourth games will never be played. If a case is a completed tournament, then the first four outcomes could be construed as a single case. This is a result of the fact that after playing two games, where player 1 won both, there would be no need to continue playing. The first four outcomes would not be distinguished from each other because games 3 and 4 would never get played. Based on such reasoning we would get the following table:

				Outcome					
Game	1	2	3	3	5	6	7	8	9	10
1	W	W	W	W	L	L	L	L	L	L
2	W	L	L	L	W	W	W	L	L	L
3		W	L	L	W	L	L	W	W	L
4			W	L		W	L	W	L

According to this table player 1 wins in outcomes 1, 2, 3, 5, 6, 8, giving him a $\frac{6}{10}$ chance of winning the tournament. According to this view of what counts as a case, we should divide the stakes by giving $\frac{6}{10}$ of the stake to player 1 and $\frac{4}{10}$ of the stake to player 2.

What is the matter with this solution? We have counted cases, and counted favorable cases, and divided the latter by the former. But we have neglected the subtle "equipossible" in the classical definition. These 10 cases cannot plausibly be regarded as equipossible.

Why not? Because it is natural to suppose, given the description of the circumstances, that (a) each player has the same chance of winning a game, and (b) that chance remains the same, whatever has happened on the preceding game. This analysis renders the cases of the first table equipossible, but the cases of the second table are not equipossible. For example, in the second table the first case has a $\frac{1}{4}$ chance of occurring, while the third has only a $\frac{1}{16}$ chance of occurring.

In effect, the classical interpretation of probability can provide us with answers to a wide range of problems provided that we have an accurate method for enumerating the total number of outcomes and the total number of successful outcomes, and provided that we are careful to ensure that the outcomes are equipossible. As a *definition* of probability, however, the classical view is clearly defective: The only way to characterize equipossibility is in terms of probability (or chance, or some other synonym of probability), as we just did in the problem of points.

3.2.2 The Probability Calculus

In this section, we will provide a brief introduction to the probability calculus. We use set-theoretical notation for our axioms, theorems, and definitions. We consider probability to be a function from *events* to the real numbers in the closed interval [0,1]. Nothing profound should be imputed to the term "events". We have taken the domain of the probability function to consist of events because it is traditional and because events can be given the appropriate structure. (We could take the domain to be lottery tickets, outcomes of experiments, measurements, etc.) The domain of the probability function is called the *sample space*.

A sample space S is a set of individual objects or events: the tosses of a coin, the balls in an urn, the recruits in an army at a certain time, telephone calls received by a switchboard, the sentences of a language, propositions, worlds, etc. *Probability* is relative to a sample space. The mathematical properties of probability are very simple: it is a function, defined on a field of subsets of the sample space, that (a) is additive and (b) has a range equal to the closed interval [0, 1].

Whereas this characterization of probability is very simple, it may not be very enlightening. In particular, it is not a *definition* of probability, because it does not specify how the *value* of the probability function for a given argument (set of elements in the field of subsets of S) is determined. For that matter, we have not specified how the field of subsets of S is determined. Both of these are nontrivial philosophical issues; we will explore some of the possible suggestions in the next chapter.

Let us begin at the beginning. A *field* of subsets of S is a set $\mathcal{F}$ of subsets of S closed under union and complementation. That is, it satisfies the following simple axioms:

Axiom 3.1. $\emptyset \in \mathcal{F}$.

Axiom 3.2. $A \in \mathcal{F} \rightarrow \overline{A} \in \mathcal{F}$.

Axiom 3.3. $A \in \mathcal{F} \wedge B \in \mathcal{F} \rightarrow A \cup B \in \mathcal{F}$.

Of course, this means that $\mathcal{F}$ is closed under intersection as well:

Theorem 3.1. $A \in \mathcal{F} \wedge B \in \mathcal{F} \rightarrow A \cap B \in \mathcal{F}$.

   Proof: $A \cap B = \overline{\overline{A} \cup \overline{B}}$. ∎

In such a structure the axioms for probability are equally simple:

Axiom 3.4 (Total Probability). $\mathcal{P}(S) = 1$.

Axiom 3.5 (Nonnegativity). $A \in \mathcal{F} \rightarrow \mathcal{P}(A) \geq 0$.

Axiom 3.6 (Additivity).

$$A, B \in \mathcal{F} \wedge A \cap B = \emptyset \rightarrow P(A \cup B) = P(A) + P(B).$$

For some writers, this is the proper axiomatization of probability; probabilities satisfying only these axioms are "finitely additive" probability measures. The grounds for this nomenclature are given by the following theorem:

Theorem 3.2.

$$(\forall i)(1 \leq i \leq n \rightarrow A_i \in \mathcal{F}) \wedge (\forall i)(\forall j)(i \neq j \rightarrow A_i \cap A_j = \emptyset)$$

$$\rightarrow P\left(\bigcup_{i=1}^{n} A_i\right) = \sum_{i=1}^{n} P(A_i).$$

Surprisingly, however, there are circumstances in which we would like to be able to consider the union of infinitely many sets. For example, we might be interested in the probability that the first head in a series of coin tosses will occur on the nth toss. It strikes many as implausible to impose any finite bound on n. (Of course, it also strikes many as implausible to consider an *infinite* sequence of tosses.)

How can we handle this case? Let FH_n be the event of the coin landing heads for the first time on the nth toss. A natural probability value for this is $(\frac{1}{2})^n$, since the coin will land heads for the first time on the nth toss only if it lands tails on each of the first $n-1$ tosses, and heads on the nth toss. A consequence is that for any finite n, $\sum_{i=1}^{n} P(\mathrm{FH}_i) < 1$.

In order to satisfy the axiom of total probability, we must consider the infinite sum $\sum_{i=1}^{\infty} P(\mathrm{FH}_i) = 1$. This in turn requires that the field $\mathcal{F}$ be closed under countable unions. Such a structure is called a σ-*field*. We say that $\mathcal{F}$ is a σ-field if it is a field (satisfies Axioms 3.1–3.3) and in addition it satisfies the condition that the union of any countable set of its members is also a member of it.

Axiom 3.7. $(\forall i)(A_i \in \mathcal{F}) \rightarrow \bigcup_{i=1}^{\infty} A_i \in \mathcal{F}.$

We must then also strengthen the third axiom for probability: We must say that if we have a set of pairwise disjoint subsets of S, the probability of their union is the (infinite) sum of their probabilities.

Axiom 3.8 (Countable Additivity).

$$(\forall i)(\forall j)((A_i, A_j \in \mathcal{F}) \wedge (i \neq j \rightarrow A_i \cap A_j = \emptyset)) \rightarrow P\left(\bigcup_{i=1}^{\infty} A_i\right) = \sum_{i=1}^{\infty} P(A_i).$$

These axioms provide us with relations among probabilities, but they do not tell us how to assign values to the probabilities of particular events (except for the event S and the event $\emptyset$). To guide our intuitions, we may think of $P(A)$ as the ratio of the number of objects of type A to the number of all objects in S. This only makes sense if the sample space S is finite, but as a guide to intuition it will not generally lead us astray.

Example 3.5. *Imagine that we have an urn which contains five red chips, three blue chips, and two green chips. Let C be the set of all chips, and let R, B, and G stand for the set of red chips, the set of blue chips, and the set of green chips, respectively.*

The field underlying our algebra consists of the sets C, R, G, B, R ∪ G = $\overline{B}$, R ∪ B =
$\overline{G}$, G ∪ B = $\overline{R}$, and, of course, ∅ = $\overline{C}$.

The intended probability measure is clearly given by $P(G) = 0.2$, $P(R) = 0.5$, and
$P(B) = 0.3$. Note that $P(G \cup B) = 0.5$; why?

Of course, the fact that half the chips in the urn are red does not at all guarantee
that half of the *draws* of chips will result in a red chip. If it is the case that *draws* from
the urn of Example 3.5 are, in the long run, 60% red, 10% blue, and 30% green, we
can take account of this in our model.

Example 3.6. *Let D be the set of draws from the urn of the previous example; let DR
be the set of red chips drawn, DG the set of green chips drawn, and DB the set of blue chips
drawn. We would assign probabilities $P(DR) = 0.6$, $P(DG) = 0.3$, and $P(DB) = 0.1$.*

These two examples illustrate two important ways of thinking of probability. One
insight is based on the image of drawing balls from an urn. The proportion of each
kind of ball is known, and the story is told in such a way that it is plausible to suppose
that each ball has the same chance of being drawn. If we consider several draws, there
are two alternatives: the balls are drawn by replacing each ball as it is drawn, mixing
the contents of the urn, and drawing again. The draws are *independent*—a concept
we shall explain shortly. Alternatively, the balls are not replaced, and each draw alters
the proportions in the urn; but it does so in a determinate and known way. This image
is a nice one, because, under either alternative, the cases are well understood, and
they remain finite and easy to count.

The other insight is based on rolls of a die, or flips of a coin. In this way of looking
at things, the basic condition is that one toss or flip should not influence the next: the
sequences are characterized by independence. Having flipped a coin, there is no way
in which the flip can be returned to the set of flips. Hence there is no way in which
we can reflect the kind of dependence we can reflect in the urn image.

These are simply two aids to the imagination. Each can serve the function of the
other: to emulate flips of a coin, you can consider an urn with two balls, and draws
(designed to ensure that each ball has the same chance of being selected) that are
made with replacement, so that each is made from an urn with the same contents. To
emulate draws without replacement from an urn containing six distinct balls, you can
consider tosses of a die, and simply disregard outcomes that have already occurred.
Each image is more useful to some people, less useful to others. They are both to be
thought about only if they are useful.

3.2.3 Elementary Theorems

There are a number of elementary theorems that will be useful for our investigations
into uncertain inference.

Theorem 3.3. $P(\overline{A}) = 1 - P(A)$.

Theorem 3.4. $P(A) = P(A \cap B) + P(A \cap \overline{B})$.

Theorem 3.5 (General Addition Theorem).

$$P(A \cup B) = P(A) + P(B) - P(A \cap B).$$

Theorem 3.6 (Theorem of Total Probability).

$$(\forall i)(\forall j)\left(i \neq j \to B_i \cap B_j = \emptyset \wedge \bigcup_{i=1}^{n} B_i = S\right) \to \mathcal{P}(A) = \sum_{i=1}^{n} \mathcal{P}(A \cap B_i).$$

Example 3.7. *To illustrate these theorems, consider an urn containing triangular (T) and square (S) chips that are colored red (R), green (G), and blue (B). The following table gives the number of each. Note that we provide the number of each kind of chip answering to a most detailed specification. This is essential to the characterization of a probability function if we are to be concerned with both shape and color. (If we were just concerned with color, we could take the field that is the domain of P to be $\{\emptyset, T, S, T \cup S\}$.)*

Type	Number	Probability
$S \cap R$	10	0.10
$S \cap G$	15	0.15
$S \cap B$	20	0.20
$T \cap R$	20	0.20
$T \cap G$	30	0.30
$T \cap B$	5	0.05

To illustrate the first theorem, we note that the probability of $\overline{T \cap R}$ is $1 - \mathcal{P}(T \cap R) = 0.80$. To illustrate the second theorem, we note that the probability of R is $\mathcal{P}(R \cap T) + \mathcal{P}(R \cap \overline{T}) = \mathcal{P}(R \cap T) + \mathcal{P}(R \cap S)$. Finally, we note that the probability of a red or triangular chip is $\mathcal{P}(R \cup T) = \mathcal{P}(R) + \mathcal{P}(T) - \mathcal{P}(R \cap T) = 0.65$.

3.3   Conditional Probability

In the examples and explanations we have provided so far, we have not worried much about what it means for an event to happen and what impact that event, or our knowledge of that event, has on our sample space. Once we have drawn the chip from the urn then what? Do we return it to the urn (called drawing with replacement) or do we keep it out of the urn (called drawing without replacement)? What is the impact of drawing with and without replacement?

Given an urn which has 100 chips in it of which 14 are blue, assuming that probability is determined by the proportion in the urn, what happens to the probabilities if we assume that we have previously drawn a blue chip?

Initially the probability of drawing a blue chip was $\frac{14}{100}$. If we draw a chip and then return it to the urn, we have no effect upon the probability of drawing a blue chip, since the contents of the urn are still the same. $\mathcal{P}(B_2)$ will still be $\frac{14}{100}$, where B_2 is the set of pairs of draws in which the second chip is blue.

If we draw a chip and do not return it to the urn, we shall affect the sample space appropriate to the next draw (i.e., the chips in the urn) and consequently the probability of drawing a blue chip. If the first chip was blue, then $\mathcal{P}(B_2) = \frac{13}{99}$. If the first chip was not blue, then $\mathcal{P}(B_2) = \frac{14}{99}$.

What has happened is that the occurrence of one event (or the knowledge that that event has occurred) has had an effect on the probability of another event. When we seek to calculate the probability of one event occurring, *given* that another event has occurred, we are seeking to calculate the conditional probability of that event.

Suppose that the original sample space for one draw of a chip is C. It is natural to consider $C^2 = C \times C$ as the appropriate space for the draw of two chips. Relative to this sample space, the appropriate reference class determining the probability that the first chip is blue is $\{\langle x, y \rangle : C(x)\}$. The appropriate reference class determining the probability that the second chip is blue is $\{\langle x, y \rangle : C(y)\}$. These subsets of C^2 of course have the same cardinality: they are the same size. These subsets are appropriate for determining probabilities when we are drawing two chips *with replacement*.

When we are drawing chips *without replacement* the appropriate subset of C^2 for determining the probability that the second chip is blue is not $\{\langle x, y \rangle : C(y)\}$, but $\{\langle x, y \rangle : B(x) \wedge C(y)\}$ if the first chip is blue, and $\{\langle x, y \rangle : \overline{B}(x) \wedge C(y)\}$ if the first chip is not blue. These two sets have different cardinalities and lead to different proportions.

When we calculate the probability of $B(y)$ *given* $B(x)$—represented symbolically by $P(B(y)|B(x))$—we are using $B(x)$ to restrict or alter the sample space over which our probabilities are calculated. In the case of drawing from an urn without replacement, the sample space is reduced.

Example 3.8. *Suppose there are five chips in the urn, and that three of them are blue. The set of pairs of chips has 25 members, and this would be the right reference class for draws with replacement. But draws* without *replacement exclude the possibility of getting the same chip twice, so that there are only 20 possibilities. Among these, given that the first chip is blue, only $\frac{2}{4} = \frac{1}{2}$ are draws in which the second chip is also blue. Of course, this is symmetrical: Given that the second chip is blue, the probability that the first chip is blue is $\frac{1}{2}$.*

3.3.1 The Axiom of Conditional Probability

Often the conditional probability function, $P(B|A)$, is introduced by a "definition" of the form: If $P(A) > 0$, then $P(B|A) = P(A \cap B)/P(A)$. This is not a logically adequate definition, since a proper definition, in logical terms, always allows the replacement of the defined term by the expression defining it in all contexts. What we must do is to think of the conditional probability function, $P(B|A)$ as a *primitive* expression. It satisfies the axiom:

Axiom 3.9. $P(A \cap B) = P(A)P(B|A)$.

From a logical point of view, it is most elegant to take the single primitive locution as $P(A|B)$, and to interpret $P(A)$ as $P(A|S)$, where S is the entire sample space. Axioms 3.4 to 3.9 are then taken to apply to $P(A|S)$. This approach leaves open the question of what you say when $P(B) = 0$, and that can be useful philosophically, as we shall see in the next chapter. Among those who have taken advantage of the flexibility that this approach allows are Karl Popper [Popper, 1959], Hugues Leblanc [Leblanc & Morgan, 1983], and Charles Morgan [Morgan, 1998].

The most important aspect of conditional probability is given by Axiom 3.9: it allows us to calculate the probability of the intersection (or meet) of two elements of the field of subsets of the sample space S that constitutes the domain of the probability function. It also allows us to define a concept that is very important in many applications of probability, *independence*.

We say that two elements A and B in the domain of the probability function $\mathcal{P}$ are independent under $\mathcal{P}$, provided that the probability of $A \cap B$ is the same as the probability of A multiplied by the probability of B:

Definition 3.2. $\text{Ind}_P(A, B)$ if and only if $\mathcal{P}(A \cap B) = \mathcal{P}(A)\mathcal{P}(B)$.
$\text{Ind}_P(A_1, \ldots, A_n)$ if and only if $(\forall I \subseteq \{1, \ldots, n\})(\prod_{i \in I} \mathcal{P}(A_i) = \mathcal{P}(\bigcap_{i \in I} A_i))$.

Note that A is independent of B if and only if B is independent of A. Independence is symmetrical.

Example 3.9. *Suppose that the contents of a 100 chip urn conform to the data in the table in Example 3.7.*

(a) *If we draw a green chip, what is the probability that it is also triangular? That is, what is the value of $\mathcal{P}(T|G)$?*

$$\mathcal{P}(G) = 0.45. \ \mathcal{P}(G \cap T) = 0.30. \ \mathcal{P}(T|G) = \mathcal{P}(T \cap G)/\mathcal{P}(G) = 0.67.$$

(b) *If we have drawn a square blue chip without replacement and then draw a green chip, what is the probability that the chip is also square ?*

$$\mathcal{P}(G) = \tfrac{45}{99}. \ \mathcal{P}(S \cap G) = \tfrac{15}{99}. \ \mathcal{P}(S|G) = \tfrac{15}{45}.$$

A large class of examples that are useful to contemplate in understanding probability is drawn from gambling apparatus: roulette wheels, dice, cards, and the like. An important aspect of such devices is *repeatability*: a coin or a die (even a biased die) can be tossed over and over again, and will be characterized by the same probabilities of outcomes on each toss. This is just to say that the trials or tosses are *independent*.

Now there is clearly no way of putting a toss of a coin back into its reference set so that it can be made again. There is, thus, another reason for taking the sample space representing the outcomes of a gambling system to be infinite. When a die is rolled and yields a one, that does not decrease the size of the set of ones in the sample space relative to the set of threes. Having got a one on the first toss, the probability of getting a three on the second toss is still $\tfrac{1}{6}$; it has not increased.

The use of infinite sample spaces represents an idealization: no one supposes that an infinite number of tosses will be made with a particular die, or even with any die. The statistical behavior of dice conforms nicely to the statistical model, and that's all we need be concerned with. Independence is a key notion in these models.

Example 3.10. *If we are tossing a six-sided die:*

(a) *What is the probability that the side which lands face up is less than 4, given that the value of that face is odd ?*

Let L4 stand for the die landing with a face up having a value less than 4. Let O stand for a face up having an odd-numbered value. $\mathcal{P}(O) = \tfrac{3}{6}. \ \mathcal{P}(L4 \cap O) = \tfrac{2}{6}. \ \mathcal{P}(L4|O) = \tfrac{2}{3}.$

(b) *What is the probability that the side which lands face up is less than 4, given that the previous toss yielded an odd value?*

Let O stand for the odd outcome of the first toss, and L4 the outcome of the second toss being less than 4. Because the tosses are regarded as independent, $\mathcal{P}(L4|O) = \mathcal{P}(L4) = \tfrac{3}{6}.$

We may state the following elementary theorems concerning conditional probability.

Theorem 3.7. *If A is independent of B, then B is independent of A, $\overline{A}$ is independent of $\overline{B}$, and B is independent of $\overline{A}$.*

Theorem 3.8. *If $P(B) > 0$, then A is independent of B if and only if $P(A|B) = P(A)$.*

Theorem 3.9. *If $P(A) > 0$, then $P(B|A)$ is a probability function defined on the field of sets in $S - \overline{A}$, where S is the universal set of the original field of sets.*

3.3.2 Bayes' Theorem

Bayes' theorem is an elementary theorem involving conditional probability. It has been thought to solve all the problems of uncertain inference, and like any other universal solvent has been the focus of considerable controversy, which we will look at in a later chapter. Its significance depends on how we *interpret* probability. Note that its validity as a theorem is completely independent of how we interpret probability—a topic to be covered in Chapter 4—just as the validity of our other theorems depends only on the formal properties of probability, and not at all on what probability *means*.

Theorem 3.10 (Bayes' Theorem).

$$P(H|E) = \frac{P(H)P(E|H)}{P(E)} = \frac{P(H)P(E|H)}{P(H)P(E|H) + P(\overline{H})P(E|\overline{H})}$$

Proof: $P(H \cap E) = P(H)P(E|H) = P(E)P(H|E)$. ∎

The letters "H" and "E" are used suggestively to bring to mind "hypothesis" and "evidence". In induction we may be interested in the probability of a hypothesis *given* some evidence. Bayes' theorem, even in this simple form, is suggestive. It captures some intuitions: How probable the hypothesis is depends on how probable it was before the evidence was collected. The less probable the evidence, given the falsity of the hypothesis, the more impact the evidence has.

It shows that we can compute the probability of a hypothesis, given some evidence, if (a) we can compute the probability of the evidence given the hypothesis, (b) we can compute the probability of the evidence given the denial of the hypothesis, and (c) we have a probability for the hypothesis.

In cases in which the hypothesis *entails* the evidence, $P(E|H) = 1$, so that the theorem boils down to

$$P(H|E) = \frac{P(H)}{P(H) + (1 - P(H))P(E|\overline{H})}.$$

Often, when the hypothesis H is a statistical hypothesis, $P(E|H)$, called the *likelihood* of E, is a quantity that people can agree on.

We will explore these connections later.

For the present, we should note that not only is Bayes' theorem a straightforward theorem of the probability calculus, but that it has applications that are quite uncontroversial whatever interpretation of probability we adopt. The following theorem slightly generalizes Theorem 3.10.

Theorem 3.11. *Let $B_1, B_2, \ldots, B_n$ partition the sample space S (i.e., their union is equal to the sample space, and they are disjoint). Then*

$$P(B_i|E) = \frac{P(B_i)P(E|B_i)}{\sum_{j=1}^{n} P(B_j)P(E|B_j)}$$

Example 3.11. *Suppose that in a certain lab in the hematology department of a major hospital there are three technicians who prepare 17%, 39%, and 44% of the tissue plasminogen activator (TPA—a thrombolytic agent that helps dissolve blood clots); call the technicians Ms. Jones, Mr. Smith, and Ms. Cardano respectively. Let us say that 0.010%, 0.003%, and 0.006%, respectively, of the TPA they make is faulty. If a batch of TPA comes in, and it is faulty, what is the probability that it was prepared by Mr. Smith?*

The sample space consists of ordered pairs: a technician paired with a batch of TPA. Let B_1 be the set of pairs in which the technician is Ms. Jones, B_2 the set of pairs in which the technician is Mr. Smith, and B_3 the set of pairs in which the technician is Ms. Cardano. B_1, B_2, and B_3 constitute a partition of the sample space, B_1 accounting for 17% of it, B_2 for 39% of it, and B_3 for 44% of it. Let the set of pairs in which the sample is faulty be D. Within B_1, 0.010% of the TPA is faulty, and similarly for the other subsets of the sample space.

We may now calculate simply in accord with Theorem 3.11:

$$P(B_2|D) = \frac{P(B_2)P(D|B_2)}{P(B_1)P(D|B_1) + P(B_2)P(D|B_2) + P(B_3)P(D|B_3)} = 0.2123.$$

3.4   Probability Distributions

A *random quantity*, often called a "random variable," is a function from a sample space to real numbers.[2] For example, in a sample space representing the outcomes of the throw of a single die, the random quantity N, the number of dots on top, is a function that gives, for each member of the sample space (i.e., each possible roll of the die), the number of spots uppermost when the die comes to rest. It has six values, corresponding to the six sides of the die. If the sample space consists of army recruits at a certain time, the weight in pounds, W, of a recruit is a random quantity: For each element in the sample space (each recruit) the value of the random quantity is the weight in pounds of that recruit. Other quantities are "length of," "number of siblings of," etc. We may also have vector valued quantities: "⟨length, width⟩," for example, defined over a space of rectangles.

A random quantity yields a *partition* of the sample space. If there are a finite or countable number of values of the random quantity, we can associate with each value of the quantity a probability: the probability associated with that subset of the sample space. Thus, in the case of a sample space corresponding to the roll of a die, the relevant random quantity will have six possible values, $\{1, 2, 3, 4, 5, 6\}$. The subset of the sample space for which the value of the random quantity N is 1 is just the set of rolls that yield that number of spots uppermost.

This leads to the temptation to replace the sample space by the minimum space that supports the probability we want to discuss. In the case of dice, rather than

---

[2]Or some other set; for example, a set of properties like {red, white, blue}; of course, these can be coded as real numbers.

Table 3.1: Frequency table for dice

Sum	Probability
2	0.0278
3	0.0556
4	0.0833
5	0.1111
6	0.1389
7	0.1667
8	0.1389
9	0.1111
10	0.0833
11	0.0556
12	0.0278

individual rolls, we could adopt a space each of whose points is an equivalence class of points in our original space based on the number of dots showing. This is a natural temptation to succumb to, and only has the drawback that when we wish to consider a different random quantity we may have to consider a different sample space even if the underlying set of events is exactly the same. It also has the drawback that it is hard to associate the probability of a subset of the sample space with its relative cardinality, for example for a biased die.

It should be clear that any real-valued function of random quantities is a random quantity. If X and Y are random quantities, so are $aX + b, X + Y, XY, X^2, e^X, \sin X$, etc.

3.4.1 Frequency Functions and Distribution Functions

A *frequency function* f_N gives, for each x, the probability that a member y of the sample space has the value x. Thus, in the case of the normal die, the value of $f_N(x)$ is $\frac{1}{6}$ if $x \in \{1, 2, 3, 4, 5, 6\}$, and 0 otherwise. The frequency function gives us the same information as the original probability function.

Example 3.12. *If the event is the tossing of two dice and we are concerned with the probability of the outcomes of the occurrence of the sum of the dots showing (2, ... , 12), we can represent this information (for a normal die) in Table 3.1. It can also be represented by a graph, as in Figure 3.1.*

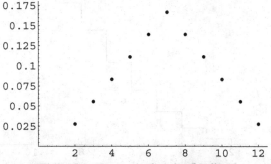

Figure 3.1: Frequency function for dice.

Table 3.2: Distribution function for dice

Sum	Probability
2	0.0278
3	0.0833
4	0.1667
5	0.2778
6	0.4167
7	0.5833
8	0.7222
9	0.8333
10	0.9167
11	0.9722
12	1.000

A *distribution function*, $F_N(x)$, gives for each x the probability that a member of the sample space has a value less than or equal to x. A distribution function also tells us the way in which the available supply of probability is "spread out" over the various possible outcomes. A distribution may be represented as a table or as a graph. The graph offers the possibility of dealing with continuous distributions, which cannot be represented as a table.

Example 3.13. *The rolls of the two dice of the previous example can be characterized by a distribution function, given in a table (Table 3.2), or in a graph (Figure 3.2).*

The random quantities that arise in gambling apparatus are discrete; but we can also consider random quantities that are continuous. For example, we can consider choosing a point at random from the unit line, or consider the distribution of weights among a hypothetical infinite population of army recruits to allow every real-valued weight between the minimum and maximum weight allowable for recruits. Such random quantities offer the advantages that are often offered by the transition from the discrete to the continuous: We don't have to worry about choosing a granularity, mathematically continuous functions are often easier to specify than discontinuous functions, and so on.

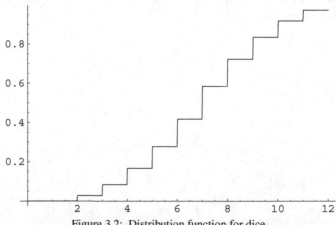

Figure 3.2: Distribution function for dice.

Example 3.14. *Consider a testing apparatus for fishing line. A 2 meter length of the line is inserted into the machine, and the line is put under tension. The tension is increased until the line breaks. The number X represents the distance from the left hand end at which the line breaks. Thus, X can have any value from 0.0 to 2.0 meters (we ignore stretch). The sample space consists of an arbitrarily large set of trials of this apparatus. A natural assumption, for good line, is that the breaking point is as likely to be at any one point as at any other point, or, more constructively, that the likelihood that the line breaks in any interval is proportional to the length of the interval. Thus, the probability that X is less than or equal to the real number x is 0 if x is less than 0, is 1.0 if x is greater than 2, and is x/2 if x is between 0 and 2.*

Because there are an uncountable number of points at which the line may break— representing every real number between 0.00 and 2.00—this probability distribution cannot be represented by a table. The distribution function—a function of real values x that gives the probability that an outcome will produce a value of X less than or equal to x—can be given. Indeed, $F_X(x)$, the probability that the string will break at a point whose coordinate is less than or equal to x, is given by x/2 in the relevant region:

$$F_X(x) = \begin{cases} 0 & \text{if } x < 0 \\ x/2 & \text{if } 0 \le x \le 2 \\ 1 & \text{if } x > 2 \end{cases}$$

For example, the probability that the line breaks at a point to the left of the point marked 0.84 is 0.42; the probability that it breaks at a point to the left of the point marked 1.64 is 0.82. Note that intuitively the probability that the line breaks at exactly the point marked 0.42 is 0.0. This distribution is easily represented by the graph of Figure 3.3.

In the case of a continuous distribution, as we have just observed, the probability that the random quantity X takes on any particular value is 0.0. There is thus no frequency function that can represent the probability. But just as in the case of a discrete probability function the value of the distribution function at the point x can be represented as the *sum* of the probabilities of all the possible values of $X \le x$, so we may consider a function whose *integral* over all possible values of $X \le x$ is $F_X(x)$:

$$F_X(x) = \int_{-\infty}^{x} f_X(x)\,dx$$

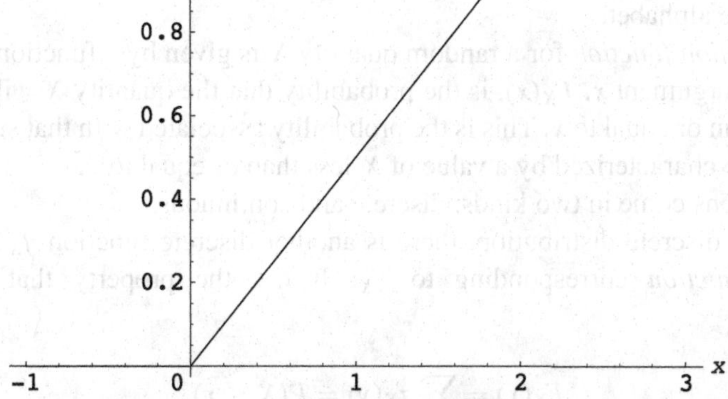

Figure 3.3: Fishing line distribution function.

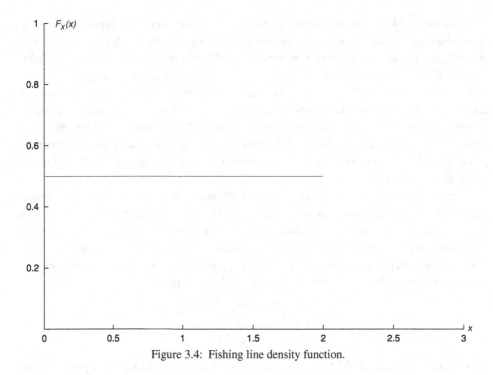

Figure 3.4: Fishing line density function.

The function f_X with this property is called a *density function* corresponding to the distribution function F_X.

Example 3.15. *In Example 3.14, the distribution function was continuous. The corresponding density function, illustrated in Figure 3.4, is*

$$f_X(x) = \begin{cases} 0 & if \quad x < 0 \\ 0.5 & if \quad 0 \le x \le 2 \\ 0 & if \quad x > 2. \end{cases}$$

Note that $\int_{-\infty}^{1.64} f_X(x)\, dx = 0.82$, as required.

Let us review all of this new terminology.

It is convenient to characterize probabilities with the help of *random quantities*.

Random quantities will be denoted by uppercase italic letters, like X, usually from the end of the alphabet.

A *distribution function* for a random quantity X is given by a function F_X whose value for an argument x, $F_X(x)$, is the probability that the quantity X will take on a value less than or equal to x. This is the probability associated with that subset of the sample space characterized by a value of X less than or equal to x.

Distributions come in two kinds: discrete and continuous.

If F_X is a discrete distribution, there is another discrete function f_X, called the *frequency function* corresponding to F_X. It has the property that for every real x,

$$F_X(x) = \sum_{-\infty}^{x} f_X(y) = \mathcal{P}(X \le x).$$

Note that there can only be a finite or countable number of values of y for which $f_X(y)$ is other than 0.

A continuous distribution has a continuum of possible values. If F_X is a continuous distribution, there is another function f_X, called the *density* function corresponding to F_X, with the property that

$$F_X(x) = \int_{-\infty}^{x} f_X(x)\,dx = \mathcal{P}(X \leq x).$$

Note particularly that $f_X(x)$ is *not* a probability in this case. A density function is just a handy auxiliary function.

3.4.2 Properties of Distributions

The *mathematical expectation* of a random quantity X is, loosely speaking, the average of X over a long run of trials. More precisely, we define:

Definition 3.3 (Expectation). *If X is a discrete random quantity, then the expectation of X is*

$$E(X) = \sum_i x_i f_X(x_i),$$

where the summation extends over all values of x_i such that $P(X = x_i) > 0$. If X is continuous, then the expectation of X is

$$E(X) = \int_{-\infty}^{\infty} x f_X(x)\,dx,$$

where f_X is the density function of X.

Example 3.16. *If X is the sum of the points on a roll of two dice, as in Example 3.12, then the expectation of X, $E(X)$, is*

$$E(X) = 2 \times \frac{1}{36} + 3 \times \frac{2}{36} + 4 \times \frac{3}{36} + 5 \times \frac{4}{36} + 6 \times \frac{5}{36} + 7 \times \frac{6}{36} + 8 \times \frac{5}{36}$$
$$+ 9 \times \frac{4}{36} + 10 \times \frac{3}{36} + 11 \times \frac{2}{36} + 12 \times \frac{1}{36} = 7.$$

Example 3.17. *Let Y represent the distribution of breaking points on the fishing line test described in Example 3.14. The expectation of Y is*

$$E(Y) = \int_{-\infty}^{\infty} x f_Y(x)\,dx = \int_{0}^{2} x \times \frac{1}{2}\,dx = 1.$$

There is obviously an important difference between a distribution in which almost all values of X are close to $E(X)$ and one in which the values of X tend to be spread widely around the mean $E(X)$. The most common measure of how spread out the values of X are is the *variance* of X, denoted by $D^2(X)$.

Definition 3.4 (Variance).

$$D^2(X) = E[(X - E(X))^2].$$

Example 3.18. *The tosses of a pair of dice described in Example 3.12 will provide a simple example. Recall that $E(X) = 7$; the possible values of $(X - E(X))^2$ are therefore*

25 *with probability* $\frac{2}{36}$, 16 *with probability* $\frac{4}{36}$, 9 *with probability* $\frac{6}{36}$, 4 *with probability* $\frac{8}{36}$, 1 *with probability* $\frac{10}{36}$, *and* 0 *with probability* $\frac{6}{36}$. *Thus,* $D^2(X) = \frac{210}{36}$.

Example 3.19. *An example of the variance of a continuous variable is provided by the fishing line of Example 3.14:* $D^2(Y) = \int_0^2 (x - E(Y))^2 \times \frac{1}{2}\, dx = \int_0^2 x^2 \times \frac{1}{2}\, dx - \int_0^2 2x \times \frac{1}{2}\, dx + \int_0^2 \frac{1}{2}\, dx = \frac{4}{3} - 2 + 1 = \frac{1}{3}$.

We know what it is for two subsets of the sample space to be independent: S and T are independent if $P(S \cap T) = P(S)P(T)$—or [equivalently if $P(S) > 0$], if $P(T) = P(T|S)$. We also speak of random quantities as independent. The random quantities X and Y are independent, provided the subsets of the sample space they pick out are independent, that is, for every real x_i and y_j such that $P(X \le x_i) > 0$ and $P(Y \le y_j) > 0$, one has $P(X \le x_i \cap Y \le y_j) = P(X \le x_i)P(Y \le y_j)$. If we think of X, Y as a two dimensional random quantity with distribution function $F_{X,Y}(x, y) = P(X \le x \cap Y \le y)$, this amounts to the claim that the distribution function $F_{X,Y}(x, y)$ is the product of the distribution functions $F_X(x)$ and $F_Y(y)$ for every x and y.

Let us take this as our definition of independence for random quantities:

Definition 3.5 (Independence). *X and Y are independent just in case for every real x and y,* $F_{X,Y}(x, y) = F_X(x)F_Y(y)$.

A number of elementary but useful theorems concerning independence, expectation, and variance follow:

Theorem 3.12. *If X and Y are random quantities, and both are discrete or both are continuous, then* $f_{X,Y}(x, y) = f_X(x)f_Y(y)$, *if and only if X and Y are independent.*

Theorem 3.13. $E(X + Y) = E(X) + E(Y)$.

Theorem 3.14. $E(aX + b) = aE(X) + b$.

Theorem 3.15. *If X and Y are independent, then* $E(XY) = E(X)\,E(Y)$.

Theorem 3.16. $D^2(X) = E(X^2) - [E(X)]^2$.

Theorem 3.17. $D^2(aX + b) = a^2 D^2(X)$.

Theorem 3.18. *If X and Y are independent, then* $D^2(X + Y) = D^2(X) + D^2(Y)$.

The two dimensional *joint* distribution of the quantities X and Y, $F_{X,Y}(x, y)$, is perfectly well defined when X and Y are not independent. It is just the probability that $X \le x \wedge Y \le y$. It may also be given by the joint density function (joint frequency function) $f_{X,Y}(x, y)$. Then $F_X(x)$ is the *marginal X* distribution function of $F_{X,Y}(x, y) = \int_{-\infty}^{\infty} f_{X,Y}(x, y)\, dy$. Similarly, $\int_{-\infty}^{\infty} f_{X,Y}(x, y)\, dy = f_X(x)$ is the marginal density of the quantity X.

It is important to note that whereas we can obtain the marginal distribution or density functions from the joint distribution or density function, we can only obtain the joint distribution function from the marginal distribution function in a special case: That in which the quantities involved are *independent*.

3.5   Sampling Distributions

Often we are interested in the properties of *samples* from a population. The sample space S may often be taken to be the set of equinumerous samples of that population. Let X be a random quantity defined for the population, and let X_i be the random quantity defined for sequences of n members of that population, whose value is $X(a_i)$, where a_i is the ith item in the sequence. If we are sampling with replacement, the random quantities X_i are independent.

Suppose we are drawing from an urn containing white and black balls. We draw a ball, observe its color, and replace it and remix the contents of the urn before drawing a second ball. Suppose we do this n times. A natural model takes as its sample space the n-fold product of the sample space consisting of the balls in the urn. The algebra of the original sample space consists of the empty set, the sample space itself, the set of black balls, and the complement of the set of black balls.

Let the random quantity X be the characteristic function of black balls: That is, it has the value 1 if its argument is black, and the value 0 otherwise. In the enlarged problem, the random quantity X_i has the value of X for the ith object. In particular, if s is a sample of n draws, $X_i(s)$ has the value 1 if the ith draw is black, and the value 0 otherwise.

Whether or not the quantities X_i are independent, the expected value of their sum is the sum of their expected values: $E(\sum_{i=1}^{n} X_i) = \sum_{i=1}^{n} E(X_i) = nE(X)$. Because the quantities are independent, the variance of the sum is n times the variance of each quantity: $D^2(\sum_{i=1}^{n} X_i) = \sum_{i=1}^{n} D^2(X_i) = nD^2(X)$.

Example 3.20. *Suppose the urn contains ten balls, of which three are black. An experiment consists of selecting five balls from the urn, one at a time, and replacing each ball after we have observed its color. Let B_i be the function whose value is 1 if the ith ball is black, and 0 otherwise. We suppose the chance of selecting each ball is the same. $E(B_i) = 0.3$; $D^2(B_i) = (0.3)(0.7) = 0.21$.*

Let FB(s) be the number of black balls among the five drawn, where s is a member of the new sample space. FB $= \sum_{i=1}^{5} B_i(s)$. The expected value of FB is $5(0.3) = 1.5$. (Note that the actual number of black balls on any run of this experiment must be integral: We can no more draw 1.5 black balls than an American family can have 2.2 children.)

The variance of the quantity FB is $\sum_{i=1}^{5} D^2(B_i) = 5(0.21) = 1.05$, making use of Theorem 3.18.

Intuitively, the smaller the variance of a random quantity, the more closely we may expect its values to be packed around its mean value. This last example suggests a possible way of making use of this. To obtain a quantitative handle on the fact that a smaller variance indicates a distribution that is more tightly packed about its mean, we turn to an important and powerful theorem. It is particularly relevant to inductive logic, as we shall see in later chapters. This theorem makes use of the *standard deviation* of a quantity.

Definition 3.6 (Standard Deviation).   *The standard deviation $D(X)$ of a quantity X is the positive square root of its variance: $D(X) = \sqrt{D^2(X)}$.*

Theorem 3.19 (Chebychef's Theorem). *Let X be a random quantity with* E(X) *and* D²(X) *finite. Then*

$$P(|X - E(X)| \geq kD(X)) \leq \frac{1}{k^2}.$$

Proof: We will use integrals ambiguously for sums (in case X is discrete) and integrals (in case X is continuous). Dividing the integral that gives the value of variance into three parts, we have

$$D^2(X) = \int_{-\infty}^{E(X)-kD(X)} (x - E(X))^2 f_X(x)\, dx \tag{3.1}$$

$$+ \int_{E(X)-kD(X)}^{E(X)+kD(X)} (x - E(X))^2 f_X(x)\, dx \tag{3.2}$$

$$+ \int_{E(X)+kD(X)}^{\infty} (x - E(X))^2 f_X(x)\, dx \tag{3.3}$$

Throw away the second term of the sum. Note that when x is less than $E(X) - kD(X)$, $x - E(X)$ is less than $-kD(X)$, so that $(x - E(X))^2$ is greater than $k^2D^2(X)$. Similarly, when x is greater than $E(X) + kD(X)$, $x - E(x)$ is greater than $kD(X)$, so that $(x - E(X))^2$ is greater than $k^2D^2(X)$. In the two terms remaining, replace $(x - E(X))^2$ by the smaller quantity $k^2D^2(X)$. We then have

$$\int_{-\infty}^{E(X)-kD(X)} k^2D^2(X) f_X(x)\, dx + \int_{E(X)+kD(X)}^{\infty} k^2D^2(X) f_X(x)\, dx \leq D^2(X).$$

Dividing by $k^2D^2(X)$, we have

$$\int_{-\infty}^{E(X)-kD(X)} f_X(x)\, dx + \int_{E(X)+kD(X)}^{\infty} f_X(x)\, dx \leq \frac{1}{k^2}$$

But the two integrals on the left give exactly the probability that the quantity X is more than a distance $kD(X)$ from $E(X)$.   ∎

Example 3.21. *Consider the urn of the Example 3.20. Suppose that instead of 5 draws, we considered 100 draws. Furthermore, rather than the random quantity* FB *we consider the quantity* RFB *(relative frequency of black), whose value is* $\frac{1}{100}$ FB. *The expected value of* RFB, *E(RFB), is 0.30; the variance is* $D^2(\text{RFB}) = \frac{1}{100}(0.21)$. *The standard deviation is 0.046.*

Applying Chebychef's theorem, we may infer that the probability that the relative frequency of black balls in a hundred draws will fail to be within 2 standard deviations of the mean is no more than a quarter: $P(0.288 \leq \text{RFB} \leq 0.392) \leq \frac{1}{4}$. *Similarly, the probability that the relative frequency of black balls in a hundred draws will fail to be within 4 standard deviations of the mean is no more than* $\frac{1}{16}$: $P(0.116 \leq \text{RFB} \leq 0.484) \leq \frac{1}{16}$.

In later chapters, we will look at the ways in which various writers have exploited the consequences of Chebychef's theorem.

3.6   Useful Distributions

Whereas there are many probability distributions that have found application in the sciences, there are two that are of particular importance from the point of view of the analysis of induction. One of these is discrete, the binomial distribution, and one is continuous, the normal distribution.

The binomial distribution arises naturally when we are concerned with a repeatable experiment that may yield success or may yield failure. Calling the experiment repeatable suggests that the probabilities of the outcomes are the same on various trials of the experiment. The experiment is similar to the urn situation described in Example 3.21.

More formally, a binomial distribution is characterized by two parameters: p, the probability of success on a single trial (in the example, the probability of getting a black ball on a single draw), and n, the number of trials being considered (in the example, the number of balls drawn).

The sample space consists of sequences of n trials. Again, let the quantity X_i take the value 1 if the ith trial results in success, and the value 0 otherwise. The outcome of n trials can be represented by a sequence of n zeros and ones. Thus, if n is 5, $\langle 0, 1, 0, 1, 1 \rangle$ represents a sequence of five trials of which the first and third were failures, and the second, fourth, and fifth successes.

In view of the fact that the quantities are independent, we can calculate the probability of any particular outcome. For example $\langle 0, 1, 0, 1, 1 \rangle$ has the probability $(1 - p)(p)(1 - p)(p)(p)$, because the probability of failure on the first trial is $1 - p$; the probability of success on the second trial is p, because success on the second trial is independent of what happened on the first trial; and so on. In general, the probability of a sequence of r successes and $m = n - r$ failures in a specified order will be the product of r p's and m $(1 - p)$'s, or $p^r(1 - p)^{n-r}$.

In a binomial distribution we care only about the numbers of successes and failures on the n trials, rather than the particular order of the successes and failures. Thus, the probability of r specified successes and $n - r$ specified failures must be multiplied by the number of ways of choosing r out of n objects, $\binom{n}{r}$.

Definition 3.7 (Binomial Distribution). *A random quantity X has a binomial distribution, characterized by the parameters p and n, if and only if for every integer x_i,* $0 \leq x_i \leq n$,

$$f_X(x_i) = \binom{n}{x_i} p^{x_i}(1 - p)^{n-x_i}.$$

It is easy to check that this is indeed a probability distribution.

Theorem 3.20.

$$\sum_{x_i=0}^{n} f_X(x_i) = 1.$$

Proof: Let $q = 1 - p$. Expand $(p + q)^n$ by the binomial theorem. We get exactly all the terms of the form $\binom{n}{x_i} p^{x_i}(1 - p)^{n-x_i}$, and their sum, being 1^n, is 1. (Hence the name "binomial".) ∎

The normal distribution is the distribution whose density function is represented by the famous bell curve, or error curve, or Gaussian curve (Figure 3.5). The normal distribution has two parameters, μ and σ, representing the mean and standard deviation of the normally distributed quantity. The density function defines the distribution.

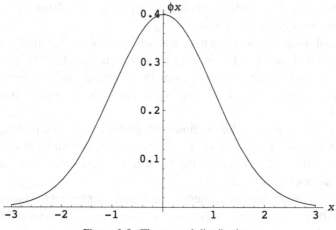

Figure 3.5: The normal distribution.

Definition 3.8 (**Normal Distribution**). *The quantity X is normally distributed if and only if the distribution has a density function of the form*

$$\phi(x) = \frac{1}{\sqrt{2\pi}} e^{-(x-\mu)^2/2\sigma^2}.$$

We present without proof:

Theorem 3.21. *If X has a normal distribution with parameters μ and σ, then $E(X) = \mu$ and $D(X) = \sigma$.*

The pervasiveness of the normal distribution is partly explained by a theorem that we will also state without proof, the *central limit theorem*:

Theorem 3.22 (**Central Limit Theorem**). *If X is the sum of a large number of independent and identically distributed random quantities, then the distribution of X will be approximately normal.*

Example 3.22. *If X has a binomial distribution, with parameters $n > 15$ and $p > 0.1$, then the distribution of X is very nearly normal.*

The theorem can be strengthened in various ways; roughly speaking, almost any sum of random quantities, provided only that they are not very dependent and none of them dominates the sum, will be approximately normally distributed.

3.7 Summary

Formally, probability is an additive function with the range [0, 1] defined on a field. It is characterized by a few simple axioms. If the field is a σ-field, i.e., contains the union of a countable number of its members, the probability axioms may include a corresponding axiom of countable additivity.

It is convenient to take conditional probability as primitive, and to define unconditioned probability as the value of a probability conditioned on the universe of the field.

Random quantities provide a convenient way to talk about many probabilities. They are either discrete or continuous. In the discrete case, probabilities may be given by either frequency functions or distribution functions. In the continuous case, probabilities may be given by either density functions or distribution functions.

It is useful to consider the expectation and variance of random quantities. The standard deviation of a random quantity is the positive square root of its variance.

The expectation of a sum of random quantities is the sum of their expectations. The variance of a sum of *independent* random quantities is the sum of their variances.

Chebychef's theorem shows that it is very probable that the value of a random quantity will be near its expected value, where nearness is measured in units of its standard deviation.

There are relatively few standard distribution functions that are of interest to us in what follows: mainly the binomial distribution and the normal distribution.

3.8   Bibliographical Notes

A very large number of fine texts in probability theory are available. Among the classic texts we may mention Cramér [Cramér, 1955] and Neyman [Neyman, 1950]. Goldberg [Goldberg, 1960] is a short, clear, elementary exposition of the basic ideas. For obvious reasons [Kyburg, 1969] exhibits a close fit to the ideas presented in this chapter. There is nothing very controversial about the subject matter itself. When we come to the application and interpretation of the probability calculus, on the other hand, we shall encounter serious controversies.

3.9   Exercises

(1)   Prove Theorem 3.2.

(2)   Prove Theorem 3.3.

(3)   Prove Theorem 3.4.

(4)   Prove Theorem 3.5.

(5)   Prove Theorem 3.6.

(6)   Prove Theorem 3.7.

(7)   Prove Theorem 3.8.

(8)   Prove Theorem 3.9.

(9)   A coin is tossed three times. What is the sample space? What probabilities would you assign to each point?

(10)   Two balls are drawn (without replacement) from an urn containing three red, two green, and five blue balls. What is the probability that the first is red and the second is blue? What is this probability if the draws are made with replacement?

(11)   A die has two sides painted red, two yellow, and two green. It is rolled twice. What is the sample space? What probability should be assigned to each point?

(12)   What is the probability of rolling a 4 with one die? What is the probability of rolling an 8 with two dice?

(13)   One urn contains two white and two black balls; a second urn contains two white and four black balls. (a) If one ball is chosen from each urn, what is the probability that they will be the same color? (b) If an urn is selected at random, and one ball drawn from it, what is the probability that the ball will be white? (c) If an urn is

selected at random, and two balls are drawn from it (without replacement), what is the probability that they will be the same color?

(14) Assume that the sex of a child is determined by a random process, and that the proportion of males is 0.5. In a family of six children: (a) What is the probability that all will be of the same sex? (b) What is the probability that the four oldest will be girls, and the two youngest boys? (c) What is the probability that exactly half the children will be girls?

(15) A coin is tossed until a head appears. What is the sample space? What is the probability that a coin will first appear on the fourth toss? What is the probability $f(x)$ that heads will appear for the first time on the xth toss?

(16) Six dice are rolled. Find an expression $f(x, y)$ for the probability of getting x ones and y twos.

(17) One die is rolled. A second die is rolled a number of times corresponding to the points showing on the first die. What is the probability that the total number of points is 4? Find a general expression for the probability of getting z points.

(18) Assume a person is equally likely to be born on any day of the year. How many people in a group do there have to be in order to have a better than even chance that two of them have the same birthday?

(19) Let p be the probability that a man will die within a year. Assume that the deaths of A, B, C, and D are independent. Find the probability that

(a) A will die within the year.

(b) A and B will die within the year, but C and D will not die within the year.

(c) Only A will die within the year.

(20) Suppose a sharpshooter has probability $\frac{1}{2}$ of hitting the bull's-eye. Assuming the shots are independent, how many shots does he have to fire for the probability to be at least 0.99 that the bull's-eye is hit at least once?

(21) One urn contains two white and two black balls; a second urn contains two white and four black balls.

(a) If one ball is chosen from each urn, what is the probability that they will be the same color?

(b) If an urn is selected at random, and one ball drawn from it, what is the probability that the ball will be white?

(c) If an urn is selected at random, and two balls are drawn from it (without replacement), what is the probability that they will be the same color?

(22) Prove Theorem 3.12.

(23) Prove Theorem 3.13.

(24) Prove Theorem 3.14.

(25) Prove Theorem 3.15.

(26) Prove Theorem 3.16.

(27) Prove Theorem 3.17.

(28) Two dice are rolled. Find an expression $f(x, y)$ for the probability of getting x ones and y twos. Calculate $E(N_1)$, where N_1 is the number showing on the first die. Calculate $E(N_1 + N_2)$, where N_2 is the number showing on the second die.

(29)   One die is rolled. A coin is tossed a number of times corresponding to the dots showing on the die. Every head counts as one point. What is the probability that the total number of points is 4? Find a general expression for the probability of getting z points.

(30)   Find the expectation and variance of Z, the number of points in the preceding problem.

(31)   Let D be the set of throws of one die. Y is defined as the function that takes an element of D into the square of the number of dots showing uppermost in that element of D. Find the frequency function and distribution function of Y, on the usual assumptions.

(32)   Find the distribution function of the random quantity Z whose frequency function is 0 for $x \leq 0$, x for $0 < x \leq 1$, $2 - x$ for $1 < x \leq 2$, and 0 for $x > 2$. (You might start by showing that this is, indeed, a frequency function.) Compute $E(Z)$ and $D^2(Z)$.

(33)   Suppose X is uniformly distributed in $[2, 4]$ and Y is uniformly distributed in $[2, 5]$, and that X and Y are independent. Compute $\mathcal{P}(X < 3 \wedge Y < 3)$. Find the joint frequency and distribution functions of X, Y. Find the frequency function of $W = X + Y$.

(34)   Under the circumstances of the previous problem, find the probability that $XY < 6$. More generally, find the distribution function of $Z = XY$.

(35)   Let $F_X(x) = k(x - a)^2$ for x in $[a, b]$, 0 for $x < a$, 1 for $x > b$. Compute the expectation of X^3 and the variance of X. (By the way, what is k?)

(36)   Find k so that $F_{X,Y}(x, y) = kx(1 - y)$ for x and y in $[0,1]$ is a frequency function. What are the marginal frequency functions? Are X and Y independent? Compute the mean and variance of $X + Y$ and XY.

(37)   Let $f_{X,Y}(x, y) = 3x$ for $0 \leq y \leq x \leq 1$. What are the marginal frequency functions? Are X and Y independent? Compute the mean and variance of $X + Y$ and XY.

(38)   Show that when X is nonnegative, with mean m_X, the probability that X exceeds km_X is less than $1/k$.

Bibliography

[Bernoulli, 1713] Jacob Bernoulli. *Ars Conjectandi.* 1713.

[Cramér, 1955] Harald Cramér. *The Elements of Probability Theory.* Wiley, New York, 1955.

[Goldberg, 1960] Samuel Goldberg. *Probability: An Introduction.* Prentice Hall, Englewood Cliffs, NJ, 1960.

[Kyburg, 1969] Henry E. Kyburg, Jr. *Probability Theory.* Prentice Hall, Englewood Cliffs, NJ, 1969.

[Laplace, 1951] Pierre Simon Marquis de Laplace. *A Philosophical Essay on Probabilities.* Dover Publications, New York, 1951.

[Leblanc & Morgan, 1983] Hugues Leblanc and Charles Morgan. Probability theory, intuitionism, semantics, and the Dutch book argument. *Notre Dame Journal of Formal Logic*, 24:289–304, 1983.

[Morgan, 1998] Charles Morgan. Non-monotonic logic is impossible. *Canadian Artificial Intelligence Magazine*, 42:18–25, 1998.

[Neyman, 1950] Jerzy Neyman. *First Course in Probability and Statistics.* Henry Holt, New York, 1950.

[Popper, 1959] K.R. Popper. *The Logic of Scientific Discovery.* Hutchinson, London, 1959.

4

Interpretations of Probability

4.1 Introduction

In Chapter 3, we discussed the axioms of the probability calculus and derived some of its theorems. We never said, however, what "probability" *meant*. From a formal or mathematical point of view, there was no need to: we could state and prove facts about the relations among probabilities without knowing what a probability *is*, just as we can state and prove theorems about points and lines without knowing what they are. (As Bertrand Russell said [Russell, 1901, p. 83] "Mathematics may be defined as the subject where we never know what we are talking about, nor whether what we are saying is true.")

Nevertheless, because our goal is to make use of the notion of probability in understanding uncertain inference and induction, we must be explicit about its interpretation. There are several reasons for this. In the first place, if we are hoping to follow the injunction to believe what is probable, we have to know what is probable. There is no hope of assigning *values* to probabilities unless we have some idea of what probability means. What determines those values? Second, we need to know what the *import* of probability is for us. How is it supposed to bear on our epistemic states or our decisions? Third, what is the domain of the probability function? In the last chapter we took the domain to be a field, but that merely assigns *structure* to the domain: it doesn't tell us what the domain objects are.

There is no generally accepted interpretation of probability. The earliest view, often referred to as the "classical" interpretation of probability, was that mentioned in Chapter 3 as involving the counting of "equipossible" cases. We dismissed that view there as not germane to the formal concerns of Chapter 3, but we promised to reconsider the view here.

4.2 The Classical View

The idea that probability could be interpreted as a ratio of favorable cases to a totality of cases is generally attributed to Laplace [Laplace, 1951] but may, according to some writers [Hacking, 1975], go back to Leibniz. The idea makes good sense when we are thinking (as all the early probabilists were thinking) of games of chance. The cases in such games are well defined: a die comes to rest on a level surface with one of six sides uppermost. A coin, tossed in the usual way, lands with the head up or with the tail up. These are *cases*. More complicated cases can also be considered: the number of cases constituting the partition of the sample space appropriate to hands of bridge is enormous—the number of ways of dividing 52 cards into four hands of 13 cards each, or

$$\binom{52}{13\ \ 13\ \ 13\ \ 13} = \frac{52!}{13!\ 13!\ 13!\ 13!}.$$

Even in these complicated circumstances, the cases are easy to identify, and, what is equally to the point, they tend to occur with roughly equal frequencies.

Of course, people can gamble on anything, and not merely on well-regimented games of chance, but there are intermediate examples in which counting cases does not seem to be quite right. For example, as popular as craps or roulette are horse races. But nobody (at least nobody who is solvent) would propose that the correct probabilities in a horse race are to be obtained by counting the various permutations of the order in which the racers might finish as "equally possible cases".

But this does not mean that it is impossible to imagine that a set of cases exists in an abstract sense from which we can generate a sample space that has the structure we think appropriate, whatever that structure may be. For example, consider a race between two horses, Gallagher and Tweeny. Suppose that we want to say that the probability that Gallagher wins is 0.35. One possibility is to imagine that the sample space S consists of a very large set of similar "trials" of this race, under similar conditions. This large number of trials is partitioned into 100 equipossible cases, of which 35 are favorable to Gallagher winning. Of course, we don't know how to describe those cases intuitively, but it could be maintained that the very fact that we say that the probability is 0.35 entails that we are assuming that these cases exist. Put more generally, when we say that the probability is 0.35 that Gallagher wins, we are assuming that there are N *similar kinds* of race, of which 35% have the result that Gallagher wins.

Thus, the view that probability is the ratio of favorable cases to the total number of equipossible cases is not entirely implausible. Due to its historical popularity, this view has become known as the *classical interpretation* of probability.

The classical interpretation has had an enormous influence over the development of probability theory; it was the definition of probability provided in probability textbooks right into the twentieth century, in part due to its simplicity. Despite its great influence, it has been generally rejected as an adequate interpretation of the probability calculus.

As we noted in the previous chapter, the classical interpretation suffers from an apparent circularity. When we define probability as the ratio of favorable cases to the total number of *equally possible* cases, we need to explain what we mean by "equally possible." At first blush we might say that we mean equally probable, but such an approach immediately leads to circularity. This circularity can be overcome if a definition of equally possible is given which is independent of the definition of probability. To do so the classical probabilists offered the *principle of indifference*. According to that principle, any set of possibilities is equally possible if we have "no reason" to prefer one possibility over the others.

According to the classical definition, probability is the ratio of successful outcomes to the total number of outcomes where each outcome is equally possible. For example, given a six sided die, it is equally possible that on a given roll of that die any particular side will land facing up. The classical definition of probability clearly covers such events. But what happens if we shave a piece off of one of the corners of the die?

It is precisely the symmetry of the die which gives us grounds for claiming that it is "equally possible" that any of the six sides will come up. With the elimination of that symmetry we destroy that equipossibility. It is no longer the case that we have six sides which are all equally possible. The classical definition of probability seems not to be able to handle this situation, for it gives us no way to arrive at a different distribution of equipossibility. Let us put this case to one side, and ask whether, if we do have clear-cut cases, the probability calculus should apply. Even in this limited domain the classical interpretation of the probability calculus will run into difficulties.

We may argue that this principle is implausible simply because it is not acceptable to use ignorance as a basis for knowledge. Of course, many writers take the application of the principle of indifference to be based on a symmetry of reasons, rather than mere lack of reasons. But this is not a complete answer to the difficulties of applying the principle without a complete specification of how strong these reasons that are balanced must be.

A more pressing problem with the principle of indifference is that it can lead to outright inconsistency. Imagine that we know there is life on Vulcan but we are ignorant as to whether it is carbon based (like us), silicon based, or germanium based. If we ask what is the probability that carbon based life exists on Vulcan we get, by the principle of indifference,

$$P(Lc) = 0.5$$
$$P(\neg Lc) = 0.5$$

where "Lc" stands for carbon based life existing on Vulcan.

We can also ask what is the probability that there is either carbon based life (Lc), silicon based life (Ls), or germanium based life (Lg) on Vulcan. By the principle of indifference we get

$$P(Lc) = 0.333$$
$$P(Ls) = 0.333$$
$$P(Lg) = 0.333.$$

These different queries about the probabilities of the existence of types of life on Vulcan leads to a quandary. By the first approach we get $P(Lc) = 0.5$, but by the second approach we get $P(Lc) = 0.333$. Clearly this is a contradiction. It is not logically possible for the probability of the same event to be both 0.333 and 0.5.

The principle of indifference can lead to contradiction in another way. Here is an example due to Nagel [Nagel, 1939].

Suppose you have a sample of liquid, and you know that its specific gravity lies between 1 and 2 grams per cubic centimeter. The principle of indifference would suggest that the probability is 0.5 that the liquid has a specific gravity of less than 1.5 g/cc. But put the same problem another way: You know the specific volume of the sample lies between 0.5 cc/g and 1.0 cc/g. The principle of indifference suggests that

It has a specific volume greater than 0.75 cc/g if and only if it has a specific gravity of less than 1.33 g/cc. But the claim that the probability is 0.5 that the liquid has a specific gravity of less than 1.5 g/cc is inconsistent with the claim that the probability is 0.5 that it has a specific gravity of less than 1.33 g/cc unless the probability is 0.0 that the specific gravity lies between 1.33 g/cc and 1.5 g/cc.

Despite the fact that these difficulties can often be circumvented (if only by limiting the scope of its application), the principle of indifference on which the classical interpretation of probability came to be based was neither so clear-cut nor so compelling as to lead to universal agreement.

4.3 Empirical Interpretations of Probability

Writers from the middle of the nineteenth century on have noticed these difficulties, and attempted to provide interpretations of probability that would meet them. Several interpretations of probability have been given in the twentieth century; they fall into three groups. These are (i) empirical interpretations, (ii) logical interpretations, and (iii) subjective interpretations. None has been thought to be wholly satisfactory, but elements of each are still vigorously defended by contemporary writers.

There are several ways in which we can treat probability as an empirical concept. One proposal is to construe probability directly as a relative frequency.

The uses of probability associated with insurance, social statistics, and scientific experiments in biology and agriculture as these began to be performed in the nineteenth and twentieth centuries were generally based on counting instances and looking at relative frequencies. The great insurance companies like Lloyd's looked to historical frequencies to guide them in assessing premiums, and not to the principle of indifference or the counting of possible cases.

The *relative frequency* of an event of kind A among a set of events of kind B is just the ratio of the cardinality of events of kind $A \cap B$ to the cardinality of events of kind B. This assumes that these cardinalities are finite. Given our current cosmological notions, it makes good sense to suppose that empirically given sets are finite: In the history of the universe the number of tosses of coins will be finite. The number of crows and the number of people will be finite (though we may not be able to put sharp bounds on either).

For example, we may take the probability of a head on a toss of a particular coin to be the relative frequency with which that coin comes up heads on all the times (past and future) that it is tossed. A difficulty of this view that must be faced is that it undermines the usual applications of the probability calculus.

Usually we calculate the probability of two heads in a row by assuming that the outcomes of the tosses are independent. But if we suppose that the probability of heads is the relative frequency of heads among the tosses of the coin and that the number of tosses is finite, the outcomes of the tosses are not independent. If there are to be a total of N tosses, then the probability of heads on the second toss is the relative frequency of heads on the remaining $N - 1$ tosses, which cannot be the same as the relative frequency among the original N tosses, however large N may be. The fact that we do not know the value of N is irrelevant: We do know that the event of getting heads on the second toss is *not* independent of getting heads on the first toss.

There are two ways in which this annoying difficulty can be dealt with. One is to world, in the sense that it is to be interpreted as the limiting relative frequency in an unbounded class of events. In that case when we have observed a head, the relative frequency of heads among the remainder of the events of coin tossing can remain unchanged. The other is to take probability to be a physical *propensity* of a chance setup [Hacking, 1965]. On this latter view, the probability is not to be identified with the relative frequency we observe (or might observe) but with a property of the physical arrangement—the coin and the tossing apparatus—that gives rise to that empirical frequency.

4.3.1   The Limiting Frequency Interpretation

The limiting frequency view of probability was extensively developed by von Mises [von Mises, 1957]. It is still highly regarded in many circles, particularly among the majority of statisticians, though most no longer worry about von Mises's concern with randomness. On this view, probability only makes sense when it is referred (usually implicitly) to a hypothetical *infinite* sequence of trials. The probability of a *kind* of event in an infinite sequence of trials is defined as the limit of the relative frequency of events of that kind in the sequence. Remember, though, that the whole construction is designed to be an *idealization* of the processes we find in the world.

The formal objects with which we are to be concerned in this interpretation of probability are infinite sequences and infinite subsequences of them. The sample space consists of the infinite sequences; the σ-algebra is determined by constraints on the sequences. For example, let S consist of infinite sequences of rolls of a die. The constraint that the first roll yielded a 1 picks out an infinite number of those sequences—viz., those whose first element has the property of yielding a one.

Let $E_{i \in \omega}$ be an infinite sequence of events of a certain sort, for example, tosses of a coin. Some of these events may have the property of yielding heads; let H be a predicate denoting this property. We say that $H(E_i)$ is true if (and only if) the ith event yields heads.

Consider a specific sequence of tosses $E = \{E_1, E_2, \ldots\}$. Clearly there is nothing problematic about defining the relative frequency of heads among the first n members of E, $\mathrm{relf}_{H,E}(n)$: it is simply $1/n$ times the number of true statements of the form $H(E_i)$ with i less than or equal to n. Let p be a real number (for example 0.5). Using the standard mathematical conception of a limit, the limit, as n approaches infinity, of $\mathrm{relf}_{H,E}(n)$ is p if and only if, for every δ, there is an N such that for any m, if $m > N$, then $|\mathrm{relf}_{H,E}(m) - p| \leq \delta$. In symbols we define probability as the limit of the relative frequency of H in the infinite (ideal) sequence E:

Definition 4.1. $\mathcal{P}(H, E) = p$, *if and only if*

$$(\forall \delta)(\exists N)(\forall m)(m > N \rightarrow |\mathrm{relf}_{H,E}(m) - p| \leq \delta).$$

For the probability to exist, the limit must exist. This need not necessarily be the case. For example, the sequence $0, 1, 1, 0, 0, 0, 1, 1, 1, 1, 1, 1, 0, \ldots$, in which the relative frequency of 1's varies from $\frac{1}{3}$ to $\frac{2}{3}$ and back again, has no limiting frequency of 1's.

Von Mises imposes an additional condition, beyond the existence of a limit, on sequences that are to qualify as *collectives* (*Kollektives*), that is, as sequences supporting with respect to a property (or subsequence determined by a property) if the relative frequency is invariant under place selection. The limit of the relative frequency of heads in E should be the same in the subsequence consisting of the odd-numbered tosses, in the subsequence consisting of the prime-numbered tosses (there are an infinite number of them), in the subsequence consisting of every third toss starting with the second, in the subsequence consisting of tosses made after a toss resulting in heads, and in the subsequence consisting of tosses made while the tosser is smoking a cigar, under the idealizing assumption (always in the back of von Mises's mind) that there are an infinite number of tosses made while the tosser is smoking a cigar.

The intuition behind this requirement is that the sequence $0, 1, 0, 1, 0, 1, 0, 1, \ldots$ does not fit our idea of a random sequence. It would be easy to devise a foolproof gambling system for betting on such a sequence: bet in favor of 1 after observing 0, and on 0 after observing 1. A gambling system is a systematic way of deciding, on the basis of the part of the sequence that has been observed, how to bet on the next event. As von Mises states the requirement: No gambling system should be possible in a collective. Whatever subsequence of the collective one chooses, the limiting frequency of the target property H should be the same.

We shall later see that this is a problematic constraint. For the time being, let us suppose that it makes sense, and continue with von Mises's construction.

It is easy to see that if the limit of the relative frequency in a collective exists, it must be between 0 and 1. Furthermore, if the limiting frequency of events of kind H exists and is p, the limiting frequency of events of kind $\overline{H}$ exists and is $1 - p$. To go beyond this requires constructing new collectives from old ones. For example, von Mises introduces a mixing operation on collectives that represents disjunction. Consider a collective that represents the limiting frequency of ones on throws of a die; consider a collective that represents the limiting frequency of threes on throws of a die. We may naturally consider a collective constructed by "mixing" these two collectives—namely, one in which we consider the relative frequency of getting either a one or a three. It is possible to show that if the limit of the relative frequency of ones in the first collective is p_1 and the limit of the relative frequency of threes in the second collective is p_3, then the limit of the relative frequency of ones or threes, in the third collective (a) exists, and (b) is equal to $p_1 + p_3$. (This depends on the fact that the outcome cannot be both a one and a three.)

There is also another way of constructing a collective from two given collectives. Let one collective, K_1, represent the limiting frequency of the property H_1, and another, K_2, represent the limiting frequency of the property H_2. We may construct a new collective consisting of those events in K_1 that correspond to the events in K_2 in which H_2 occurred. For example, let K_1 represent the limiting frequency of rain, among days in general, and let K_2 represent the limiting frequency of clouds on the horizon among days in general. We may be able to construct a new collective representing the limiting frequency of rain on days on which there are clouds on the horizon. This is not assured (though if the two events are independent, it is assured), because there may be regularities that preclude the sequence of days on which there are clouds on the horizon from being a collective with respect to rain—there may be a place selection that does not yield the same limiting frequency. But if it exists, the new collective represents the *conditional* probability of rain, given clouds.

Von Mises thus can capture the classical probability calculus with his limiting fruitful source of mathematical results. The requirement that there be no subsequence of a collective K in which the limiting frequency of a property H differs from that in K as a whole is clearly impossible to meet except in sequences in which every member has the property or every member lacks the property. The reason is that the modern conception of a function as a set of ordered pairs entails that there is a function f from the natural numbers to 0, 1 such that the event E_i has the property H if and only if $f(i) = 1$. The function f thus constitutes a perfect gambling system. The only problem is that we don't know what f is; nevertheless, we know it exists, and thus we know that there is a subsequence of any putative collective K that prevents it from being a collective.

One natural response to this observation is to propose that we restrict our attention to "natural" subsequences. If the sequence of tosses of a coin were $HTHTHTHTHT\ldots$, it would not be hard to devise a gambling system that would guess correctly more than 50% of the time. One proposal anent randomness was to require that collectives be "after-effect free"—that is, that the frequency of H after events with the property H be the same as the frequency of H in general, that the frequency of H after any fixed pattern of H's among n successive events be the same as that of H in general, etc. It is possible to show that infinite sequences exist in which the limiting frequency of H exists, and in which the occurrences of H exhibit this property of aftereffect freedom.

But what other constraints are possible? This question has led to the study of what functions are practically computable that would help the gambler beat the house. And this study is part of the modern theory of computability. A rough modern answer to what constitutes randomness in von Mises's sense is that a finite sequence is random with respect to a property H if any program that told you which objects in the sequence had the property H would be as long as the sequence itself.

An alternative approach is to deny the importance, except from a pragmatic point of view, of the existence of subsequences that have different limiting frequencies of the target property. Hans Reichenbach [Reichenbach, 1949] adopted this point of view. He provided a limiting relative frequency interpretation of probability in which von Mises's formal requirement of randomness is rejected. For Reichenbach, a probability sequence is given empirically. That collectives can be constructed he takes to be of mathematical interest, and that these constructions provide a model of the probability calculus is philosophically interesting primarily because it demonstrates the consistency of the limiting frequency interpretation.

For Reichenbach, any empirically given sequence can be a basis for a probability. It may be finite; it need not exhibit disorder. What serves the place of von Mises's requirement of randomness is the selection of an appropriate reference sequence for each particular application of the theory. The appropriateness of a particular choice is a pragmatic matter.

For both Reichenbach and von Mises, probability statements, because they concern the properties of infinite sequences of events, are theoretical. They cannot be verified in the way that observation statements like "the cat is on the mat" can be. The question, "How do you know the probability of heads is a half?" is, for both von Mises and Reichenbach, incomplete. A probability only makes sense if it includes reference to the collective, or reference sequence. The correct form of the question is "How do you know that the probability of heads *in the sequence of coin tosses T* is a half?"

But how do we answer this last, correctly formulated question? For von Mises the question is analogous to any other question about theoretical knowledge, e.g., "How do you know that gravitational force varies inversely with the square of the distance?" The idea of using probability to help us to understand scientific inference is, for von Mises, to put the cart before the horse: probability represents just one of the various kinds of scientific conclusions we would like to be able to justify.

This point of view is bolstered by reference to the *scientific* use of probability: in kinetic theory and quantum mechanics, as well as in characterization of gambling apparatus and in demographics. There is something to be said about how we decide what probability statements to accept; von Mises said little about this in detail, but, as we shall see, the influential British–American school of statistics adopts just this point of view. What we need are pragmatic rules, such as those provided by statisticians, for deciding when to accept a probability statement. This is not to say the rules are "merely pragmatic" in a pejorative sense of that phrase; we can explore the consequences of using one rule or another and find good reasons for using the rules that we use. The rules themselves, however, are pragmatic, rather than logical; they represent actions rather than arguments.

Reichenbach's view is similar, except that he explicitly proposes a particular general rule: having observed n As, of which m have been Bs, *project* that the limiting frequency of Bs among As is m/n. This simple rule, combined with the fact that Reichenbach imposes no constraints on the infinite sequences that can serve as the reference sequences for probabilities, creates a formidable problem, to which we will return repeatedly in this work: the problem of the reference class. Any event E_i belongs to a large (or even infinite) number of *kinds*. Each of these kinds may have exhibited a different historical frequency of events with property H. Each of these kinds may therefore be thought of as forming an infinite sequence to which E_i may belong. To which kind should we refer the probability that event E_i has the property H?

Example 4.1. *Let E_i be the toss of a coin I shall next perform. It is the toss of a quarter; it is the toss of a coin of U.S. mintage; it is the toss of a coin by me; it is the toss of a coin by a member of my family; it is the toss of a coin by a college professor; it is the toss of a coin performed in Lyons during December; .... Let these classes be $\{C_1, C_2, \ldots\}$. I have some information about each of them: m_1/n_1 of C_1 have yielded heads, m_2/n_2 of C_2 have yielded heads, etc. Each of these historical frequencies, according to Reichenbach, can (correctly) be projected into the future as a long run or limiting frequency.*

In a sense each projection can be taken to correctly characterize a probability concerning heads on the next toss. But if we want to use probability as a guide to belief, or as an ingredient in a decision theory, we can only make use of one value of the probability.

This leads us to the *problem of the single case*—the problem of applying probability to specific events, such as the next toss of the coin, or the next person to take out insurance with our company. The contrast between von Mises's more mathematical approach to the frequency interpretation of probability and Reichenbach's more pragmatic approach to the same problem can be best appreciated if we consider how they look at applying probabilities in the real world.

For neither Reichenbach nor von Mises can we literally apply probability to a single event. Probability is a property of sequences, not individual events within sequences of events. To talk of the *probability* of heads on the next toss, the *probability* of suffering a loss on the next applicant for insurance, is, for both von Mises and Reichenbach, just plain mistaken. There is no such probability. We can only talk of the limiting frequency of heads in a *sequence* of tosses, the limiting frequency of losses in a *sequence* of applicants, for every probability is the limit of a relative frequency in a sequence. Thus, von Mises deals with the single case simply by saying that it is not his concern; his concern is with science, and the concern of science is the general, not the particular.

For von Mises, the probability of a single event, for example the probability of death of a particular individual, is not a subject for the probability calculus. He writes, "The theory of probability [is] a science of the same order as geometry or theoretical mechanics . . . just as the subject matter of geometry is the study of spatial phenomena, so probability deals with mass phenomena and repeatable events" [von Mises, 1957, p. vii]. He also writes ". . . the solution of a problem in the theory of probability can teach us something definite about the real world. It gives a prediction of the result of a long sequence of physical events; this prediction can be tested by observation" [von Mises, 1957, p. 63].

Example 4.2. *We can calculate the limiting frequency with which sets of n draws from an urn containing equal numbers of red and black balls yield samples in which the sample frequency of red balls is 0.50, on the assumption that the draws are binomial with parameters $r = 0.50$ and n. Suppose this frequency is 0.72. What this tells us about the world is that if our model is correct, and if we perform the experiment of making n draws from the urn a great many—indeed, for von Mises, an infinite number of—times, then the limiting frequency with which such experiments yield samples consisting of 50% red draws is 0.72.*

Notice that, in conformity with the assertion that probability deals with long sequences of repeatable events, this assertion about the world says nothing about the outcome of any particular experiment of drawing n balls, however large n may be. The experiment of drawing n balls is a single (complex) experiment.

Although von Mises contends that probability theory cannot say anything about the probability of a single event, we do use statements of probability that employ the indefinite article "a" to make claims about what appears to be a single event: "the probability of getting a six in *a* throw of a die is $\frac{1}{6}$," "the probability that *a* birth is the birth of a girl is 0.48", etc. Von Mises construes these statements as elliptical versions of probability statements concerning the general case.

But Reichenbach does, where von Mises does not, seek to come to terms with the ordinary language use of "probability" in which we talk of the probability that a specific lottery ticket will win the lottery, the probability of getting heads on the next toss of a coin, the probability that Mary's baby will be a girl, and so on. In the time-honored fashion of philosophers, he does so by inventing new terminology: what we apply to the next toss is a *posit*. It is not a relative frequency, because one cannot find a meaningful relative frequency in a class containing only one member— the class whose only member is the next toss. But clearly some frequencies are more relevant to the next toss than others, and if, as a practical matter, we can pick out the most relevant frequency, we can use that as a basis for a posit concerning a particular event—e.g., the next toss. The doctrine works like this.

Although probabilities can only literally apply to aggregates—the sequence of to allow us to relate the general relative frequency and the single case. A posit is a guess at a truth value. In the simplest case, we might posit the occurrence of a kind of event that happens more often than not in the long run. A more sophisticated procedure would lead us to assign weights to posits: and these weights, on Reichenbach's view, are exactly the probabilities of the event in question in the *appropriate* reference class. These weighted posits are reflected in ordinary language in talk about the probability of the single case. Literally, it is nonsense to talk of the "probability" of the next draw from the urn producing a red ball; but the doctrine of posits gives us a way to make sense of the common language claim that the probability of the next draw being that of a red ball is *p*. Although for Reichenbach the claim is not correct, we must strive to make sense of our ordinary intuitions.

Consider an example given by Wesley Salmon [Salmon & Greeno, 1971, p. 40] to illustrate a "fundamental difficulty" related to the frequency interpretation of probability. Suppose there are two urns on a table. The one on the left contains only red balls, while the one on the right contains equal numbers of red, white, and blue balls. The reference class R_1 might be the sequence of draws from the urn on the right (with replacement to ensure independence). The reference class L_1 might be the sequence of draws made alternately from the left and right urns.

Suppose the event in question is the drawing of a red ball on the next trial in the setup described. Should it be one-third (the presumed relative frequency of draws of red balls from the right hand urn), or 1.0 (the presumed relative frequency of draws of red balls from the left hand urn), or two-thirds (the presumed relative frequency of draws of red balls on trials of the experiment)? Again, that event—the drawing of a red ball on the next trial—may also be the drawing of a red ball by Alice on the next trial, or the drawing of a red ball by means of a pair of tweezers on the next trial, or any one of a large number of *descriptions* of the event constituting the next draw.

Given the attribute (the drawing of a red ball), we must devise a method for choosing the right reference class. Reichenbach recommends selecting "the narrowest class for which reliable statistics can be compiled" as the reference class. He writes, "If we are asked to find the probability holding for an individual future event, we must first incorporate the case in a suitable reference class. An individual thing or event may be incorporated in many reference classes .... We then proceed by considering the narrowest reference class for which reliable statistics can be compiled" [Reichenbach, 1949, p. 374].

Reichenbach does not deal with the problem of choosing the correct attribute class to represent the event under consideration (nor, for that matter, does he explain the phrase "reliable statistics"). This may be a matter of oversight, or, more likely, reflect the fact that the choice of the reference class is typically much more controversial than the choice of the attribute class.

Philosophers like Reichenbach regard the problem of selecting the correct reference class on which one will base an assertion of probability as merely a pragmatic matter, rather than a serious philosophical one. They argue that the job of the scientist is over when he has provided us with a bunch of statistical hypotheses that he considers acceptable. Taking one hypothesis rather than another to guide our actions, according to this view, is not the scientist's problem. Reichenbach's advice, to choose the narrowest reference class for which we have reliable statistics, is purely pragmatic.

However, often we have data about the potential reference class *A* (say, the set by Natasha), but though the set of bets made by Natasha on throws of a coin is a perfectly good reference class, we may have no knowledge about it at all. The advice to choose the narrowest class is not always applicable.

Even when it is applicable, this piece of advice is not unambiguous, as we shall see later. Whether or not we have reliable statistics about a reference class depends on what we mean by "having statistics." If we take "having statistics" to mean being in a position to accept a precise limiting frequency statement ("The limiting frequency of heads in tosses of coin C is 0.50132"), then we so rarely have statistics as to make the application of probability useless. On the other hand, if we allow vague or approximate statistical statements ("The limiting frequency of heads in tosses of coin C is between 0.4 and 0.6"), we have statistics about every possible reference class—if only the information that the limiting frequency of A lies between 0.0 and 1.0. If we construe that as "having statistics," it is clear that the narrowest reference class will be characterized exactly by those limits, 0.0 and 1.0, since the narrowest reference class will be the class having as its only member the object or event in question.

As we shall see later, it is a complicated matter to adjudicate the conflict between knowledge of reference classes that apply more specifically to the instance at hand and knowledge of reference classes about which our knowledge is more precise. It is also both crucial and central to the application of statistics.

4.3.2   The Propensity Interpretation

Like the frequency approach, the propensity approach takes probabilities to be objective features of the empirical world. However, whereas the frequency view takes probability to belong collectively to the sequence that constitutes the reference class or the collective, the propensity approach takes probability to belong distributively to the individual trials in that class.

Thus, on the former view, the probability of heads on a toss of this coin is $\frac{1}{2}$, because $\frac{1}{2}$ is the limit of the relative frequency of heads in a (hypothetical) infinite sequence of tosses of the coin. This probability characterizes the whole *sequence* of coin tosses. On the latter view, the limit of the relative frequency is (with probability 1—see the statement of Chebychef's theorem in the last chapter) $\frac{1}{2}$, because each toss has a propensity of magnitude $\frac{1}{2}$ to land heads, and the sequence of tosses is independent. This probability is a propensity that each toss has distributively, as opposed to the collective property that holds of the sequence of tosses as a whole. Furthermore, it is this propensity, this *disposition* of the coin (or coin plus tossing apparatus), that *explains* the frequencies we observe.

This is not at all to deny that samples from the sequence of tosses may constitute the best *evidence* we have concerning the propensity of tosses to yield heads. More important, the propensity of a certain kind of trial to yield a certain result depends on the circumstances of the trial—what Hacking [Hacking, 1965, p. 13] calls the chance setup under which the trial is conducted.

Karl Popper puts the matter this way: "The probabilities . . . may be looked upon as properties of this arrangement. They characterize the disposition, or the propensity, of the experimental arrangement to give rise to certain characteristic frequencies when the experiment is often repeated" [Popper, 1959, p. 67].

To see what this comes to, let us consider a sequence of throws of a die by means the limiting frequency with which the six comes up is $\frac{1}{4}$. Now suppose that three of these throws were made with a different apparatus which is, as we would put it, "fair". Popper argues that despite the fact that the limiting frequency of six for all throws is $\frac{1}{4}$, the correct probability of six in the three *fair* throws is $\frac{1}{6}$. The claim is that although there are only three of these throws, and their results are swamped by the long run results, the *propensity* of yielding a six, or the *causal tendency* to yield a six, on each of these three throws is $\frac{1}{6}$, whatever their actual outcome.

One difficulty faced by the frequency theorist was that of finding it hard to account for the attribution of probability to a single case: for the frequency theorist probability is only an attribute of the collective constituted by the sequence of trials. The propensity theorist seems to have no such problem. His interpretation of probability is exactly focused on the single case. No matter if only one of the tosses of the die was performed in such a manner as to be fair; that toss had a propensity of $\frac{1}{6}$ to yield a six [Fetzer, 1977; Giere, 1973].

On the other hand, the propensity theorist faces his own problems with ordinary language. The propensity of the roll of the die to produce a six is $\frac{1}{4}$, in Popper's example. But that applies to every roll of the die in that apparatus, including the one just performed. If I have seen that the die resulted in a two, it seems strange to ask me to attribute a probability of $\frac{1}{4}$ to its yielding a six.

The only way to make sense of this is to construe probabilities as intensional; we can offer an interpretation of probability in terms of possible worlds, in which a single event is said to have an attribute with probability 0.4 in this world, just in case in the worlds accessible from this world that same event has the attribute 40% of the time.

That still gives us no way to understand the difference in epistemic attitude we have toward past events whose attributes we happen to know and future events of the same sort. There is a sense in which we would not want to attribute a probability of a half to heads on a toss of a coin we have already observed, though the propensities surrounding that toss may be just as well known as the propensities surrounding the next toss.

In addition, a given event may fall under a number of descriptions, just as it could fall into a number of collectives in the frequency interpretation. We must still face all the problems entailed by the multiplicity of descriptions, and the coordinated multiplicity of propensities, that we faced in the frequency case. Specifically, coins tossed by me have a certain propensity to land heads. Coins tossed on Tuesdays have a certain propensity to land heads, which may be different. Coins tossed by me, on Tuesdays, have a certain propensity to land heads, which may match neither of the preceding propensities. A single toss, by me, on Tuesday, will partake of all three propensities, which by no means need be the same. How do I decide which propensity to call the probability?

Furthermore, as Wesley Salmon has pointed out, we use probabilities in both directions: we argue from the nature of the chance setup to the probability of the result of a trial, but we also argue from the results of trials to the probable nature of the chance setup. From the observation that a quarter of the tosses of the die in the apparatus have resulted in a six, are we to infer that there is in fact a propensity for the procedure to be biased? No. But we will quite reasonably say (as we observed in the last chapter) that the probability that the procedure is biased, given the evidence, is high. How are we to interpret *this* probability?

The empirical interpretations of probability certainly have a lot going for them, when ordinary language uses "probability" to express some epistemic notion: to express the weight of evidence, or the degree of rational belief, concerning some eventuality. The most clear-cut illustration of the conflict is that, with regard to the draw just completed that produced a red ball, we feel the impulse both to say that the probability *was* one-third that the ball *would* turn out to be red, and to say that the probability that that particular draw did result in a red ball is 1.0. Surely no nonextreme odds would be fair as a basis for betting (now) on the color of the ball produced by that draw.

How do these empirical interpretations of probability influence our investigation into uncertain and inductive inference? They make probability statements into empirical hypotheses that are themselves the conclusions of uncertain inference. Logically, the most important thing to observe is that these statements are strictly general: they do not concern individual events, individual balls, individual statements. The propensity view may suggest otherwise, because on that view probability may properly be attributed to a single event. But this suggestion is misleading, because it is not the individual event as such that is the subject of the propensity claim, but the individual event as *type*. The very same event may fall under two descriptions, and thus fall under two types, and thus partake of two different propensities to produce result R. In any case, as most writers recognize, these views of probability provide us with no normative handle on the logic of uncertain inference. A conclusion is true or false, as the case may be. That is as far as an empirical view of probability takes us. Both Reichenbach and von Mises understood this, as do the modern holders of frequency views of probability among statisticians.

4.4   Logical Interpretations of Probability

A logical interpretation of probability takes probability to measure something like the degree of validity of an argument. From "the coin is tossed" we may infer, with cogency measured by $\frac{1}{2}$, that the coin lands heads. Rational belief is the degree to which one ought to believe a statement. Probability is to be *legislative* for rational belief. Knowing what I know, the validity of the argument from what I know to "the die lands one up" is $\frac{1}{6}$. That is the degree to which I ought rationally to believe "the die lands one up".

The logical approach was explicitly adopted by John Maynard Keynes in his 1921 book, *A Treatise on Probability* [Keynes, 1952]. Keynes's system was proposed precisely to provide a formal logic for uncertain inference. "Part of our knowledge we obtain direct; and part by argument. The Theory of Probability is concerned with that part which we obtain by argument, and it treats of the different degrees in which the results so obtained are conclusive or inconclusive" [Keynes, 1952, p. 3].

Keynes's view has been unjustly neglected until recently. An important feature of his view, but one that makes his probabilities hard to compute with, is the claim that probabilities are only partially ordered: two probabilities may be incomparable. The first may be neither greater, nor less than, nor yet equal to the third. This theory of "imprecise probabilities" has only recently attracted significant attention.[1] Keynes

---

[1] The first conference devoted to imprecise probabilities was held in 1999.

himself had little to offer by way of a systematization of his logical probabilities, and this may be one of the reasons for the neglect of his work.

The most widely known approach along logical lines is that of Rudolf Carnap [Carnap, 1950], whose theory of probability is directed toward constructing a formal inductive logic in the same spirit in which we have a formal deductive logic. Probability is taken to be a logic of rational belief in the sense that, given our total body of evidence, the degree of partial belief in a given statement that is *rationally justified* by that evidence is to be determined on logical grounds alone. Probability is to be legislative for rational degrees of belief.

Carnap assumes a logical language in which both deductive and inductive relations can be expressed. In deductive logic, if a statement e entails another statement h, then if e is true, h must be true. In the same vein, if e is evidence for a hypothesis h, then the relation we see between the evidence e and the hypothesis h is what Carnap calls a relation of partial entailment. Carnap views this relation of partial entailment between the evidence e and hypothesis h as a logical or necessary relation. In this he is following Keynes, whose book defended the same idea.

What does it mean to call the relation "necessary"? It means that when e is our total evidence, and e confirms the hypothesis h to the degree r, then the statement expressing this fact, $c(h, e) = r$, is true, not in virtue of any contingent fact, but in virtue of the very concept of confirmation. Of course, the statements e and h may both be empirical.

Carnap develops his conception of probability step by step, starting with very simple languages. To begin with, he construes the degree of confirmation of h relative to e, $c(h, e)$, as a conditional probability. It is the probability of the proposition h, given the total evidence expressed in e. The requirement of total evidence is important: given some more evidence, the probability of h may go either up or down. Because $c(h, e)$ is a conditional probability, it can be expressed as the ratio of two unconditional probabilities: $m(e \wedge h)/m(e)$. The problem of defining confirmation for pairs of sentences h and e has been reduced to the problem of defining a *measure function* m over the sentences of a language.

Carnap restricts his attention to first order languages with a finite number of one-place predicates, without functions or relations, and with a finite number of proper names, assumed to name uniquely each individual in the domain of discourse: each individual in the domain is to have one and only one name.

Furthermore, he assumed that the languages were descriptively complete (so that we need take account of no phenomena that cannot be described in the language), and that the predicates were *logically independent* of each other in the sense that there is no logical difficulty in supposing that any combination of primitive predicates can be instantiated in a single object. (In contrast, we have for example color predicates such as "red" and "green", which are often assumed to be logically, and not merely empirically, exclusive.) All of these constraints have subsequently been lifted, though the procedure about to be described becomes more complicated as we consider richer and more general languages.

Carnap assumes the standard logical machinery in all of his languages: the usual sentential connectives and quantifiers of first order logic.

In the simple language just described, we can say everything there is to say about an individual i by means of a sentence that, for each of the set of primitive predicates,

$P_1, \ldots, P_n$, says whether or not the individual i has the property denoted by that predicate. There are 2^n compound predicates we can define on the basis of n primitive predicates. These compound predicates are what Carnap calls Q-predicates. If there are N individual names in the language, a complete description of the world (for that language) consists of the assignment of one of the 2^n Q-predicates to each individual in the world. There are thus $(2^n)^N$ distinguishable descriptions of the world—the number of functions from N to the set of 2^n Q-predicates. Each such description is called a *state description*.

Two state descriptions are *isomorphic* if and only if one could be turned into the other by a permutation of the individual constants of the language: thus $Q_4(a) \wedge Q_6(b)$ is isomorphic in a two name language to $Q_6(a) \wedge Q_4(b)$. It is useful to group isomorphic state descriptions together. A maximal group of isomorphic state descriptions is called a *structure description*: a structure description is the set of all state descriptions isomorphic to a given state description. The structure descriptions of a language constitute a *partition* of the set of state descriptions.

It will be useful in seeing how this all works to consider an extremely simple language. Suppose this language has three individual terms (names), "a," "b," and "c," and two descriptive predicates "F" and "G". In this language there are four Q-predicates:

$$Q_1(x) \quad Fx \wedge Gx$$
$$Q_2(x) \quad Fx \wedge \neg Gx$$
$$Q_3(x) \quad \neg Fx \wedge Gx$$
$$Q_4(x) \quad \neg Fx \wedge \neg Gx$$

The number of state descriptions is $4^3 = 64$. (In a language with two individuals and two predicates the number of state descriptions is $(2^2)^2 = 16$.) The first eight of them are

$$Q_1 a \wedge Q_1 b \wedge Q_1 c \qquad Q_1 a \wedge Q_2 b \wedge Q_1 c$$
$$Q_1 a \wedge Q_1 b \wedge Q_2 c \qquad Q_1 a \wedge Q_2 b \wedge Q_2 c$$
$$Q_1 a \wedge Q_1 b \wedge Q_3 c \qquad Q_1 a \wedge Q_2 b \wedge Q_3 c$$
$$Q_1 a \wedge Q_1 b \wedge Q_4 c \qquad Q_1 a \wedge Q_2 b \wedge Q_4 c$$
$$\vdots \qquad\qquad \vdots$$

The first thing to observe is that every consistent sentence of the language can be expressed as a disjunction of state descriptions. "$Fa \wedge Ga \wedge Fb \wedge Gb$" is equivalent to the disjunction of the four state descriptions in the first column. The statement "There is at least one thing that has the property F" can be translated into a disjunction of state descriptions. The statement "Everything has the property F and the property G" is equivalent to the first state description or the universal quantification of $Q_1(x)$.

It should be observed that, but for the constraint that each name in the language names a different entity, state descriptions are similar to *models* of the language, having a given domain. Indeed, that is the way Carnap thought of them: not as linguistic entities, but as propositions, or "possible worlds", in the sense in which we are using that term.

The state descriptions in any disjunction are called the *range* of that statement: they constitute the set of the state descriptions that make the sentence true. An inconsistent statement is equivalent to the denial of all state descriptions. Its range is empty.

Noting that state descriptions are logically exclusive, we observe that if the measure will be the sum of the measures assigned to the state descriptions in its range. We have achieved another level of reduction. If we assign each state description or possible world a measure that is a positive real number, such that the sum of all these numbers is 1.0, then the measure of every sentence in the language has been determined. (Any inconsistent statement, having a range of zero state descriptions, will get the measure zero.) But if $\mathbf{m}$ is defined for every sentence in the language, then the degree of confirmation $\mathbf{c}$ is defined for every pair of sentences of the language, provided the second—the evidence—has positive measure.[2]

Carnap's problem has been reduced to that of finding a measure $\mathbf{m}$ for state descriptions.

A natural possibility is to assign each state description in the language the same measure. One might even motivate this assignment by a principle of indifference; restricted to this language, the principle of indifference is even demonstrably consistent. This measure function is denoted by $\mathbf{m}^{\dagger}$.

The problem with this measure is that it precludes learning from experience in a natural and desirable sense. We may see this as follows: In our example language, we can calculate the measure $\mathbf{m}^{\dagger}(Fc)$; it is $\frac{1}{2}$—its range consists of 32 state descriptions of the original 64, each of which has measure $\frac{1}{64}$. Suppose now that we explore our world, and gather evidence. We look at objects a and b, and discover that they are both F. That surely should raise our confidence that c, too, is F. But the confirmation function $\mathbf{c}^{\dagger}$ that corresponds to $\mathbf{m}^{\dagger}$ has the following value for this case:

$$\mathbf{c}^{\dagger}(Fc|Fa \wedge Fb) = \frac{\mathbf{m}^{\dagger}(Fa \wedge Fb \wedge Fc)}{\mathbf{m}^{\dagger}(Fa \wedge Fb)} = \frac{1}{2}.$$

Taking account of our evidence has not changed our logical probability. The statement $F(c)$ is *probabilistically independent* of the evidence provided by other statements involving F, $F(a)$, and $F(b)$. That everything else is F has not increased the probability that the last thing in the universe is F. In a certain sense, we have not learned from experience.

Carnap next suggests a measure function $\mathbf{m}^*$, which assigns equal weights to each *structure description*, and then divides that weight equally among the state descriptions in that structure description. (Recall that a structure description is a set of isomorphic state descriptions.) The idea can be best understood if we look at Table 4.1. There are 20 structure descriptions that are equally weighted. The weight of a structure description is divided among the state descriptions that belong to it to obtain the measure assigned to each state description.

The corresponding confirmation function, $\mathbf{c}^*$, satisfies the intuition that we should expect the future should be like the past: if all observed objects have been F, then we should have an increased expectation that the next object should be F also:

$$\mathbf{c}^*(Fc \mid Fa \wedge Fb) = \frac{\mathbf{m}^*(Fa \wedge Fb \wedge Fc)}{\mathbf{m}^*(Fa \wedge Fb)} = \frac{2}{3} = 0.67$$

Because $\mathbf{m}^*(Fc) = \frac{1}{2}$, we have learned from experience.

There are, of course, other ways of distributing weights (or logical measures, or logical probabilities) among the state descriptions of a language. However it is done,

---

[2]Carnap in [Carnap, 1950] restricted his attention mainly to what he called *regular* languages—languages in

Table 4.1: Values of $\mathbf{m}^*$

Structure	Weight	State description	Measure
All Q_1	$\frac{1}{20}$	$Q_1a \wedge Q_1b \wedge Q_1c$	$\frac{1}{20}$
$\frac{2}{3}Q_1, \frac{1}{3}Q_2$	$\frac{1}{20}$	$Q_1a \wedge Q_1b \wedge Q_2c$	$\frac{1}{60}$
		$Q_1a \wedge Q_2b \wedge Q_1c$	$\frac{1}{60}$
		$Q_2a \wedge Q_1b \wedge Q_1c$	$\frac{1}{60}$
$\frac{2}{3}Q_1, \frac{1}{3}Q_3$	$\frac{1}{20}$	$Q_1a \wedge Q_1b \wedge Q_3c$	$\frac{1}{60}$
		$Q_1a \wedge Q_3b \wedge Q_1c$	$\frac{1}{60}$
		$Q_3a \wedge Q_1b \wedge Q_1c$	$\frac{1}{60}$
$\frac{2}{3}Q_1, \frac{1}{3}Q_4$	$\frac{1}{20}$	$Q_1a \wedge Q_1b \wedge Q_4c$	$\frac{1}{60}$
		$Q_1a \wedge Q_4b \wedge Q_1c$	$\frac{1}{60}$
		$Q_4a \wedge Q_1b \wedge Q_1c$	$\frac{1}{60}$
$\frac{1}{3}Q_1, \frac{2}{3}Q_2$	$\frac{1}{20}$	$Q_1a \wedge Q_2b \wedge Q_2c$	$\frac{1}{60}$
		$Q_2a \wedge Q_1b \wedge Q_2c$	$\frac{1}{60}$
		$Q_2a \wedge Q_2b \wedge Q_1c$	$\frac{1}{60}$
$\frac{1}{3}Q_1, \frac{2}{3}Q_3$	$\frac{1}{20}$	$Q_1a \wedge Q_3b \wedge Q_3c$	$\frac{1}{60}$
		$Q_3a \wedge Q_1b \wedge Q_3c$	$\frac{1}{60}$
		$Q_3a \wedge Q_3b \wedge Q_1c$	$\frac{1}{60}$
$\frac{1}{3}Q_1, \frac{2}{3}Q_4$	$\frac{1}{20}$	$Q_1a \wedge Q_4b \wedge Q_4c$	$\frac{1}{60}$
		$Q_4a \wedge Q_1b \wedge Q_4c$	$\frac{1}{60}$
		$Q_4a \wedge Q_4b \wedge Q_1c$	$\frac{1}{60}$
$\frac{1}{3}Q_1, \frac{1}{3}Q_2, \frac{1}{3}Q_3$	$\frac{1}{20}$	$Q_1a \wedge Q_2b \wedge Q_3c$	$\frac{1}{120}$
		$Q_1a \wedge Q_3b \wedge Q_2c$	$\frac{1}{120}$
		$Q_2a \wedge Q_1b \wedge Q_3c$	$\frac{1}{120}$
		$Q_2a \wedge Q_3b \wedge Q_1c$	$\frac{1}{120}$
		$Q_3a \wedge Q_1b \wedge Q_2c$	$\frac{1}{120}$
		$Q_3a \wedge Q_2b \wedge Q_1c$	$\frac{1}{120}$
$\frac{1}{3}Q_1, \frac{1}{3}Q_2, \frac{1}{3}Q_4$	$\frac{1}{20}$	$Q_1a \wedge Q_2b \wedge Q_4c$	$\frac{1}{120}$
		$Q_1a \wedge Q_4b \wedge Q_2c$	$\frac{1}{120}$
		$Q_2a \wedge Q_1b \wedge Q_4c$	$\frac{1}{120}$
		$Q_2a \wedge Q_4b \wedge Q_1c$	$\frac{1}{120}$
		$Q_4a \wedge Q_1b \wedge Q_2c$	$\frac{1}{120}$
		$Q_4a \wedge Q_2b \wedge Q_1c$	$\frac{1}{120}$
$\frac{1}{3}Q_1, \frac{1}{3}Q_3, \frac{1}{3}Q_4$	$\frac{1}{20}$	$Q_1a \wedge Q_3b \wedge Q_4c$	$\frac{1}{120}$
		$Q_1a \wedge Q_4b \wedge Q_3c$	$\frac{1}{120}$
		$Q_3a \wedge Q_1b \wedge Q_4c$	$\frac{1}{120}$
		$Q_3a \wedge Q_4b \wedge Q_1c$	$\frac{1}{120}$
		$Q_4a \wedge Q_1b \wedge Q_3c$	$\frac{1}{120}$
		$Q_4a \wedge Q_3b \wedge Q_1c$	$\frac{1}{120}$
All Q_2	$\frac{1}{20}$	$Q_2a \wedge Q_2b \wedge Q_2c$	$\frac{1}{20}$
$\frac{2}{3}Q_2, \frac{1}{3}Q_3$	$\frac{1}{20}$	$Q_2a \wedge Q_2b \wedge Q_3c$	$\frac{1}{60}$
		$Q_2a \wedge Q_3b \wedge Q_2c$	$\frac{1}{60}$
		$Q_3a \wedge Q_2b \wedge Q_2c$	$\frac{1}{60}$

(cont.)

Table 4.1: *Continued*

Structure	Weight	State description	Measure
$\frac{2}{3}Q_2, \frac{1}{3}Q_4$	$\frac{1}{20}$	$Q_2a \wedge Q_2b \wedge Q_4c$	$\frac{1}{60}$
		$Q_2a \wedge Q_4b \wedge Q_2c$	$\frac{1}{60}$
		$Q_4a \wedge Q_2b \wedge Q_2c$	$\frac{1}{60}$
$\frac{1}{3}Q_2, \frac{2}{3}Q_3$	$\frac{1}{20}$	$Q_2a \wedge Q_3b \wedge Q_3c$	$\frac{1}{60}$
		$Q_3a \wedge Q_2b \wedge Q_3c$	$\frac{1}{60}$
		$Q_3a \wedge Q_3b \wedge Q_2c$	$\frac{1}{60}$
$\frac{1}{3}Q_2, \frac{2}{3}Q_4$	$\frac{1}{20}$	$Q_2a \wedge Q_4b \wedge Q_4c$	$\frac{1}{60}$
		$Q_4a \wedge Q_2b \wedge Q_4c$	$\frac{1}{60}$
		$Q_4a \wedge Q_4b \wedge Q_2c$	$\frac{1}{60}$
$\frac{1}{3}Q_2, \frac{1}{3}Q_3, \frac{1}{3}Q_4$	$\frac{1}{20}$	$Q_2a \wedge Q_3b \wedge Q_4c$	$\frac{1}{120}$
		$Q_2a \wedge Q_4b \wedge Q_3c$	$\frac{1}{120}$
		$Q_3a \wedge Q_2b \wedge Q_4c$	$\frac{1}{120}$
		$Q_3a \wedge Q_4b \wedge Q_2c$	$\frac{1}{120}$
		$Q_4a \wedge Q_2b \wedge Q_3c$	$\frac{1}{120}$
		$Q_4a \wedge Q_3b \wedge Q_2c$	$\frac{1}{120}$
All Q_3	$\frac{1}{20}$	$Q_3a \wedge Q_3b \wedge Q_3c$	$\frac{1}{20}$
$\frac{2}{3}Q_3, \frac{1}{3}Q_4$	$\frac{1}{20}$	$Q_3a \wedge Q_3b \wedge Q_4c$	$\frac{1}{60}$
		$Q_3a \wedge Q_4b \wedge Q_3c$	$\frac{1}{60}$
		$Q_4a \wedge Q_3b \wedge Q_3c$	$\frac{1}{60}$
$\frac{1}{3}Q_3, \frac{2}{3}Q_4$	$\frac{1}{20}$	$Q_3a \wedge Q_4b \wedge Q_4c$	$\frac{1}{60}$
		$Q_4a \wedge Q_3b \wedge Q_4c$	$\frac{1}{60}$
		$Q_4a \wedge Q_4b \wedge Q_3c$	$\frac{1}{60}$
All Q_4	$\frac{1}{20}$	$Q_4a \wedge Q_4b \wedge Q_4c$	$\frac{1}{20}$

though, the sum of the measures assigned to the state descriptions or models of the language should be 1.0. In addition, Carnap imposes the requirement of regularity: This is the requirement that no state description be assigned a measure of 0. This is to reflect the fact that before we have gathered any evidence about the world, no state description should be regarded as impossible.

An immediate consequence of this is that the degree of confirmation of any universal statement of the form $\ulcorner(\forall x)\phi\urcorner$, in a language having an infinite domain, is 0, and remains 0 regardless of what evidence we have, unless the condition ϕ is tautologous.

Many of the restrictions within which Carnap worked have been relaxed by other writers. The restriction to languages containing only one-place predicates was lifted by Carnap himself in later work [Carnap, 1971; Carnap, 1980]. Other writers [Hintikka, 1966; Niiniluoto, 1976] have provided systems in the spirit of Carnap that allow universal generalizations to be given finite probabilities (and thus to be increasingly confirmed by their instances).

Logically, there is no need to restrict the language. Consider a fixed countable domain. The models of a first order language with this domain (possible worlds) are countable. These models may play the role of Carnap's state descriptions. Because there are only a countable number of them, it is possible to assign a discrete measure to them. (Order them, assign the first a measure $\frac{1}{2}$, the second a measure $\frac{1}{4}$, etc.) Every sentence picks out the set of these models in which it is true. The sum of the measures of those models in which the sentence is true is the measure **m** of that sentence. This measure is a probability measure. The sample space S is just the set of models. The subsets picked out by sentences of the language form an algebra or a σ-algebra. (The set of models picked out by $\ulcorner \neg\phi \urcorner$ is just the complement of the set of models picked out by ϕ, and the set of models picked out by $\ulcorner \phi \vee \psi \urcorner$ is the union of the set of models picked out by ϕ and by ψ. Carnap's languages, having a finite domain, have no need of countable additivity, but a language having a countably infinite number of constants or implying the existence of a countably infinite number of objects would require countable additivity.)

The standard by which we judge such logical systems of probability as that offered by Carnap seems to be the intuitive appeal of their results: **m***, for example, is more appealing than **m**†. There thus seems to be an unavoidable arbitrariness in the assignment of measures to the state descriptions of the language; the fundamental step in the design of logical probabilities seems to be immune to critical assessment. Many writers [Bacchus, 1992; Halpern, 1990; Morgan, 1998; Leblanc & Morgan, 1983] seem to think that from a logical point of view our concern should be with the *logical implications* of an arbitrary probability assignment, rather than what particular probability assignment we start with.

But if it is arbitrary anyway, we might as well bite the bullet and say that probability is a matter of personal beliefs rather than a matter of logic. It is a remarkable fact that such an approach can have plausible and interesting results. We shall explore this approach in the next section.

We may also ask what relevance these logical probabilities have to uncertain *inference*. Carnap's view is that properly speaking there is no such thing as uncertain inference. When we speak casually, as of course we do, of inferring a consequence from data or premises that do not entail that consequence, we should strictly be speaking of assigning a *probability* to that consequence, relative to the premises, together with what else we take ourselves to know. Because degree of confirmation represents a necessary relation, the *inference* that leads to the assignment of a conditional probability is a straightforward deductive inference. Thus, induction is the deductive computation of probabilities of statements that go beyond the entailments of our evidence. The same is true of the subjective view of probability: science consists in updating the probabilities that we assign to hypotheses and predictions that are not *entailed* by the data we have.

If there were an agreed-on assignment of probabilities to the propositions of a canonical language of science, this would be a live option as an approach to induction. There is no such agreement, and there seems to be little prospect of such an agreement. If there were agreement about the measure function **m**, it could still be questioned whether or not there is anything more to say. There are some who would claim that there are good reasons for supposing that there is a rationally defensible reconstruction of inductive inference that takes the full acceptance of uncertain conclusions as central. A number of the arguments for this point of view will be considered later, when we consider the positive and negative aspects of the acceptance of uncertain conclusions.

4.5 Subjective Interpretations of Probability

According to the logical approach to probability, probability is to be legislative for rational belief. This means that if your total body of evidence is e, the degree of belief it would be *rational* for you to have in h is exactly $P(h \mid e)$. But what is rational belief? Or belief in general? For that matter, how can we get at this elusive set of formulas that we simplistically call "your total body of evidence"?

Many writers would construe degree of rational belief as a disposition to act. The classic challenge to someone who claims an unusual belief is to invite him to put his money where his mouth is. Less crudely, the philosopher F. P. Ramsey [Ramsey, 1931] suggested that the degree of one's belief in a proposition could be measured by the odds that one is willing to take in a bet on that proposition. For example, if I am just willing to bet at even odds on the truth of the proposition that the next toss of this coin will yield heads, but unwilling to offer any better odds, then that is evidence that my degree of belief in the proposition is a half.

Once we construe probability as concerned with actual degrees of belief, we have also solved the problem of the representation of "total evidence". Our total evidence is whatever led us to our actual degrees of belief; it need not be given any explicit representation.

4.5.1   Dutch Book

Ramsey shows that if you are willing to take any bet on a proposition at odds corresponding to your degree of belief in it, then your degrees of belief conform to the laws of the probability calculus if and only if it is impossible for a clever bettor to devise a set of bets under which you will lose, no matter how the facts turn out.

For example, if you are always willing to bet at odds of two to one on heads (that is, you will bet two dollars against one dollar that the next toss lands heads) and always willing to bet at odds of two to one on tails, the clever bettor will simply take both bets. If the toss yields tails, the bettor wins two dollars on the first bet but loses a dollar on the second. If the toss yields heads, the bettor loses a dollar on the first bet but wins two dollars on the second. In either case, you have lost a dollar.

In general, the relation between odds and probabilities (on this view) is this: The probability you attribute to a proposition S is $P(S)$ if and only if you are willing to bet in favor of S at odds of $P(S)$ to $1 - P(S)$, but at no greater odds. If you are unfamiliar with "odds," another way of putting the matter was offered by the statistician L. J. Savage [Savage, 1966]. Take $P(S)$ to be the price you would regard as fair to pay for a ticket that will return \$1.00 if S turns out to be true, and nothing otherwise. This comes to the same thing, if we are willing to suppose that you will buy or sell any number of these tickets at the fair price.

We can now state Ramsey's result as a theorem.

Theorem 4.1 (Dutch Book Theorem). *If F is a finite field of propositions (i.e., closed under finite unions, and containing the negation of any proposition that occurs in it), and P is my price function for the propositions in that field, in the sense due to Savage, then I will be protected against certain loss if and only if P is an additive probability function.*

Proof: The price function $\mathcal{P}$ must have a range in [0, 1], because if there is a proposition S for which $\mathcal{P}(S)$ is less than 0 or more than 1, I will be selling tickets for less than nothing, or buying them for more than they could possibly be worth no matter what happens. Suppose that S and T are logically exclusive, and that the values of my price function for each of them and their disjunction are $\mathcal{P}(S)$, $\mathcal{P}(T)$, and $\mathcal{P}(S \vee T)$. We show that if $\mathcal{P}(S) + \mathcal{P}(T) < \mathcal{P}(S \vee T)$, I face a certain loss. My opponent[3] buys a ticket on S and a ticket on T, and sells me one on the disjunction. Because S and T are exclusive, only three things can happen: S alone, T alone, or neither. In each of the first two cases, I must pay out a dollar on one ticket, but I make a dollar on another. In the third case, none of the tickets pays off. But in *every* case, I have received $\mathcal{P}(S) + \mathcal{P}(T)$, and paid out $\mathcal{P}(S \vee T)$. My opponent has made money. I have suffered a certain loss. The argument is similar, of course, if $\mathcal{P}(S) + \mathcal{P}(T) > \mathcal{P}(S \vee T)$. The only way to avoid certain loss is to have $\mathcal{P}(S) + \mathcal{P}(T) = \mathcal{P}(S \vee T)$. But this is all that is required to have $\mathcal{P}$ be a probability function when we identify the domain of the function with the field of propositions described. ∎

A somewhat more powerful theorem can be stated that is alleged to justify countable additivity as well as finite additivity. This was first stated by Shimony [Shimony, 1955] and Lehman [Lehman, 1959].

Theorem 4.2. *If $\mathcal{F}$ is a σ-field of propositions (i.e., closed under countable unions, and containing the negation of any proposition that occurs in it), and $\mathcal{P}$ is my price function for the propositions in that field, in the sense due to Savage, then I shall be protected against the certainty of no gain combined with the possibility of loss if and only if P is a countably additive probability function.*

Proof: The proof is similar. ∎

Ramsey claimed that he could find no principle, other than conformity to the probability calculus, that could plausibly be imposed on a person's degrees of belief. "...we do not regard it as belonging to formal logic to say what should be a man's expectation of drawing a white or a black ball from an urn; his original expectations may within the limits of consistency [the limits imposed by the probability calculus] be any he likes" [Ramsey, 1931, p. 189].

On this view one man's belief is as good as another's. If two people disagree, their disagreement may be treated merely as a datum. At the same time, in many areas people's degrees of belief are in fairly close agreement—we all assign a degree of belief of about a half to heads on the next toss. On this account, we can simply accept agreement or disagreement as a given datum and proceed from there. Of course, that leaves the prevalence of agreement, where it exists, as something to be explained.

In this sense, the subjective view is maximally tolerant, and tolerance is generally regarded as a virtue. Another reason for the current popularity of the subjective theory is that it seems to be minimally committal. The subjective theory implies only relatively weak constraints on the degrees of belief that man has. It only requires our degrees of belief to satisfy the axioms of probability calculus. The theory is maximally tolerant and minimally restrictive.

---

[3]The opponent in these arguments is often referred to as "the bookie," and it is in allusion to bookies' slang

There are arguments other than the Dutch book argument for normative role of the probability calculus in governing degrees of belief. Howson and Urbach [Howson & Urbach, 1993, pp. 75–76], for example, find sufficient motivation in the somewhat different notion of *fairness* of bets: if I think fair odds on rain are $2:3$, then I am rationally obliged to think that odds of $3:2$ against rain are fair.

4.5.2   Conditionalization

If this were as far as the subjective view of probability went, it wouldn't be very interesting. As a static theory of idealized degrees of belief, it could be psychologically interesting (though easy to falsify, if construed literally), but would be of little interest to logicians, epistemologists, and statisticians, not to mention philosophers of science and computer scientists. It would certainly cast little light on the issues of inductive inference. The importance of the theory stems largely from its employment of conditionalization in *revising* degrees of belief. In fact, what is called "Bayesianism" is characterized mainly by the importance it attributes to Bayes' theorem and conditionalization.

Recall that the conditional probability of S given T, $P(S \mid T)$, when $\mathcal{P}(T) > 0$, is taken to be $P(S \wedge T)/\mathcal{P}(T)$. The subjectivist interpretation of probability claims not only that our degrees of belief *ought* to conform to the probability calculus in a static sense, but that our degrees of belief should be modified, in the light of new evidence, by conditionalization. That is to say, when we observe T, we should replace our old probability function $\mathcal{P}$ by the new one $\mathcal{P}_T$, where for every S, $\mathcal{P}_T(S) = P(S \wedge T)/\mathcal{P}(T) = \mathcal{P}(S \mid T)$.

We must be a bit careful here with our talk of "old" and "new" probability functions. The "new" probability function we are exhorted to adopt is not really new: it is simply the "old" *conditional* probability. As Keynes pointed out [Keynes, 1952], it is an error to think of new evidence as "correcting" an old probability. What new evidence does is make a different *argument* or a different *measure* relevant. If our belief in S is $\mathcal{P}(S \mid B)$, where B represents our background evidence, and we obtain new evidence T, our belief in S should become $\mathcal{P}(S \mid B \wedge T)$. This is the *same* two place probability function we had before, but with a different argument in the second place: $B \wedge T$ rather than B. We have, and need, only one probability function: $\mathcal{P}$. This is true, of course, of logical probabilities as well as subjective probabilities.

There is an argument for the claim that we should update our probabilities by conditionalization, too, but it is not as straightforward as the argument for static conformity to the rules of the probability calculus. Let $\mathcal{P}_T$ be my conditional price function—that is, the price I would take to be fair for a ticket that returns a dollar if T is true and its argument also turns out to be true, but whose price is refunded if T turns out to be false. Let us call such a ticket a ticket *conditional* on T; it is a ticket that is refundable if T fails to be true. This definition allows us to prove the following theorem.

Theorem 4.3. *If F is a field of propositions, $\mathcal{P}$ and $\mathcal{P}_T$ are my price functions as defined above, $\mathcal{P}$ is a probability function, $\mathcal{P}(T) > 0$, and the truth of T is settled before that of S, then I shall be protected against certain loss if and only if $\mathcal{P}_T(S) = P(S \wedge T)/\mathcal{P}(T)$.*

Proof: Assume $\mathcal{P}_T(S) > P(S \wedge T)/\mathcal{P}(T)$. My opponent the bookie sells me

$\mathcal{P}_T$. If T fails to occur, the first bets are called off, but he keeps the price of the ticket on T. If T does occur, he has to pay me a dollar, but then he buys N tickets on S at the price of $\mathcal{P}(S \wedge T)/\mathcal{P}(T)$. I must regard this price as fair, since it is $\mathcal{P}(S \wedge T)/[\mathcal{P}(S \wedge T) + \mathcal{P}(\neg S \wedge T)]$, and that is just the original distribution of my belief between S and $\neg S$ in the case where T is true. Whether or not S occurs, we break even on the payoffs; but since $N\mathcal{P}_T(S)$ is greater than $N(\mathcal{P}(S \wedge T)/\mathcal{P}(T))$, he is ahead in virtue of the original tickets bought and sold. A similar argument works for the case in which $\mathcal{P}_T(S)$ is smaller than $\mathcal{P}(S \wedge T)/\mathcal{P}(T)$. ∎

The import of this theorem for the subjectivist is that I should change my degrees of belief by conditionalization as I obtain more evidence.

But how can one take seriously the proposal that all we can do with probability is to protect ourselves against ingenious bettors? Probability plays a role in statistics and in science generally; can we account for that role on the basis of an interpretation of probability as mere personal opinion? The claim that probability can be useful even when it is interpreted as personal degrees of belief can be buttressed by a remarkable theorem due to the Italian statistician and probabilist Bruno de Finetti [de Finetti, 1980]. The upshot of the theorem is that under certain circumstances, as two individuals gather more evidence about a certain sequence of events, their personal probabilities regarding a future occurrence of an event in that sequence will converge. This is an important result, and though the theorem is complex, it will be worthwhile following its steps.

A preliminary notion is that of exchangeability. We say that a sequence of values 0 and 1 of the random quantity X is *exchangeable* according to a probability function $\mathcal{P}$ if and only if (a) $\mathcal{P}(X_i = 1) = \mathcal{P}(X_j = 1)$ for all i and j, and (b) the probability of any conjunction of values, $\mathcal{P}(X_{i_1} = v_1 \wedge \cdots \wedge X_{i_m} = v_m)$, is the same for every permutation of the values $v_1, \ldots, v_m$. The order of the occurrence of 0's and 1's is irrelevant—only the number of 0's and the number of 1's count.

An example that appeals to intuition is the sequence generated by draws from an urn, with replacement, where a 0 corresponds to a white ball and a 1 corresponds to a black ball. We don't know what the frequency of 0's is, but before sampling, we would assign the same probability to 0 on the sixth draw as to 0 on the tenth draw. Furthermore, we take the order of the draws to be irrelevant, and so would assign the same probability to the outcome sequence $\langle 0, 0, 1, 0, 1, 1 \rangle$ as to $\langle 0, 1, 0, 1, 1, 0 \rangle$.

Theorem 4.4 (de Finetti's Theorem). *Let a sequence of 0's and 1's represent a sequence of trials on which a certain kind of event may occur, with 0 representing nonoccurrence and 1 representing occurrence. Two probability functions $\mathcal{P}_A$ and $\mathcal{P}_B$ represent the opinions of two people about the probability of the kind of event in question. Both individuals take the sequence of events to be exchangeable. Provided neither $\mathcal{P}_A$ nor $\mathcal{P}_B$ is extreme (assumes independence, or assigns probability 1 or probability 0 to all events in the sequence), for any ϵ there is an N such that if A and B have shared a sample of length N as evidence, the probabilities they assign to the next trial resulting in a 1 will differ by less than ϵ, with probability greater than $1 - \delta$ for each of A and B.*

Proof: Let X_i be the ith member of the sequence. Let $\mathcal{P}_A(X_i = 1) = m_A = E_A(X_i)$, $\mathcal{P}_B(X_i = 1) = m_B$, $\mathcal{P}_A(X_i = 1 \wedge X_j, = 1) = s_A = E_A(X_i X_j)$ for $i \neq j$, and $\mathcal{P}_B(X_i =$

n trials, we will show that after n trials both A and B take the difference between the relative frequency of the event on the observed trials and the relative frequency of the event on a long subsequent sequence to be almost certainly less than $\epsilon/2$. In view of the exchangeability of the events, the expectation of the relative frequency in the subsequent sequence must have the same value as the conditional probability of the $n + 1$st event occurring, conditional on the observation of n trials. The strategy is to show that $E(|rf(E_n) - rf(E_k)| < \epsilon/2$ for each of A and B, where $rf(E_n)$ is the relative frequency of the event among the first n trials and $rf(E_k)$ is the relative frequency of the event on a subsequent sequence of k trials.

We compute both $E(rf(E_n) - rf(E_k))$ and $D^2(rf(E_n) - rf(E_k))$ for each of A and B, where E_k is a subsequent sequence of k trials. We obtain $E_A(rf(E_n)) = (1/n)(E_A(X_1) + E_A(X_2) + \cdots + E_A(X_n)) = m_A$ and $E_A(rf(E_n) - rf(E_k)) = 0$. Similarly $E_B(rf(E_n) - rf(E_k)) = 0$. Both A and B expect to find a relative frequency close to their prior probability of $X_i = 1$.

But

$$D_A^2(rf(E_n) - rf(E_k)) = E_A((rf(E_n) - rf(E_k))^2) - (E_A(rf(E_n) - rf(E_k)))^2$$
$$= E_A((rf(E_n) - rf(E_k))^2)$$

and

$$E_A((rf(E_n) - rf(E_k))^2) = E_A\left[\frac{1}{n}\sum_{i=1}^{n}X_i - \frac{1}{k}\sum_{i=n+1}^{n+k}X_i\right]^2$$

$$= E_A\frac{1}{n^2}\left(\sum_{i=1}^{n}(X_i)^2\right) + E_A\frac{1}{k^2}\left(\sum_{i=n+1}^{n+k}(X_i)^2\right)$$

$$- E_A\left(\frac{2}{nk}\sum_{i=1,j=n+1,i\neq k}^{i=n,j=n+k}X_iX_j\right)$$

$$= \frac{1}{n^2}[nm_A + n(n-1)s_A] + \frac{1}{k^2}[km_A + k(k-1)s_A] - \frac{2}{nk}nks_A$$

$$= \left(\frac{1}{k} + \frac{1}{n}\right)m_A + \left(\frac{n-1}{n} + \frac{k-1}{k} - 2\right)s_A$$

$$= \left(\frac{1}{n} + \frac{1}{k}\right)(m_A - s_A)$$

Having calculated the variance of the quantity $rf(E_n) - rf(E_k)$, we apply Chebychef's theorem:

$$P_A\left(|rf(E_n) - rf(E_k)| \geq \frac{\epsilon}{2}\right) \leq \frac{4}{\epsilon^2}\left(\frac{1}{n} + \frac{1}{k}\right)(m_A - s_A)$$

The length k of the subsequent sequence may be arbitrarily large; we are only using that sequence to pass from expected frequency to probability. We can thus make the probability, for A, that the difference between the two relative frequencies is less than $\frac{\epsilon}{2}$ as large as we want by choosing n large enough; in particular we can make it larger than $1 - \delta$.

The same can be said for B. If we take N to be the size of the larger of the two samples, we can say that both A and B can be practically certain that after a sample of this size, they will assign probabilities to the occurrence of the event on the $N + 1$st

This important fact (and more general versions of it) went a long way toward making the subjective interpretation of probability respectable among statisticians, though it is fair to say that most statisticians do not accept the subjective interpretation. The appeal to the phenomenon of the "washing out of priors" has been especially common among philosophers of Bayesian bent.

We should notice the conditions of the theorem, however. It is required that both A and B hold the sequence in question to be exchangeable. This means roughly that the probability assigned to every ordering of the results in a sample is the same: 50 heads followed by 50 tails must have the same probability as any other ordering of 50 heads and 50 tails. It is required that neither A nor B hold an extreme opinion—for example, one, embodying independence, that takes $P(X_i = 1 \wedge X_j = 1)$ to be equal to $P(X_i = 1)P(X_j = 1)$, or one that takes $P(X_i = 1)$ to be equal to zero or equal to one. The theorem also does not specify how large N must be in general, though the proof gives a way of calculating it for the case under discussion. It could turn out to be impracticably large, and anyway we do not always have the freedom to gather samples large enough to come to agreement about probabilities before making a decision; there always comes a time when we must make a decision based on the evidence we have *at that time*.

Furthermore, the theorem assumes that both A and B have patterns of belief that satisfy the probability calculus. This is unlikely to be true for any actual person, and the subjective view does not assume that it is true. The subjective view assumes merely that a rational agent, when he discovers an "incoherence" among his degrees of belief, will change his beliefs so that they more closely reflect the probability calculus. However, there are many ways in which those changes could be made. It is unclear that in the face of such changes the import of the theorem would hold.

Nevertheless, the subjective view of probability offers enough to enough people that it is now one of the mainstream interpretations of the probability calculus, par-ticularly among philosophers. The strongest argument for adopting that view is a negative one, reflecting Ramsey's position: No principles, other than those embodied in the probability calculus, seem (to subjectivists) to be defensible as principles of rationality.

For quite different reasons the subjective view is often adopted in artificial intel-ligence. Many of those who work in artificial intelligence are concerned to get things done, to accomplish certain practical goals. Often these goals involve the manipulation of uncertainty. The subjective interpretation of probability allows them to move ahead on the basis of opinion. This works out, since the "opinions" that are taken as basic in a given domain are often the opinions of experienced experts in that domain, and these "opinions" are often based on extensive objective data. It is questionable whether the adjective "subjective" is an appropriate one in these contexts.

In *Scientific Reasoning: The Bayesian Approach* [Howson & Urbach, 1993] Howson and Urbach defend the twofold thesis that (1) scientific inference consists of assigning probabilities to hypotheses, just as the logical theorists would have it, but (2) probabilities represent "ideal" personal degrees of belief, and are not to be constrained by anything beyond the probability calculus, just as Ramsey said. De Finetti's theorem provides an *explanation* of why responsible parties with access to all the evidence tend (roughly) to agree about their assessments of probability. What to do when the conditions of de Finetti's theorem are not satisfied is left open.

In this connection, it is interesting to note some recent results [Seidenfeld & Wasserman, 1993; Seidenfeld & Wasserman, 1994] that show there are also circumstances under which an expansion of evidence leads the agents toward more *extreme* differences of opinion.[4]

The Bayesian view is currently quite popular in philosophy, for the reasons already mentioned: It is maximally permissive, and yet demands conformity to a defensible and useful structure. The important question is what it has to tell us about uncertain inference. It is easy to say that it tells us nothing about uncertain inference: no inferring, according to this point of view, is in the least uncertain. What the Bayesian view gives us is a number of constraints that our inferences *about* uncertainty should satisfy. But that is another matter altogether; these inferences themselves are certain and deductive.

That the subjective view tells us nothing about uncertain inference may not be to the point. It does purport to tell us what uncertainty is and how to manipulate it. Like the logical view, it regards the very idea of uncertain inference as a mistake. What we should be looking at is the way in which we modify our uncertainties in the light of evidence. As Isaac Levi pointed out long ago [Levi, 1967], however, it is hard to justify the prohibition against starting over with a new coherent probability function whenever you find that conditionalization has led you to a state you regard as unreasonable. What he calls "temporal conditionalization" is hard to defend.

Nothing we have said in criticism of the subjective point of view establishes that there is anything better. It may be that there *are* no rules or principles for uncertain inference. If not, and if there are no general principles for the assignment of prior probabilities of the sort that the logical interpreters suppose there are, then, as Earman says [Earman, 1992] subjectivistic Bayesianism is the only game in town. If that is the case, then both classical statistical inference and the commonsense view of inductive reasoning are in fundamental error, for both of these views embody the idea of the acceptance (or rejection) of hypotheses going beyond our current data. If all we need do is make decisions, and all we are concerned with is logical and probabilistic consistency, then perhaps the subjectivistic Bayesian view can give us all we deserve.

4.6 Summary

Each of these broad standard interpretations of the probability calculus has its advantages and its difficulties. The empirical interpretation of probability is clearly the kind of thing that most scientists have in mind when they propose probabilistic laws, when they report probabilistic regularities. It is the concept of probability that underlies almost all the work in statistical inference and, therefore, the use of probability in assessing experiments and reporting results in both the social and the physical sciences. The main difficulty with the empirical interpretations is that they do not provide us with a way of measuring uncertainty in any particular case. If what probability *means* is a frequency in a class, or a chance in a chance setup, it cannot depend on what we know or don't know. But this relativity is clearly required of any

---

[4]It is worth observing that these circumstances are pretty rare, though they can arise in the case of almost any distribution. Essentially they amount to the case when one agent takes a kind of evidence to be favorable, and the other takes it to be unfavorable, to a given type of event.

measure of uncertainty in inference from known evidence. In general the empirical interpretations of probability were designed, not for measuring the uncertainty of inference, but for expressing stochastic facts about the world, and for this they are admirably suited. The issue of which empirical interpretation is the most appropriate is an issue—and a deep issue—in what might be called the metaphysics of science, and this issue lies beyond the scope of our concerns in this volume. On the other hand, there may be a way in which empirical probabilities impinge on the uncertain inferences we make. We will come back to this possibility in a later chapter.

The logical approach to probability was designed explicitly for dealing with the uncertainty of inferences we make from empirical data. Many scholars think it fails, for several reasons.

(1) It presupposes a formal (usually first order) language, in which the sentences concerning a certain domain of science can be expressed. Few parts of science have been given an explicit formalization.

(2) Given a formal language, the assignment of logical measures to its sentences is to some degree arbitrary. Carnap's original work, in which c^* and $c^\dagger$ were given as examples of confirmation functions, was succeeded by a more general study in which a whole continuum of inductive methods, parameterized by a real-valued quantity λ, was offered. In Carnap's last work two parameters were thought necessary. In related work by other authors [Hintikka, 1966], three, four, or five arbitrary parameters have been suggested.

(3) Finally, one may question what the enterprise has to do with uncertain *inference*. It clearly has to do with uncertainty: we can measure the degree of confirmation of one sentence by another sentence or a set of sentences representing our total evidence. We infer the degree of confirmation, if you will, but that inference (given that we have accepted a particular logical measure function) is not at all uncertain. Even if the evidence e gives the hypothesis h a very high degree of confirmation, that does not entitle us to make the inductive leap of accepting h.

The subjective interpretation of probability is a natural one for expressing the uncertainty we feel about some statements; it provides a natural basis for decision theory, to which it is very closely related. But, in addition to suffering the same awkwardness with regard to inference that the logical theory does, it suffers also from arbitrariness. We said that it was an interpretation of probability that was maximally liberal and minimally constraining. And so it is. De Finetti's theorem speaks directly to that difficulty, by pointing out that as evidence accumulates for two reasoners, their personal probabilities will become closer, *provided* certain conditions are met. But the theorem does not always apply—its conditions may not be met. Furthermore, even if the theorem does apply, the two reasoners may not have opinions that are close enough at a given time for them to agree on a choice or a course of action at that time.

All three of these major types of interpretation of probability are found in the contemporary literature, and all three are ably defended by numerous scholars. It seems clear, as Carnap was the first to point out in detail, that there is room for more than one interpretation of probability. In subsequent chapters we will examine a number of the ways in which various of these views have entered into the theory of

4.7   Bibliographical Notes

John Venn [Venn, 1866] was the first to suggest regarding probability as the limit of a relative frequency, but he was not clear about how you could approach a limit. Richard von Mises [von Mises, 1957] is generally credited with being the father of the modern frequency theory, and, more or less accidentally, he started the modern theory of randomness. D. H. Mellor [Mellor, 1971] holds a propensity interpretation of probability; other varieties have been offered by Giere [Giere, 1973] and Fetzer [Fetzer, 1977]. The most common statistical attitude is clearly presented by Harald Cramér [Cramér, 1951], who takes probability simply to be the "abstract counterpart" of frequencies, and forgoes any attempt to provide it with an explicit definition.

The earliest efforts at a logical interpretation of probability were those of von Kries [von Kries, 1886]. John Maynard Keynes, inspired by the work in logic of Russell and Whitehead, wrote *A Treatise on Probability* in 1921 [Keynes, 1952]. Keynes took the probability concept to be a logical one, but did not offer a definition. The most famous logical approach is that of Carnap, who systematically defined measures on formal languages to serve as a foundation for probability [Carnap, 1950; Carnap, 1952]. Following him in this were Hintikka [Hintikka, 1966], Hilpinen [Hilpinen, 1966], Tuomela [Tuomela, 1966], Niiniluoto [Niiniluoto, 1976], and others. Contemporary work along the same lines is pursued in computer science by Aleliunas [Aleliunas, 1990], Halpern [Halpern, 1990], and Bacchus [Bacchus, 1992].

The two (independent) founders of the subjective interpretation of probability were the statistician Bruno de Finetti [de Finetti, 1980] and the philosopher Frank P. Ramsey [Ramsey, 1931]. In statistics the interpretation owes its success largely to the work of L. J. Savage [Savage, 1954]. A relatively elementary exposition of the theory, from a philosophical point of view, will be found in Skyrms's *Choice and Chance* [Skyrms, 1966]. Richard Jeffrey, in *The Logic of Decision* [Jeffrey, 1965], exploited the relation between subjective probability and utility theory. The most recent book (1993) defending the subjective view is that of Howson and Urbach [Howson & Urbach, 1993]. Another philosopher, whom one might not want to call a subjectivist, but who has much in common with that point of view, is Isaac Levi [Levi, 1980].

4.8   Exercises

(1)   Give a formal specification of "mixing" collectives, and show that if K_1 and K_2 are collectives, in which the limiting relative frequencies of H_1 and H_2 exist and are equal to p_1 and p_2, then the limit of the relative frequency of $H_1 \vee H_2$ exists and is equal to $p_1 + p_2$ in the mixture of K_1 and K_2.

(2)   "The probability is at least 0.95 that the probability is between 0.6 and 0.7 that a person who receives a commendation in this organization is a man." Discuss the two occurrences of the word "probability" (do they mean the same thing or two different things?) and the role of the indefinite article "a" in "a person." What different interpretations of probability might be appropriate for these two occurrences? Can we make sense of the sentence using only a single interpretation?

(3)   There are three classical interpretations of "probability" in current use: limiting or finite frequencies; logical relations; subjective degrees of belief. Each has advantages and drawbacks. Evaluate these virtues and deficiencies when it comes to discussing problems of uncertain inference. Is there a role for more than one interpretation?

(4)   Richard von Mises and most British and American statisticians would argue that probabilistic inference is not "uncertain inference". Explain. This is also true of many subjectivists. Explain.

(5)   In a simple language Carnap assigns probability $\frac{1}{2}$ to any consistent atomic statement S before any evidence is gathered bearing on it. This probability is an a priori prior probability. Is this assignment of probability $\frac{1}{2}$ to S based on some principle similar to the principle of indifference? If so, why so? If not, why not?

(6)   Why do many statisticians reject the idea of prior probability? Do you agree with them? Why or why not?

(7)   What is the point of a "logical interpretation" of probability, as promoted by Carnap and others? If probability statements are, like theorems of mathematics, empty of empirical content, then what are they good for?

(8)   "It is probable that Caesar crossed the Rubicon in March, 55 B.C." Give a frequency, a logical, and a subjective interpretation of this statement, and, in each case, evaluate its truth or explain why you can't evaluate its truth.

(9)   "The probability is about a half that Mary's child will be a boy." Give a frequency, a logical, and a subjective interpretation of this statement.

(10)   "It is highly probable that a sequence of 100 tosses of a coin will yield between 40 and 60 heads." "It is highly probable that the next sequence of tosses of this coin will yield between 40 and 50 heads." Use these two statements to draw a contrast between a logical and a frequency interpretation of probability.

Bibliography

[Aleliunas, 1990] Romas Aleliunas. A new normative theory of probabilistic logic. In Henry E. Kyburg, Jr., Ronald P. Loui, and Greg N. Carlson, editors, *Knowledge Representation and Defeasible Reasoning*, pp. 387–403. Kluwer, Dordrecht, 1990.

[Bacchus, 1992] Fahiem Bacchus. *Representing and Reasoning with Probabilistic Knowledge*. The MIT Press, 1992.

[Carnap, 1950] Rudolf Carnap. *The Logical Foundations of Probability*, University of Chicago Press, Chicago, 1950.

[Carnap, 1952] Rudolf Carnap. *The Continuum of Inductive Methods*. University of Chicago Press, Chicago, 1952.

[Carnap, 1971] Rudolf Carnap. A basic system of inductive logic, Part I. In Rudolf Carnap and Richard C. Jeffrey, editors, *Studies in Inductive Logic and Probability I*, pp. 33–165. University of California Press, Berkeley, 1971.

[Carnap, 1980] Rudolf Carnap. A basic system for inductive logic: Part II. In R. C. Jeffrey, editor, *Studies in Inductive Logic and Probability*, pp. 7–155. University of California Press, Berkeley, 1980.

[Cramér, 1951] Harald Cramér. *Mathematical Methods of Statistics*. Princeton University Press, Princeton, 1951.

[de Finetti, 1980] Bruno de Finetti. Foresight: Its logical laws, its subjective sources. In Henry E. Kyburg, Jr., and Howard Smokler, editors, *Studies in Subjective Probability*, pp. 53–118. Krieger, Huntington, NY, 1980.

[Earman, 1992] John Earman. *Bayes or Bust?* MIT Press, Cambridge, MA, 1992.

[Fetzer, 1977] James H. Fetzer. Reichenbach, reference classes, and single case 'probabilities.' *Synthese*, 34:185–217, 1977.

[Giere, 1973] Ronald M. Giere. Objective single-case probabilities and the foundations of statistics. In Patrick Suppes, Leon Henkin, Athanase Joja, and GR. C. Moisil, editors, *Logic Methodology and Philosophy of Science IV*, pp. 467–484. North Holland, Amsterdam, 1973.

[Hacking, 1965] Ian Hacking. *Logic of Statistical Inference*. Cambridge University Press, Cambridge,

[Hacking, 1975] Ian Hacking. *The Emergence of Probability*. Cambridge University Press, 1975.

[Halpern, 1990] Joseph Y. Halpern. An analysis of first-order logics of probability. *Artificial Intelligence*, 46:311–350, 1990.

[Hilpinen, 1966] Risto Hilpinen. On inductive generalization in monadic first order logic with identity. In Jaakko Hintikka and Patrick Suppes, editors, *Aspects of Inductive Logic*, pp. 133–154. North Holland, Amsterdam, 1966.

[Hintikka, 1966] Jaakko Hintikka. A two-dimensional continuum of inductive logic. In Hintikka and Suppes, editors, *Aspects of Inductive Logic*, pp. 113–132. North Holland, Amsterdam, 1966.

[Howson & Urbach, 1993] Colin Howson and Peter Urbach. *Scientific Reasoning: The Bayesian Approach*. Open Court, La Salle, IL, 1993.

[Jeffrey, 1965] Richard C. Jeffrey. *The Logic of Decision*. McGraw-Hill, New York, 1965.

[Keynes, 1952] John Maynard Keynes. *A Treatise on Probability*. Macmillan, London, 1952.

[Laplace, 1951] Pierre Simon Marquis de Laplace. *A Philosophical Essay on Probabilities*. Dover Publications, New York, 1951.

[Leblanc & Morgan, 1983] Hugues Leblanc and Charles Morgan. Probability theory, intuitionism, semantics, and the Dutch book argument. *Notre Dame Journal of Formal Logic*, 24:289–304, 1983.

[Lehman, 1959] E. L. Lehman. *Testing Statistical Hypotheses*. Wiley, New York, 1959.

[Levi, 1967] Isaac Levi. *Gambling with Truth*. Knopf, New York, 1967.

[Levi, 1980] Isaac Levi. *The Enterprise of Knowledge*. MIT Press, Cambridge, MA, 1980.

[Mellor, 1971] D. H. Mellor. *The Matter of Chance*. Cambridge University Press, Cambridge, U.K., 1971.

[Morgan, 1998] Charles Morgan. Non-monotonic logic is impossible. *Canadian Artificial Intelligence Magazine*, 42:18–25, 1998.

[Nagel, 1939] Ernest Nagel. *Principles of the Theory of Probability*. University of Chicago Press, Chicago, 1939, 1949.

[Niiniluoto, 1976] Ilkka Niiniluoto. On a k-dimensional system of inductive logic. In *Psa 1976*, pp. 425–447. Philosophy of Science Association, 1976.

[Popper, 1959] K. R. Popper. *The Logic of Scientific Discovery*. Hutchinson, London, 1959.

[Ramsey, 1931] F. P. Ramsey. *The Foundations of Mathematics and Other Essays*. Humanities Press, New York, 1931.

[Reichenbach, 1949] Hans Reichenbach. *The Theory of Probability*. University of California Press, Berkeley and Los Angeles, 1949.

[Russell, 1901] Bertrand Russell. Recent works on the principles of mathematics. *The International Monthly*, 4:83–101, 1901.

[Salmon & Greeno, 1971] Wesley C. Salmon and James Greeno. *Statistical Explanation and Statistical Relevance*. University of Pittsburgh Press, Pittsburgh, 1971.

[Savage, 1954] L. J. Savage. *Foundations of Statistics*. Wiley, New York, 1954.

[Savage, 1966] L. J. Savage. Implications of personal probability for induction. *Journal of Philosophy*, 63:593–607, 1966.

[Seidenfeld & Wasserman, 1993] Teddy Seidenfeld and Larry Wasserman. Dilation for sets of probabilities. *The Annals of Statistics*, 21:1139–1154, 1993.

[Seidenfeld & Wasserman, 1994] Teddy Seidenfeld and Larry Wasserman. The dilation phenomenon in robust Bayesian inference. *Journal of Statistical Planning and Inference*, 40:345–356, 1994.

[Shimony, 1955] Abner Shimony. Coherence and the axioms of confirmation. *Journal of Symbolic Logic*, 20:1–28, 1955.

[Skyrms, 1966] Brian Skyrms. *Choice and Chance: An Introduction to Inductive Logic*. Dickenson, Belmont, CA, 1966.

[Tuomela, 1966] Raimo Tuomela. Inductive generalizations in an ordered universe. In Jaakko Hintikka and Patrick Suppes, editors, *Aspects of Inductive Logic*, pp. 155–174. North Holland, Amsterdam, 1966.

[Venn, 1866] John Venn. *The Logic of Chance*. Macmillan, London, 1866.

[von Kries, 1886] J. von Kries. *Die Principien der Wahrscheinlichkeitsrechnung*. Freiburg, 1886.

[von Mises, 1957] Richard von Mises. *Probability Statistics and Truth*, George Allen and Unwin, London, 1957.

5

Nonstandard Measures of Support

5.1 Support

As Carnap points out [Carnap, 1950], some of the controversy concerning the support of empirical hypotheses by data is a result of the conflation of two distinct notions. One is the total support given a hypothesis by a body of evidence. Carnap's initial measure for this is his $\mathbf{c}^*$; this is intended as an explication of one sense of the ordinary language word "probability." This is the sense involved when we say, "Relative to the evidence we have, the probability is high that rabies is caused by a virus." The other notion is that of "support" in the active sense, in which we say that a certain piece of evidence *supports* a hypothesis, as in "The detectable presence of antibodies supports the viral hypothesis." This does not mean that that single piece of evidence makes the hypothesis "highly probable" (much less "acceptable"), but that it makes the hypothesis more probable than it was. Thus, the presence of water on Mars *supports* the hypothesis that that there was once life on Mars, but it does not make that hypothesis highly probable, or even more probable than not.

Whereas $\mathbf{c}^*(h, e)$ is (for Carnap, in 1950) the correct measure of the degree of support of the hypothesis h by the evidence e, the increase of the support of h due to e given background knowledge b is the amount by which e *increases* the probability of h: $\mathbf{c}^*(h, b \wedge e) - \mathbf{c}^*(h, b)$. We would say that e supports h relative to background b if this quantity is positive, and undermines h relative to b if this quantity is negative.

One does not have to interpret probability as a logical relation in order to find this distinction useful. It has been adopted by many Bayesians of the subjective stripe as well. There is clearly a difference between the probability that E confers on H and the *increase* of probability given by the evidence E: the difference between $\mathcal{P}(H|E)$ and $\mathcal{P}(H|E) - \mathcal{P}(H)$.

But recognizing the difference between the overall support of H and the degree to which E increases the support of H—what might also be called the *evidential support* given H by E—is not necessarily to agree that $\mathcal{P}(H|E) - \mathcal{P}(H)$ is the best measure of evidential support. A number of writers have argued that the guidance we need for handling uncertainty in inference is best given by other functions of measures on sentences.

5.2 Karl Popper

One philosopher whose views have been extensively honed against those of Carnap is Karl Popper, who has devised his own measure of evidential support.[1] Popper's views on uncertain inference stem from his general views concerning scientific discovery: for him what is most important is our ability to choose among alternative scientific hypotheses.

In *The Logic of Scientific Discovery* [Popper, 1959], Karl Popper presents the uncontroversial thesis that theories cannot be *shown*, by any finite amount of empirical evidence, to be true. Instead, theories can either be *corroborated* to some degree by evidence or they can be shown to be false by evidence. Popper bases his approach to uncertain inference on the rejection of the *principle of induction*.

According to Popper, the principle of induction is intended to provide criteria for justifying inductive inferences. Unlike deductive inference, where the conclusion follows logically from the premises, an inductive inference proceeds from a set of singular statements to a universal generalization that is not logically entailed by those singular statements. The acceptability of the statement "All swans are white" depends both on the set of singular statements and on an appropriate application of the principle of induction.

But for Popper "All swans are white" cannot be *justified* by any number of observations of swans. If I examine 50 swans and find all of them to be white, that is very good evidence that those swans are white but it does not ensure that all swans are white. There are a potentially infinite number of swans covered by the law "All swans are white"; Popper's argument is that only by examining all swans (an impossible task) could we be sure that the theory is true. He puts severe demands on the concept of justification.

Whereas we can never be sure that a theory is true, we *can* be sure that a theory is false. If we claim that "All swans are white" and then discover a black swan, our claim "All swans are white" has been refuted; we say that the theory is falsified.

Because Popper holds that all theories and laws are universal statements, he finds an infinite regress in the attempt to justify the principle of induction. Following Hume's argument, he argues that the principle of induction cannot be a logical truth, for if it were, then all inductive inferences would turn out to be merely logical transformations. These transformations would be just like deductive inferences, and that would mean that induction was simply deduction. Because inductive inferences are not (by definition) deductive inferences, the principle of induction itself must be justified.

One such justification would be this: Because the principle of induction works in applications in science and everyday life, we have reason to suppose that it works in general. This amounts to saying that there are many instances of past applications of the principle of induction that have provided successful conclusions. These instances provide justification for supposing that the principle of induction will be successful in general. If we use a finite set of instances of the success of the principle of induction as justification for the principle of induction, then we are once again using finite instances to prove a general law. We will need some justification for that proof. But that is what we wanted the principle of induction for! We are back where we started.

---

[1]This controversy is captured in the exchanges between Carnap and Popper in [Lakatos, 1968].

"Thus, the attempt to base the principle of induction on experience breaks down, since it must lead to an infinite regress" [Popper, 1959, p. 29].

Once we reject the claim that experience can tell us about the *truth* of universal laws, we might imagine that experience can give us some *degree of reliability* for the universal law, and we might think that this degree of reliability was the probability that the universal statement is true. But, like Hume before him, Popper rejects this move.

Popper accepts an *empirical* interpretation of probability, and therefore argues that this approach also leads us to use the success of previous applications of the principle of induction to justify future applications of the principle of induction. But, as we have seen, that approach leads to circularity or an infinite regress.

If evidence cannot determine that a theory is true, and cannot determine even that a theory is probable, then the purpose of science cannot be to determine that theories are true or probable. According to Popper, the purpose of scientific inquiry is to *conjecture* theories and put them to the most rigorous tests possible in an attempt to falsify them. When a theory fails a test, it becomes definitively falsified: we know, of that general theory, that it is false. Whenever a theory passes a test, the evidence (the test results) corroborates the theory to some degree, though it does not give it a probability. An indication of the degree to which a theory is corroborated is given by taking account of the "severity" of the test. Popper's approach to uncertain inference is to seek to formalize this notion of corroboration.

5.2.1 Corroboration

What does Popper mean by "corroboration"? Popper explicitly denies that corroboration is just a new name for probability [Popper, 1959, p. 270]. Popper is attempting to measure how well a theory or hypothesis "has been able to prove its fitness to survive by standing up to tests" [Popper, 1959, p. 251].

A measure of corroboration should tell us how much a law or theory is backed up by the evidence. If we have laws L_1 and L_2 and evidence E that undermines L_1 but corroborates L_2, then we surely want the corroboration of L_1 relative to E to be less than that of L_2 relative to E: $C(L_1|E) < C(L_2|E)$. Probability does not guarantee this result, because the conditional probability depends on the prior probability, as well as the evidence—we could have $\mathcal{P}(L_1) \gg \mathcal{P}(L_2)$.

The measure of corroboration should influence how we regard a given theory L. A high degree of corroboration characterizes a theory that we take to be worthy of additional tests. Science is primarily interested in informational content, and good theories should have high informational content. But the informational content of the hypothesis L is high just in case the hypothesis is *not* very probable. In essence, if a theory either says very little about the world or if it goes very little beyond what we already know is true, then the probability that the theory is true is high; while if we go very far beyond our data, then the probability that the theory is true is low, but the theory embodies a lot of content, and it is content that we seek.[2]

Because science wants high informational content, it wants a theory that has a low probability. If corroboration were simply probability, then science would want

---

[2]Popper writes in this vein even though he does not accept a logical interpretation of probability. He does, however, accept the usefulness of "probability" measures on the sentences of a language, because those measures reflect information content.

a theory that is not corroborated very much. But what science wants is theories that are highly corroborated.

Popper has developed a conception of corroboration that reflects the scientific desire for high information content. This conception takes account of informational content and explanatory power. Popper's formal definition of corroboration depends on his notion of explanatory power. Although in general Popper interprets probability empirically, in terms of propensities, in this context he allows that a useful function can be served by a probabilistic measure over the sentences of the language of science. We can use this measure to define *informational content* and *explanatory power* (Expl):

Definition 5.1 (The Explanatory Power of L with Respect to E). *If L is consistent and $P(E) > 0$, then*

$$\text{Expl}(L|E) = \frac{P(E|L) - P(E)}{P(E|L) + P(E)}$$

Note that $-1 \leq \text{Expl}(L|E) \leq 1$. Clearly Expl is not additive. In terms of Expl, Popper defines corroboration:

Definition 5.2 (The Corroboration of L by E).

$$C(L|E) = \text{Expl}(L|E)[1 + P(L)P(L|E)]$$

Whereas it must be admitted that this definition does not immediately grab the intuition, it turns out to have a number of interesting and natural properties. The following nine theorems correspond to Popper's "desiderata" [Popper, 1959, pp. 400–401] for a measure of corroboration.

Theorem 5.1.

(a) $C(L|E) > 0$ *if and only if E supports L.*
(b) $C(L|E) = 0$ *if and only if E is independent of L.*
(c) $C(L|E) < 0$ *if and only if E undermines L.*

Popper wants to set the upper and lower bounds for corroboration to 1 and −1. If a theory is corroborated to degree 1, then it is true; if it is corroborated to degree −1, then it is falsified:

Theorem 5.2. $-1 = C(\neg E|E) \leq C(L|E) \leq C(L|L) \leq 1$.

Note that $C(L) = C(L|L)$, the degree of *confirmability* of L, is simply $P(\neg L)$, and thus is additive:

Theorem 5.3. $0 \leq C(L|L) = C(L) = P(\neg L)$.

A theory cannot be corroborated by information that could not have undermined it:

Theorem 5.4. *If E entails L then $C(L|E) = C(L|L) = C(L)$.*

A theory is falsified when the evidence entails its negation:

Theorem 5.5. *If E entails ¬L then $C(L|E) = C(\neg E|E) = -1$.*

Corroboration increases with the power of the theory to explain the evidence:

Theorem 5.6. $C(L|E)$ *increases with* $\text{Expl}(L|E)$.

Corroboration is proportional to conditional probability, other things being equal—i.e., the prior probabilities being the same:

Theorem 5.7. *If* $C(L|L) = C(L'|L') \neq 1$, *then*

(a) $C(L|E) > C(L'|E')$ *if and only if* $P(L|E) > P(L'|E')$.
(b) $C(L|E) = C(L'|E')$ *if and only if* $P(L|E) = P(L'|E')$.
(c) $C(L|E) < C(L'|E')$ *if and only if* $P(L|E) < P(L'|E')$.

Proof: $P(L|E) > P(L'|E')$ if and only if
$1 + P(L)P(L|E) > 1 + P(L')P(L'|E')$; $P(L|E) > P(L'|E')$ if and only if

$$\frac{P(L|E) - P(L)}{P(L|E) + P(L)} > \frac{P(L'|E') - P(L')}{P(L'|E') + P(L')}$$

since all terms are positive.　　　　　　　　　　　　　　　　　　■

Theorem 5.8. *If L entails E, then*

(a) $C(L|E) \geq 0$.
(b) *For fixed L,* $C(L|E)$ *and* $C(E|E) = P(\neg E)$ *increase together.*
(c) *For fixed E,* $C(L|E)$ *and* $P(L)$ *increase together.*

Theorem 5.9. *If $\neg L$ is consistent and entails E,*

(a) $C(L|E) \leq 0$.
(b) *For fixed L,* $C(L|E)$ *and* $P(E)$ *increase together.*
(c) *For fixed E,* $C(L|E)$ *and* $P(L)$ *increase together.*

5.2.2　Levi's Criticism

In *Gambling with Truth* [Levi, 1967], Isaac Levi argues that Popper's notion of corroboration is not adequately tied to his conception of the aim of science. In particular, Popper does not provide any proof that adopting corroboration as a measure of a good theory will help us achieve the aims of science.

We can better understand Levi's objection if we examine it in the light of the following quotation from *Gambling with Truth* [Levi, 1967, p. vii]:

Scientific inquiry, like other forms of human deliberation, is a goal directed activity. Consequently, an adequate conception of the goal or goals of scientific inquiry ought to shed light on the difference between valid and invalid inferences; for valid inferences are good strategies designed to obtain these goals.... However, knowing what one wants does not determine the best way to obtain it.

Levi believes that we need to know where we are going before we can determine how to get there. Once we know where to go, we must determine how to get there in the best possible way, or at least in an efficient way.

It is quite clear that corroboration is intended to provide a way to achieve the ends of science. If we have a group of theories that are corroborated to different degrees, then we should choose the one that is most highly corroborated. This should lead us to Popper's goal (or to the aim of science): theories that are true and that have high explanatory power, content, and testability. But Popper has failed to provide any

Table 5.1: Corroboration vs epistemic utility

H_1, E_1		H_2, E_2	
$H_1 \wedge E_1$	0.20	$H_2 \wedge E_2$	0.02
$H_1 \wedge \neg E_1$	0.10	$H_2 \wedge \neg E_2$	0.01
$\neg H_1 \wedge E_1$	0.10	$\neg H_2 \wedge E_2$	0.10
$\neg H_1 \wedge \neg E_1$	0.60	$\neg H_2 \wedge \neg E_2$	0.87
$C(E_1|H_1) = 0.46$		$C(H_2|E_2) = 0.70$	
$U(H_1|E_1) = 0.37$		$U(H_2|E_2) = 0.14$	

Levi introduces the notion of *epistemic utility*. Popper's concerns clearly include the gain of information obtained from accepting a theory; but his corroboration function does not lead to the acceptance of the hypothesis with the highest degree of expected epistemic utility.

Levi gives us a very simple and natural epistemic or cognitive utility measure: $P(\neg x)$ is the utility of accepting x when x is true, and $-P(x)$ is the utility of accepting x when it is false. Using these as utilities, we calculate the expected cognitive utility of accepting hypothesis x on evidence y to be

$$U(x|y) = P(x|y)P(\neg x) - P(\neg x|y)P(x)$$

Choosing hypotheses to maximize corroboration does not maximize utility in this sense. Table 5.1 shows that Popper's measure of corroboration and Levi's measure of epistemic utility yield opposite results: Popper's measure ranks H_1 given E_1 below H_2 given E_2, and Levi's measure reverses the order.

In fact, Levi argues that there is no plausible sense of epistemic utility that can be maximized by maximizing corroboration.

5.3 Other Measures

Other writers than Popper have questioned the use of probability as a basis for induction. Even Carnap, responding to Popper's objections to probability or degree of confirmation [Carnap, 1950, second edition], offered a measure designed to reflect *increase* of confirmation, as opposed to confirmation: $c(h, e) - c(h)$, which represents the degree to which e *adds* to the confirmation of h.

John Kemeny and Paul Oppenheim [Kemeny & Oppenheim, 1952] follow much the same program as Popper: they lay down certain intuitively natural conditions of adequacy for a "measure of factual support". $F(H, E)$ represents the support given H by the evidence E. The "simplest" formula that satisfies their desiderata is:

Definition 5.3.

$$F(H, E) = \frac{P(E|H) - P(E|\neg H)}{P(E|H) + P(E|\neg H)}.$$

Note that this is the same as the main component of Popper's corroboration. Nicholas Rescher [Rescher, 1958] plays this game, too. He defines a notion of "degree of evidential support", des(h, e), as follows:

Definition 5.4.

$$des(h, e) = \frac{P(h|e) - P(h)}{P(e)}$$

Table 5.2: Measures of evidential support

Author	Factor 1	Factor 2	Factor 3	Factor 4
Carnap	$\mathcal{P}(e\|h) - \mathcal{P}(e)$	$\mathcal{P}(h)$	1	$\frac{1}{\mathcal{P}(e)}$
Rescher	$\mathcal{P}(e\|h) - \mathcal{P}(e)$	$\mathcal{P}(h)$	$\frac{1}{\mathcal{P}(\neg h)}$	1
Popper	$\mathcal{P}(e\|h) - \mathcal{P}(e)$	1	1	$\frac{1}{\mathcal{P}(e\|h)+\mathcal{P}(e)}$
Kemeny	$\mathcal{P}(e\|h) - \mathcal{P}(e)$	1	1	$\frac{1}{\mathcal{P}(e\|h)+\mathcal{P}(e\|\neg h)}$
Levi	$\mathcal{P}(e\|h) - \mathcal{P}(e)$	$\mathcal{P}(h)$	$\frac{1}{\mathcal{P}(\neg h)}$	$\frac{1}{\mathcal{P}(e)}$

This is just a sample of measures of evidential support that have been offered. What is interesting about the sample is that while the conditions of adequacy stipulated the various authors are different (though similar), and the notation and form of the final definition of corroboration or support vary, in the final analysis there is more in common among them than meets the eye.

In Table 5.2 I have performed some simple algebraic manipulations designed to bring out the similarities of these apparently distinct measures. The entries in a row represent factors, which, when multiplied, yield the corroborative measure suggested by the author in the first column.

There are several lessons to be learned from Table 5.2. The most important is that all of these measures boil down to functions of probability measures defined over the language with which we are concerned. Although the authors start from different places, and have apparently different intuitions, probability measures on sentences are basic.

The second lesson is that the first factor, the difference between the probability of the evidence on the hypothesis and the probability of the evidence, is a fundamental factor on each of the five different views. The second and third factors reflect a fundamental difference between two points of view: one takes the antecedent probability of the hypothesis to be a positive consideration, the other does not. The fourth factor is basically a normalizing factor.

None of these measures has achieved any significant degree of acceptance. They represent the efforts of scholars in this domain to articulate and formalize their intuitions regarding the support of hypotheses by empirical evidence. The effort to find a generally agreed-on sharp measure of this support in terms of probabilities has failed.

5.4 Dempster–Shafer Belief Functions

One of the oddities of the principle of indifference—the principle that if we have similar information about two alternatives, the alternatives should be assigned the same probability—is that it yields the same sharp probabilities for a pair of alternatives about which we know nothing at all as it does for the alternative outcomes of a toss of a thoroughly balanced and tested coin. One way of capturing this difference is offered by Glen Shafer in *A Mathematical Theory of Evidence* [Shafer, 1976]. Glenn Shafer argues that our belief in a particular proposition should be based explicitly on the evidence we have for that proposition. If we have very little evidence either way, we might want to assign a small degree of support (0.2, say) to T but also a small degree of support, say 0.1, to its denial $\neg T$.

This approach to the representation and manipulation of uncertainty has been influential in the past decade and a half in computer science, where the handling of uncertainty has been the focus of much research.

5.4.1 Belief Functions and Mass Functions

A *frame of discernment* Θ is a collection of possible worlds, or a collection of models with a given domain. It is so called "in order to emphasize the epistemic nature of the set of possibilities" [Shafer, 1976, p. 76]. The frame of discernment plays the role of the sample space S in standard probability theory. We also have a field of subsets X of Θ that consists of all the possibilities we are going to take seriously. If we are contemplating the toss of a coin, it may consist of just the four sets $\{\Theta, H, T, \emptyset\}$. If we are contemplating a sequence of n tosses of a coin, it may consist of the algebra whose atoms are the 2^n possible outcomes of tossing a coin n times.

Given a particular frame of discernment, we are interested in the function m from 2^Θ to $[0, 1]$. The quantity $m(X)$ represents the *mass* assigned to a set X in Θ, which in turn represents the amount of support allocated exactly to that subset X of Θ, and not to any proper subset of X. Of course, we may assign 0 to most of the elements of 2^Θ; thus we may focus on a small algebra rather than the whole powerset of Θ. The function m is called a *basic probability assignment*.

Definition 5.5. *If Θ is a frame of discernment, then the function $m : 2^\Theta \to [0, 1]$ is called a* basic probability assignment *provided*

(1) $m(\emptyset) = 0$ and
(2) $\sum_{X \subseteq \Theta} m(X) = 1$.

The total belief allocated to the set of possibilities X is the sum of all the measures allocated to X and its subsets. This is given by a *belief function* Bel:

Definition 5.6. $\mathrm{Bel}(X) = \sum_{A \subseteq X} m(A)$.

We can also provide axioms for Bel directly without using m:

Theorem 5.10. *If Θ is a frame of discernment, then a function* Bel *is a belief function if and only if*

(1) $\mathrm{Bel}(\emptyset) = 0$,
(2) $\mathrm{Bel}(\Theta) = 1$, and
(3) $\mathrm{Bel}(A_1 \cup \cdots \cup A_n) \geq \sum_{I \subseteq \{1,\dots,n\}} (-1)^{|I|+1} \mathrm{Bel}(\bigcap_{i \in I} A_i)$.

Clearly there is a relationship between m and Bel, given by the following theorem:

Theorem 5.11. *Suppose* Bel $: 2^\Theta \to [0, 1]$ *is the belief function given by the basic probability assignment $m : 2^\Theta \to [0, 1]$. Then*

$$m(A) = \sum_{X \subseteq A} (-1)^{|A-X|} \mathrm{Bel}(X) \quad \text{for all} \quad A \subseteq \Theta.$$

A subset X of a frame of discernment Θ is called a *focal element* of a belief function Bel if $m(X) > 0$. The union of focal elements of a belief function is called its *core*.

In a probabilistic framework, the probability we assign to X determines the probability of $\neg X$. This is not so for belief functions. The degree of belief we have in X is separate from the degree of belief we have in $\neg X$.

For example let Θ be a frame of discernment in which our beliefs about the outcome of a particular coin toss are represented. The coin looks odd, so the focal elements are heads, tails $= \neg$ heads, and Θ itself and the measures we assign to the focal elements might be $m(\text{heads}) = 0.3$, $m(\text{tails}) = 0.3$, $m(\Theta) = 0.4$. We may easily calculate $\text{Bel}(\text{heads}) = 0.3$ and $\text{Bel}(\neg \text{heads}) = 0.3$; their sum is less than 1.0.

What is $1 - \text{Bel}(\neg\text{heads}) = 0.7$, then? It is called *plausibility*, and represents an upper bound for belief in heads, as will shortly be explained.

Definition 5.7. $\text{Pl}(X) = 1 - \text{Bel}(\overline{X})$.

Example 5.1. *For a working example of this approach to uncertainty, consider the rolls of a suspicious die. The outcomes can naturally be divided into the six alternatives we will label $O_1, \ldots, O_6$, corresponding to the number of dots that land uppermost. In the case of a die we knew to be normal, we would assign mass of $\frac{1}{6}$ to each of the singleton sets corresponding to these possibilities. But let us suppose that this is a suspicious-looking die, and that we are quite sure it is biased so that the one comes up more often than the two, or vice versa, but that the other sides each occur a sixth of the time.*

We may take the long run frequency of ones to have any value between 0 and $\frac{1}{3}$, with the long run frequency of twos to be correspondingly biased. We may represent this state of knowledge by assigning a mass of $\frac{1}{6}$ to each of $\{O_3\}, \ldots, \{O_6\}$, a mass of 0.0 to each of $\{O_1\}$ and $\{O_2\}$, and a mass of $\frac{1}{3}$ to $\{O_1, O_2\}$.

One may think of the mass assigned to any set as confidence that may flow to any subset of that set. Thus, the mass of $\frac{1}{3}$ assigned to $\{O_1, O_2\}$ may potentially be assigned to its subsets in any proportion, corresponding to any bias.

The universal set Θ may also be assigned mass. Thus, to embody the belief that a die has an unknown bias, but nothing so severe as to reduce the frequency of any side below $\frac{1}{10}$, we could assign a mass of 0.10 to each singleton set $\{O_1\}, \ldots, \{O_6\}$, and a mass of 0.4 to the universal set Θ. This leads to the assignment $\text{Bel}(\{O_1\}) = 0.10$ and $\text{Pl}(\{O_1\}) = 0.5$.

5.4.2 Reduction to Sets of Probabilities

It should no longer surprise us to find that there is an intimate relation between belief functions and probabilistic measures. In fact, every belief function can be represented by a *convex set of probability functions* defined on the frame of discernment Θ [Kyburg, 1987].

A convex set of probability functions is a set of functions, each of which is a probability function (with the same domain), and that is closed: i.e., if P_1 is in the set and P_2 is in the set, then any linear combination of $\mathcal{P}_1$ and $\mathcal{P}_2$ is in the set. A linear combination of two functions with a common domain is a function whose value for any object in the domain is that linear function of the values assigned by the two initial functions. Thus, if $\mathcal{P}_1(A) = p$ and $\mathcal{P}_2(A) = q$, then $(a\mathcal{P}_1 + (1 - a)\mathcal{P}_2)(A) = ap + (1 - a)q$.

Theorem 5.12. *Let* Bel *be a belief function defined on a frame of discernment Θ. There exists a convex set of probability functions $\mathcal{F}P$ such that for every subset X of Θ,*

Proof: Let the belief function Bel have a finite or countable number of foci. Let them be ordered $A_1, A_2, \ldots$. Let the mass associated with A_1 be $m(A_1)$, and let the rest of the mass be assigned to Θ. The set of probability distributions $\{\mathcal{P} : \mathcal{P}$ is a probability distribution $\wedge \mathcal{P}(A_1) \geq m(A_1)\}$ is obviously convex, and for every subset X of Θ, $\mathrm{Bel}(X) = \min\{\mathcal{P}(X) : \mathcal{P} \in \mathcal{F}P\}$ and $\mathrm{Pl}(X) = \max\{1 - \mathcal{P}(\neg X) : \mathcal{P} \in \mathcal{F}P\}$. Now suppose the theorem holds for k foci other than Θ, and that the corresponding set of convex probability distributions is $\mathcal{P}_k$. Consider the focus A_{k+1}, and the set of probability distributions $\{\mathcal{P} : \mathcal{P}(A_{k+1}) \geq m(A_{k+1})\}$. Let the intersection of this set with $\mathcal{F}P_k$ be P. This set is convex, since the intersection of two convex sets is convex. We show that $\mathrm{Bel}(X) = \min\{\mathcal{P}(X) : \mathcal{P} \in \mathcal{F}P\}$. Consider a subset X of Θ. Let $A_{j_1}, \ldots, A_{j_h}$ be the foci among $A_1, \ldots, A_{k+1}$ included in X. For every $\mathcal{P}$ in $\mathcal{F}P$, we have $\mathcal{P}(A_{j_1} \cup \cdots \cup A_{j_h}) \geq \mathrm{Bel}(X)$. Suppose inequality obtains. Then, since the theorem holds for the first k foci, $m(A_{k+1})$ must differ from $\min\{\mathcal{P} : \mathcal{P}$ is a probability distribution $\wedge \mathcal{P}(A_{k+1}) \geq m(A_{k+1})\}$. But that is impossible. ∎

Example 5.2. *Suppose we attribute a measure of* 0.3 *to heads on a toss of an odd-looking coin, and a measure of* 0.3 *to tails, with* 0.4 *being assigned to* Θ. *This is equivalent to believing that the probability of heads lies between* 0.3 *and* 0.7. *If we think, more realistically, of the* distribution *of heads on two tosses of the coin, the simplest way to think of it, under the ordinary assumptions, is as a binomial distribution with a parameter lying between* 0.3 *and* 0.7. *The bounds on this distribution can of course be given an interpretation in terms of belief functions as well as in terms of sets of probabilities. Observe that general convexity now fails: A* 50–50 *mixture of* $\langle 0.09, 0.21, 0.21, 0.49 \rangle$ *and* $\langle 0.49, 0.21, 0.21, 0.09 \rangle$ *is not a binomial distribution.*

Belief functions are less general than sets of probability functions. It is not the case that every distribution in terms of sets of probability functions can be represented as a belief function. Thus, we may want to model states of uncertainty that can be modeled by sets of probability distributions, but that cannot be modeled by belief functions.

Example 5.3. *Consider a compound experiment consisting of either (1) tossing a fair coin twice, or (2) drawing a coin from a bag containing* 40% *two-headed and* 60% *two-tailed coins, and tossing it twice. The two parts are performed in some unknown ratio* p *so that, for example, the probability that the first toss lands heads is* $\frac{1}{2}p + 0.4(1 - p)$. *Let A be the proposition that the first toss lands heads, and B the proposition that the second toss lands tails. The sample space may be taken to be simply* $\{TT, TH, HT, HH\}$, *since* p *is unknown. The probabilities are*

$$TT: \quad \tfrac{1}{4}p + 0.6(1 - p)$$

$$TH: \quad \tfrac{1}{4}p$$

$$HT: \quad \tfrac{1}{4}p$$

$$HH: \quad \tfrac{1}{4}p + 0.4(1 - p)$$

And the set of probability distributions is

$$S_P = \left\{ \langle \tfrac{1}{4}p + 0.6(1 - p), \tfrac{1}{4}p, \tfrac{1}{4}p, \tfrac{1}{4}p + 0.4(1 - p) \rangle : p \in [0, 1] \right\}$$

$0.4 + 0.5 - 0.0$. *But belief functions are superadditive:* $\text{Bel}(A \cup B) \geq \text{Bel}(A) + \text{Bel}(B) -$ $\text{Bel}(A \cap B)$ [Shafer, 1976, p. 32].[3]

5.4.3 Combining Evidence

Quite often we have more than one item of evidence relevant to a subset of a frame of discernment. This may be a result of seeking and obtaining new evidence. One of the chief novelties of Shafer's approach to uncertainty lies in the way different items of evidence can be combined or updated.

What Shafer calls Dempster's rule of combination allows us to combine items of evidence and obtain what is called their orthogonal sum: their combined effect. This orthogonal sum is a new belief function based on the combined evidence as represented by initial belief functions.

We denote the combining operation by $\oplus$, so that the new mass function resulting from the combination of m_1 and m_2 is $m_1 \oplus m_2$ and the new belief function resulting from the combination of Bel_1 and Bel_2 is $\text{Bel}_1 \oplus \text{Bel}_2$.

Here is the formal definition:

Definition 5.8. *Suppose* Bel_1 *and* Bel_2 *are belief functions over the same frame* Θ, *with basic probability assignments* m_1 *and* m_2, *and focal elements* $A_1, \ldots, A_k$ *and* $B_1, \ldots, B_n$ *respectively. The basic probability assignment of the combined belief function* $\text{Bel}_1 \oplus \text{Bel}_2$ *is defined to be*

$$m_1 \oplus m_2(X) = \frac{\sum_{A_i \cap B_j = X} m_1(A_i) m_2(B_j)}{1 - \sum_{A_i \cap B_j = 0} m_1(A_i) m_2(B_j)}$$

provided $1 - \sum_{A_i \cap B_j = 0} m_1(A_i) m_2(B_j) > 0$.

Example 5.4. *Suppose that the frame of discernment* Θ *embodies various possible causes of my car's failure to start. I consult Expert$_1$, and after absorbing a long discourse, much of which I don't understand, my epistemic state is reflected by the mass function* m_1 *with values*

$$m_1(\text{battery}) = 0.2$$
$$m_1(\text{gasoline}) = 0.3$$
$$m_1(\text{starter}) = 0.2$$
$$m_1(\Theta) = 0.3$$

From these values we can calculate Bel_1 *and* Pl_1 *values. For example,* $\text{Bel}_1(\text{battery}) = 0.2$ *and* $\text{Pl}_1(\text{battery}) = 0.5$.

Disturbed by the uncertainty remaining, I consult Expert$_2$, who gives me another long discourse. Relative to this discourse alone, I would have an epistemic state reflected by the mass function m_2 *with values*

$$m_2(\text{battery}) = 0.2$$
$$m_2(\text{carburetor}) = 0.4$$
$$m_2(\text{starter}) = 0.1$$
$$m_2(\Theta) = 0.3$$

Again, we can calculate Bel_2 *values:* $\text{Bel}_2(\text{battery}) = 0.2$ *and* $\text{Pl}_2(\text{battery}) = 0.5$.

Table 5.3: Combined mass values

	m_2	bat $m_1 = 0.2$	gas 0.3	carb 0.0	starter 0.2	Θ 0.3
bat	0.2	0.04	0.06	0.00	0.04	0.06
gas	0.0	0.00	0.00	0.00	0.00	0.00
starter	0.1	0.02	0.03	0.00	0.02	0.03
carb	0.4	0.08	0.12	0.00	0.08	0.12
Θ	0.3	0.06	0.09	0.00	0.06	0.09

Having consulted two experts, I want to combine the results of my consultations. For this we employ Dempster's rule of combination to obtain a new belief function, $\text{Bel}_1 \oplus \text{Bel}_2$.

We form the combination by constructing a new core consisting of all the possible intersections of the core of Bel_1 and the core of Bel_2. This is represented in Table 5.3. Every nonempty set in this new core is assigned a measure equal to the product of the measures assigned to it by the original belief functions. Because there may also be some pairs of foci whose intersection is empty, we must renormalize by dividing by one minus the sum of the products of measures of pairs of foci whose intersection is empty.

Our experts are assuming, implausibly perhaps, that only one cause can be operating, for example, that the intersection of bat and gas is empty. The mass assigned to empty sets of this form is $0.06 + 0.04 + 0.02 + 0.03 + 0.08 + 0.12 + 0.08 = 0.43$. To renormalize (since a belief function can only assign 0 to the empty set), we divide each mass value by $1 - 0.43 = 0.57$. We get a combined mass function:

$$m_1 \oplus m_2(\text{battery}) = \frac{0.04 + 0.06 + 0.06}{0.57} = 0.28$$

$$m_1 \oplus m_2(\text{gasoline}) = \frac{0.09}{0.57} = 0.16$$

$$m_1 \oplus m_2(\text{carburetor}) = \frac{0.12}{0.57} = 0.21$$

$$m_1 \oplus m_2(\text{starter}) = \frac{0.02 + 0.06 + 0.03}{0.57} = 0.19$$

$$m_1 \oplus m_2(\Theta) = \frac{0.09}{0.57} = 0.16.$$

We may compute new values for belief and plausibility; for example, $\text{Bel}_1 \oplus \text{Bel}_2(\text{battery}) = 0.28$ and $\text{Pl}_1 \oplus \text{Pl}_2(\text{battery}) = 0.44$; $\text{Bel}_1 \oplus \text{Bel}_2(\text{starter}) = 0.19$ and $\text{Pl}_1 \oplus \text{Pl}_2(\text{starter}) = 0.35$.

It is worthwhile noting the similarity to Popper's views of hypothesis testing: If any piece of evidence assigns 0 mass to a hypothesis, then the only way in which combined evidence can assign positive mass to that hypothesis is through combination with Θ. In the absence of generalized uncertainty—when the mass assigned to Θ is 0—a single piece of evidence that assigns mass 0 to a hypothesis assures that it will always be assigned mass 0 when combined with any other evidence.

The combination of belief functions is commutative (it doesn't matter in which order you combine evidence) and associative (it doesn't matter how you group

Theorem 5.13.

$$\mathrm{Bel}_1 \oplus \mathrm{Bel}_2 = \mathrm{Bel}_2 \oplus \mathrm{Bel}_1$$
$$\mathrm{Bel}_1 \oplus (\mathrm{Bel}_2 \oplus \mathrm{Bel}_3) = (\mathrm{Bel}_1 \oplus \mathrm{Bel}_2) \oplus \mathrm{Bel}_3.$$

5.4.4 Special Cases

Simple support functions constitute a subclass of belief functions. The most elementary type of support that evidence provides in a frame of discernment is support for exactly one proposition A in Θ (subset of Θ) other than Θ itself. (Of course, the evidence also supports any proposition that entails A.)

Definition 5.9. Bel *is a* simple support function *focused on A if for some s the degree of support for* $B \subseteq \Theta$ *is given by*

(1) $\mathrm{Bel}(B) = 0$ *if* $B \subset A$,
(2) $\mathrm{Bel}(B) = s$ *if* $A \subseteq B \subset \Theta$,
(3) $\mathrm{Bel}(B) = 1$ *if* $B = \Theta$.

This is of interest because many collections of evidence can be construed as having the effect of the combination of simple support functions.

Another form of support that is frequently encountered is statistical support. This was given no name by Shafer (though he did discuss similar cases). Let us say that a statistical support function is one that assigns a measure s to a proposition A and a measure $1 - s$ to its complement.

Definition 5.10. Bel *is a* statistical support function *focused on A if for some s the degree of support for* $B \subseteq \Theta$ *is given by*

(1) $\mathrm{Bel}(B) = s$ *if* $A \subset B$ *and* $B \neq \Theta$,
(2) $\mathrm{Bel}(B) = 1 - s$ *if* $A \subset \neg B$ *and* $B \neq \Theta$,
(3) $\mathrm{Bel}(B) = 1$ *if* $B = \Theta$, *and*
(4) $\mathrm{Bel}(B) = 0$ *otherwise.*

A special form of conditioning applies to belief functions in which the belief function being conditioned upon assigns the measure 1 to a subset of Θ. This is called Dempster conditioning, and follows from Dempster's rule of combination. Intuitively, it corresponds to Bayesian conditioning, where the evidence we are conditioning on is assigned probability 1.

Definition 5.11. *If* $\mathrm{Bel}_2(X) = 1$ *for* $B \subseteq X$ *and* 0 *otherwise, then*

$$\mathrm{Bel}_1(A|B) = (\mathrm{Bel}_1 \oplus \mathrm{Bel}_2)(A)$$
$$\mathrm{Pl}_1(A|B) = (\mathrm{Pl}_1 \oplus \mathrm{Pl}_2)(A).$$

The following theorem provides the calculations:

Theorem 5.14.

$$\mathrm{Bel}_1(A|B) = \frac{\mathrm{Bel}_1(A \cup \neg B) - \mathrm{Bel}_1(\neg B)}{1 - \mathrm{Bel}_1(\neg B)}$$

$$\mathrm{Pl}_1(A|B) = \frac{\mathrm{Pl}(A \cap B)}{\mathrm{Pl}(B)}.$$

Recall that one way to represent belief functions is as convex sets of probabilities. We can also apply classical conditionalization to a convex set of probability functions S_P to get a new convex set of probability functions $S_{P|e}$ conditional on the evidence e: Take $S_{P|e}$ to be $\{\mathcal{P} : (\exists p)(p \in S_P \wedge p(e) > 0 \wedge (\forall s)(\mathcal{P}(s) = p(s \wedge e)/p(e)))\}$.

Theorem 5.15. *If S_P is a convex set of probability functions, so is $S_{P|e} = \{\mathcal{P} : (\exists p)(p \in S_P \wedge p(e) > 0 \wedge (\forall s)(\mathcal{P}(s) = p(s \wedge e)/p(e)))\}$.*

Finally, we may ask how convex conditionalization compares to Dempster conditioning. We have the following theorem:

Theorem 5.16. *If S_P is a convex set of probabilities, then whenever $p(A|B)$ is defined for $p \in S_P$, one has $\min_{p \in S_P} p(A|B) \leq \mathrm{Bel}(A|B) \leq \mathrm{Pl}(A|B) \leq \max_{p \in S_P} p(A|B)$. Identities hold only if certain subsets of Θ are assigned measure 0—i.e., are regarded as impossible.[4]*

5.4.5 Assessment of Belief Functions

Dempster conditioning leads to stronger constraints than does probabilistic conditionalization. This may seem like a good thing, if we view the greater precision as a matter of squeezing a bit more information out of the available evidence. If we think of the underlying probabilities as essentially subjective, this interpretation may be satisfactory.

On the other hand, if we think of probabilities as being tied to frequencies (we will later argue for this position), then Dempster conditioning is definitely misleading. You are led to believe that the conditional frequency is in the interval (p, q) when in fact it is only constrained to lie in a much broader interval. The following example illustrates this point.

Example 5.5. *Lost and hungry in the woods, I come upon some berries. Are they good to eat? I have two items of evidence: the berries are red, and I know that 60% of red berries are good to eat. I also know that the berries are soft, and that 70% of soft berries are good to eat. The frame of discernment Θ has a distinguished subset, E representing edibility. One piece of evidence is represented by a mass function m_r that assigns mass 0.6 to E and 0.4 to $\neg E$. The other piece of evidence m_s assigns mass 0.7 to E and 0.3 to $\neg E$. Combined according to Dempster's rule, we obtain $m_r \oplus m_s(E) = 0.42/0.54 = 0.78$.[5]*

There are two noteworthy facts about this example. First, note that we get useful results where mere statistics would give us nothing: It is compatible with our evidence that 0% of soft and red berries are edible, and compatible with our evidence that 100% of soft and red berries are edible. Second the combination of the two items of evidence gives us, on this theory, *more* reason to believe that the berries are edible than either item of evidence alone. This seems implausible, unless we somehow think of softness and redness interacting causally to promote edibility. We shall argue later that if we are going to go beyond what is entailed by the statistical knowledge we have, a reasonable representation for edibility is the interval [0.6, 0.7].

The theory of belief functions also has some other counterintuitive consequences. Suppose someone offered to bet 2000 dollars against my 100 dollars that a one will

---

[4][Kyburg, 1987].

[5]Due to Jerry Feldman, in conversation.

not show uppermost on the the next toss of a particular die. This seems like a good deal, but it also seems like a lot of money. I would like to have more evidence about the outcome, so that I will have a better idea as to how to allot my belief to the proposition that the toss will yield a one. Well, one thing I might do is call up a statistician. Let's assume that the statistician gives me the response: "The chance of a one is one-sixth."

I use this information as the basis for my belief about the probability of heads. We represent this as follows: Let D stand for the proposition that the die lands with a one facing up, and let $\neg D$ stand for the proposition that the die does not land with a one facing up. Let Θ be the set of all relevant possibilities; we are concerned with two proper subsets, D and $\neg D$. Thus, we have

$$m_1(\emptyset) = 0 \qquad \mathrm{Bel}_1(\emptyset) = 0$$
$$m_1(D) = \tfrac{1}{6} \qquad \mathrm{Bel}_1(D) = \tfrac{1}{6}$$
$$m_1(\neg D) = \tfrac{5}{6} \qquad \mathrm{Bel}_1(\neg D) = \tfrac{5}{6}$$
$$m_1(\Theta) = 0 \qquad \mathrm{Bel}_1(\Theta) = 1.0.$$

Because a hundred dollars is a lot of money, I decide to get a second opinion. I call up another statistician to get an independent opinion. He also tells me that the probability of getting a one is one-sixth. Using the same notation, we consider the same set of possible worlds, and would represent our uncertainty by

$$m_2(\emptyset) = 0 \qquad \mathrm{Bel}_2(\emptyset) = 0$$
$$m_2(D) = \tfrac{1}{6} \qquad \mathrm{Bel}_2(D) = \tfrac{1}{6}$$
$$m_2(\neg D) = \tfrac{5}{6} \qquad \mathrm{Bel}_2(\neg D) = \tfrac{5}{6}$$
$$m_2(\Theta) = 0 \qquad \mathrm{Bel}_2(\Theta) = 1.0.$$

Because we also have evidence from the first statistician, we should combine these two bodies of evidence. Each piece of evidence is represented by a set of mass and belief functions. We can combine them, as before, by using Dempster's rule of combination. Writing $m_{1,2}$ for $m_1 \oplus m_2$ and $\mathrm{Bel}_{1,2}$ for $\mathrm{Bel}_1 \oplus \mathrm{Bel}_2$, the new distribution is

$$m_{1,2}(\emptyset) = 0 \qquad \mathrm{Bel}_{1,2}(\emptyset) = 0$$
$$m_{1,2}(D) = \tfrac{1}{26} \qquad \mathrm{Bel}_{1,2}(D) = \tfrac{1}{26}$$
$$m_{1,2}(\neg D) = \tfrac{25}{26} \qquad \mathrm{Bel}_{1,2}(\neg D) = \tfrac{25}{26}$$
$$m_{1,2}(\Theta) = 0 \qquad \mathrm{Bel}_{1,2}(\Theta) = 1.0.$$

This result seems counterintuitive. The evidence provided by either statistician alone would lead to a belief in the occurrence of one of $\tfrac{1}{6}$, and would make the offer hard to refuse. In classical probabilistic terms, the expectation of the bet is $\$2000 \times \tfrac{1}{6} - \$100 \times \tfrac{5}{6} = \$250$. But both statisticians provide evidence against the occurrence of a one, and, combined by Dempster's rule, the combined evidence makes the occurrence of a one so doubtful that it no longer seems advantageous to take the bet. Something is surely wrong.

The answer, according to Dempster–Shafer theory, is that the two pieces of evidence represented by m_1 and m_2 are not independent, even though there is no collusion between the two statisticians. They both accept the same information—namely, that the long run relative frequency of ones is about a sixth. But this is not the stochastic dependence with which we are familiar from probability theory. It is a concept of evidential independence that, in our opinion, has not yet been clearly explicated.

There is also the question of how the theory of belief functions is to be used in making decisions. In the case of probability theory, however it is interpreted, the answer is clear: if the relevant probabilities are known, we can construe the utility of each act as a random quantity, and compute the expected utility of each act. We then choose that act with the maximum expected utility. The answer is not so clear in the case of the theory of belief functions, especially in those cases in which the beliefs assigned to X and to $\neg X$ do not add up to 1.0. The reduction of belief theory to a theory of sets of convex probabilities does provide a tie: In the case in which an act has higher expected utility than any other *whatever* probability function is used, that act is clearly preferable.[6]

5.5 Sets of Probability Functions

Sets of probability functions, or imprecise probabilities,[7] as we have already noted in connection with belief functions, provide another nonstandard interval-valued measure of evidential support. The idea of representing uncertainty by intervals is not new. The earliest proposals were [Good, 1962], [Kyburg, 1961], [Smith, 1961]. For Good, and perhaps for Smith, the underlying idea was that, although each individual had implicit degrees of belief that satisfied the probability calculus, these degrees of belief were hard to access. Thus, a reasonable and feasible approach to handling uncertainty was to characterize an individual's degrees of belief by the upper and lower odds he would offer for a bet on a proposition. The idea in [Kyburg, 1961] was that beliefs should be based on statistical knowledge, and that, since statistical knowledge is always approximate, so should the normative constraints imposed on belief also be approximate.

Levi [Levi, 1974] has proposed that epistemic probabilities be represented by convex sets of classical probability functions. This proposal provides more structure than is provided by intervals alone, but it is not entirely satisfactory. When we are faced with a bent coin, we can be quite sure that the behavior of the coin is binomial, though we can't be sure what the parameter p of the binomial distribution is. But the set of binomial distributions with $p \in [r_1, r_2]$ is not convex: the mixture of two such distributions is not a binomial distribution.

A more thorough-going probability-set treatment is proposed in [Kyburg & Pittarelli, 1996]. Indeed, the simple case in which we pull a coin from one of two bags, one consisting of coins biased 0.4 against heads, the other of coins biased 0.4 against tails, and toss it twice, seems most naturally represented as based on a pair of binomial distributions. Because this view is close to the view we shall discuss in detail later, we postpone further consideration of it here.

Another nonstandard approach to uncertainty consists in weakening the axioms that lead to the classical probability calculus. The leading proponent of this approach

---

[6] An even simpler example of the counterintuitive consequences of the theory has been offered by Zadeh [Zadeh, 1986].

[7] This terminology was introduced by Peter Walley [Walley, 1991].

is Terrence Fine [Fine, 1973]. Rather than requiring that probabilities form a complete ordered set, we may suppose that they form merely a partial order. Note that this is true also of intervals: if we regard one interval as less than another if all of its points are less than any of the points of the other, we obtain a weak order on intervals.

Thus, some of the intuitions that have led to interval-valued or imprecise probabilities have also led to partially ordered and weakly ordered probability systems. The latter, however, have more structure than is ordinarily considered in interval relations.

5.6 Summary

The fact that none of the standard interpretations of probability seem to do what we want when it comes to the weighing of evidence has led a number of writers to devise quantitative replacements for probability. Most of these measures make use of a logical probability function defined on the sentences of a formal language. Among these measures based on logical probabilities, we find Popper's "corroboration," Carnap's "increase in confirmation," Levi's measure of "expected epistemic utility," Rescher's "degree of evidential support," and the degree of "factual support" offered by Kemeny and Oppenheim. In the same general class are the "certainty factors" of Shortliffe (cited below), which have their own rules of combination. None of these measures has been widely adopted.

The Dempster–Shafer formalism continues to enjoy some popularity in artificial intelligence, despite the criticisms it has attracted. It is often combined with with other formalisms, as in the transferable belief model of Smets, or in work of Dubois and Prade. The main drawback of the Dempster–Shafer formalism is that its rule of combination can be counterintuitive. Sometimes, but not always, these difficulties can be explained in terms of a violation of the condition that the evidence to be combined must be independent. The most attractive feature of the Dempster–Shafer formalism is that it yields a distinction between the degree to which a possibility is supported by the evidence and the degree to which the evidence supports the denial of that possibility.

This feature is captured by several interval-valued formalisms for support measures. In fact there is a natural way of interpreting belief functions in terms of convex sets of probabilities, though the rule of combination of belief functions does not correspond to conditionalization. Levi has suggested that support be represented by convex sets of (presumably subjective) probability measures, and that conditionalization be the rule of combination. Kyburg and Pittarelli have followed a similar line, but with emphasis on empirical probability measures. Kyburg, Good, and Smith also have related probabilities to intervals determined by sets of probability functions.

The most thorough exploration to date of approximate probability in the sense of general-interval valued representations of uncertainty is that of Peter Walley, cited below.

5.7 Bibliographical Notes

An idea related to those mentioned in the text has been developed in artificial intelligence for expert systems. This is the formalism of certainty factors, first presented by Shortliffe [Shortliffe, 1976]. This idea has been used in a number of medical expert systems. A thorough study of the relation between certainty factors and probability measures may be found in David Heckerman's thesis [Heckerman, 1990].

The "fuzzy logic" of Lotfi Zadeh [Zadeh, 1978] has led to *possibility theory*, which has been developed extensively in Europe [Dubois & Prade, 1980].

Another view of the representation of belief and its dynamics in the light of evidence is presented by Philippe Smets; this is the transferable belief model [Smets & Kennes, 1994], which borrows both from Dempster–Shafer theory and from possibility theory.

A recent and through exploration of probabilities that are not treated as precise is to be found in [Walley, 1991]; this is done within a classical statistical framework, with very careful attention being paid to both epistemological and decision-theoretic aspects of probability.

5.8 Exercises

(1) Give an illustration—the more realistic the better—of the principle that the more surprising the evidence, the more it supports a hypothesis that predicts it.

(2) Explain why Popper thinks that a hypothesis that is a priori less probable than another should be preferred to it.

(3) It has been objected to Popper's view that it does not apply to scientific laws containing *mixed quantifiers*, such as the law "For every action, there is an equal and opposite reaction", because the denial of such a law also contains a universal component ("There is some action such that every reaction is either not equal or not opposite.") Discuss this problem, if it is a problem.

(4) Show that ⊕ is commutative and associative (Theorem 5.13).

(5) Do the calculations to show that Example 10.1 works.

(6) Prove Theorem 5.15.

(7) Suppose I know that 60% of red berries are good, but that I do not know that any red berries are not good; similarly, suppose I know that 70% of soft berries are good, but that I do not know that any are not good. Apply Dempster's rule to calculate belief and plausibility concerning berries that are both soft and red.

(8) Construct a concrete example (a model) that shows that it is possible that more than 70% of the soft berries are good, more than 60% of the red berries are good, and yet 0% of the soft red berries are good.

(9) Suppose that Θ has distinguished subsets R (red), B (blue), $R \cap C$ (red and circular), and $B \cap S$ (blue and square). These subsets of Θ are assigned masses, respectively, of 0.2, 0.3, 0.1, and 0.2. Where is the rest of the mass? What is the belief assigned to red? What is the belief assigned to not-blue? Now suppose new evidence supports redness to some degree. The impact is represented on Θ' with mass 0.5 on R, 0.2 on $\neg R$, and 0.3 on Θ'. What are the new (combined) belief and plausibility functions?

(10) Consider the distribution of heads on tosses of a coin that is known to be either biased 0.6 toward heads or biased 0.6 toward tails. We have a set consisting of two distinct probability distributions. Give an argument as to why we might want to be concerned with the convex closure of these distributions.

Bibliography

[Carnap, 1950] Rudolf Carnap. *The Logical Foundations of Probability*. Second edition. University of Chicago Press, Chicago, 1950.

[Dubois & Prade, 1980] Didier Dubois and Henri Prade. *Fuzzy Sets and Systems: Theory and Applications*. Academic Press, San Diego, 1980.

[Fine, 1973] Terrence Fine. *Theories of Probability*. Academic Press, New York, 1973.

[Good, 1962] I. J. Good. Subjective probability as a measure of a nonmeasurable set. In Patrick Suppes, Ernest Nagel, and Alfred Tarski, editors, *Logic, Methodology and Philosophy of Science*, pp. 319–329. University of California Press, Berkeley, 1962.

[Heckerman, 1990] David E. Heckerman. *Probabilistic Similarity Networks*. PhD thesis, Stanford University, 1990.

[Kemeny & Oppenheim, 1952] John G. Kemeny and Paul Oppenheim. Degree of factual support. *Philosophy of Science*, 19:307–324, 1952.

[Kyburg, 1961] Henry E. Kyburg, Jr. *Probability and the Logic of Rational Belief*. Wesleyan University Press, Middletown, 1961.

[Kyburg, 1987] Henry E. Kyburg, Jr. Bayesian and non-bayesian evidential updating. *AI Journal*, 31:271–294, 1987.

[Kyburg & Pittarelli, 1996] Henry E. Kyburg, Jr., and Michael Pittarelli. Set-based bayesianism. *IEEE Transactions on Systems, Man, and Cybernetics*, 26:324–339, 1996.

[Lakatos, 1968] Imre Lakatos, editor. *The Problem of Inductive Logic*. North Holland, Amsterdam, 1968.

[Levi, 1967] Isaac Levi. *Gambling with Truth*. Knopf, New York, 1967.

[Levi, 1974] Isaac Levi. On indeterminate probabilities. *Journal of Philosophy*, 71:391–418, 1974.

[Popper, 1959] K. R. Popper. *The Logic of Scientific Discovery*. Hutchinson, London, 1959.

[Rescher, 1958] Nicholas Rescher. Theory of evidence. *Philosophy of Science*, 25:83–94, 1958.

[Shafer, 1976] Glenn Shafer. *A Mathematical Theory of Evidence*. Princeton University Press, Princeton, 1976.

[Shortliffe, 1976] E. Shortliffe. *Computer-Based Medical Consultations: MYCIN*. North Holland, 1976.

[Smets & Kennes, 1994] Philippe Smets and Robert Kennes. The transferable belief model. *Artificial Intelligence*, 66:191–234, 1994.

[Smith, 1961] C. A. B. Smith. Consistency in statistical inference and decision. *Journal of the Royal Statistical Society B*, 23:1–37, 1961.

[Walley, 1991] Peter Walley. *Statistical Reasoning with Imprecise Probabilities*. Chapman and Hall, London, 1991.

[Zadeh, 1978] L. A. Zadeh. Fuzzy sets as a basis for a theory of possibility. *Fuzzy Sets and Systems*, 1:3–28, 1978.

[Zadeh, 1986] Lotfi A. Zadeh. A simple view of the Dempster–Shafer theory of evidence and its implications for the rule of combination. *AI Magazine*, 8:85–90, 1986.

6

Nonmonotonic Reasoning

6.1 Introduction

Classical logic—including first order logic, which we studied in Chapter 2—is concerned with deductive inference. If the premises are true, the conclusions drawn using classical logic are always also true. Although this kind of reasoning is not inductive, in the sense that any conclusion we can draw from a set of premises is already "buried" in the premises themselves,[1] it is nonetheless fundamental to many kinds of reasoning tasks. In addition to the study of formal systems such as mathematics, in other domains such as planning and scheduling a problem can in many cases also be constrained to be mainly deductive.

Because of this pervasiveness, many logics for uncertain inference incorporate classical logic at the core. Rather than replacing classical logic, we extend it in various ways to handle reasoning with uncertainty. In this chapter, we will study a number of these formalisms, grouped under the banner *nonmonotonic reasoning*. Monotonicity, a key property of classical logic, is given up, so that an addition to the premises may invalidate some previous conclusions. This models our experience: the world and our knowledge of it are not static; often we need to retract some previously drawn conclusion on learning new information.

6.2 Logic and (Non)monotonicity

One of the main characteristics of classical logic is that it is *monotonic*, that is, adding more formulas to the set of premises does not invalidate the proofs of the formulas derivable from the original premises alone. In other words, a formula that can be derived from the original premises remains derivable in the expanded premise set. For example, given that we know birds are animals, and Tweety is a bird, we can conclude deductively that Tweety is an animal. In first order logic, this can be

---

[1]We need to make the distinction between knowledge and truth here. We may not know whether "84193621 is a prime number" is a true statement or not, but its truth or falsity can certainly be deduced from a simple set of premises.

formulated as follows:

$$\begin{Bmatrix} \texttt{bird(tweety)}, \\ (\forall x)\,(\texttt{bird}(x) \rightarrow \texttt{animal}(x)) \end{Bmatrix}. \qquad (6.1)$$

The conclusion $\texttt{animal(tweety)}$, which follows from the premises, stands regardless of what else we happen to know, now or in the future. For example, if we were to know in addition that Tweety is black and white and lives in Antarctica, we can infer also that Tweety is an Antarctic black-and-white animal, but an animal nonetheless.

There is nothing we can add to the knowledge base that will lead us to retract the deduction that Tweety is an animal. Note that even if we explicitly add to the premise set (6.1) the "fact" that Tweety is *not* an animal,

$$\neg\texttt{animal(tweety)},$$

we can still derive $\texttt{animal(tweety)}$ from the amended premise set. In fact, this latter premise set is inconsistent, and every formula in the whole first order language can be derived from it, including both $\texttt{animal(tweety)}$ and $\neg\texttt{animal(tweety)}$. In other words, from the amended inconsistent premises, we can still deduce that Tweety is an animal, just as from the original premises, even though at the same time we can also deduce that Tweety is not an animal.

Thus, in classical logic, the set of conclusions can never shrink as we add more premises. The set of derivable conclusions increases *monotonically* as more formulas are added to the premise set. Formally we have the following definition of monotonicity.

Definition 6.1. *An inference relation $\rightsquigarrow$ is monotonic iff for all sets of formulas Δ and Δ', if $\Delta \rightsquigarrow \phi$, then $\Delta \cup \Delta' \rightsquigarrow \phi$.*

The monotonic property is convenient for many purposes. For example, in theorem proving we commonly make use of lemmas that have been derived earlier based on only a subset of the premises. Without monotonicity, we cannot be sure that a lemma so derived will still be valid when we consider the full set of premises.

Contrast deduction with inductive or uncertain reasoning. In this context, we would like to discover new and interesting, although not necessarily certain, conclusions. These conclusions are subject to revision or retraction in case they are later found to be not justified after all. Nonmonotonic reasoning attempts to formalize the dynamics of drawing conclusions that may have to be retracted later. This cannot be easily accomplished in classical logic. For example, given that we know Tweety is a bird, it is always the case that it is also an animal, a conclusion we have shown deductively. However, we may also want to allow the line of reasoning to the effect that since birds "typically" (or usually, or probably) fly, we can conclude (nondeductively) that Tweety flies as well unless there is a good reason not to do so. How are we to formalize this in classical logic? One thought is to add

$$(\forall x)\,(\texttt{bird}(x) \rightarrow \texttt{fly}(x)) \qquad (6.2)$$

to the premise set (6.1). We can then deduce that Tweety flies, that is, $\texttt{fly(tweety)}$.

However, formula (6.2) amounts to asserting that *all* birds fly, without exception. Suppose in addition we know that Tweety is a penguin: $\texttt{penguin(tweety)}$. Now $\texttt{fly(tweety)}$ seems dubious, even though Tweety is still a bird, and birds still typically fly. (Note that this is weaker than asserting that Tweety does not fly. We are simply refraining from drawing conclusions about Tweety's flying ability one way or another.) Because classical logic is monotonic, the new piece of information, penguin(tweety), cannot interfere with the previous conclusion, fly(tweety), and this conclusion stays even though intuitively we would want to block or retract its derivation in this case.

One way to get around this is to amend (6.2) to exclude the penguins explicitly:

$$(\forall x)\,(\text{bird}(x) \wedge \neg\text{penguin}(x) \to \text{fly}(x)). \qquad (6.3)$$

This will allow us to avoid the conclusion fly(tweety) if penguin(tweety) is derivable. We are fine for now, but there are two problems: the rule is both too weak and too strong.

- What about ostriches? And baby birds? And stone birds, dead birds, birds with clipped wings, birds soaked in tar, ...? The list is endless. Not only do we have to revise our rule (6.3) every time we want to take into account an additional exception, but it is impossible to list all the exceptions to ensure whenever the antecedent of the rule is true, the bird in question certainly can fly. This is known as the *qualification problem* in the AI literature. In some sense the rule is too weak to cover all the exception cases.
- Rule (6.3) can only be invoked for Tweety when we can *prove* that Tweety is not a penguin, that is, ¬penguin(tweety) has to be deducible from the premises. In the more typical case when all we know about Tweety is that it is a bird, we cannot conclude that it flies, as we have no information about whether Tweety is a penguin or not. Thus, rule (6.3) imposes too strong a condition: We would like to conclude that Tweety flies in the absence of contrary information, such as that Tweety is a penguin, without having to *prove* the falsity of this contrary information.

These considerations led to the study of many nonmonotonic reasoning formalisms. These mechanisms reject monotonicity, at least partially, allowing us to draw conclusions that are typical of the situation but do not follow deductively from the premises. These conclusions may have to be retracted as more information is added. The set of conclusions derivable thus does not necessarily grow monotonically with the expansion of the premises, as is the case for classical logic; hence the nonmonotonicity. We can express notions such as "birds typically (but not always) fly" as nonmonotonic inference rules that have the generic form:

> If α is true, and $\beta_1, \ldots, \beta_n$ are consistent with what we know,
> then conclude γ. $\qquad (6.4)$

Here, α is what we know about a particular situation, and the *negations* of the β's are each a piece of contrary information we do *not* want to possess. Given such an inference rule, we can draw the nonmonotonic conclusion γ from α in the absence of all of $\neg\beta_1, \ldots, \neg\beta_n$. In particular, for the birds we can utilize the following rule:

> If x is a bird, and "x flies" is consistent with what we know,
> then conclude "x flies." $\qquad (6.5)$

Note that in contrast to monotonic rules, which only depend on what is present in the knowledge base, nonmonotonic rules also depend on what is *not* present in the knowledge base. In (6.4), we can conclude γ in the presence of α only as long as the negations of the list of β's are all absent. If one of the β's becomes provably false, either from the addition of new information, or from the sudden discovery of old information, the rule can no longer be used to support the conclusion of γ, and γ may need to be retracted if it has no other supporting derivations.

For example, rule (6.5) can be used to conclude Tweety flies as long as information such as "Tweety is a penguin", "Tweety is an ostrich", and the like are not present in the knowledge base. If we add the formulas

$$\left\{ \begin{array}{l} \texttt{penguin(tweety)}, \\ (\forall x)\,(\texttt{penguin}(x) \rightarrow \neg\texttt{fly}(x)) \end{array} \right\}$$

we can then derive $\neg\texttt{fly(tweety)}$, and rule (6.5) is no longer applicable for Tweety, because x (in this case Tweety) flies is not consistent with the rest of the knowledge base. Thus, the absence or the nonprovability of $\neg\texttt{fly(tweety)}$ is critical to the applicability of the nonmonotonic rule. When $\neg\texttt{fly(tweety)}$ is brought to light, we then have to retract the original conclusion $\texttt{fly(tweety)}$ derived from the nonmonotonic rule.[2]

We will examine a number of nonmonotonic reasoning mechanisms in the following sections. The one we will spend most time on is default logic [Reiter, 1980], but the issues we discuss with respect to default logic are equally relevant to the other formalisms. We will also discuss briefly autoepistemic logic [Moore, 1985] and circumscription [McCarthy, 1980].

Before we move on to the discussion of the specific formalisms, note that although nonmonotonicity is mostly discussed in the context of logic, this characteristic is not inherently qualitative, and it is not restricted to qualitative formalisms. Statistical inference, for instance, may be considered nonmonotonic. Statistical conclusions are accepted at a certain confidence level, and on the basis of a number of procedural assumptions. New data or the choice of a new confidence level may undermine old conclusions; new information bearing on the procedural assumptions may also lead to the rejection of the conclusions. For example, we can recognize and reject "bad" samples and sampling procedures, but under most conditions we cannot obtain a sample that is certifiably random. Instead, we take a sample to be random by default, unless we can show that it violates some essential criteria. This follows the same nonmonotonic inference pattern as Tweety the bird: we take that Tweety flies by default, unless we have reasons to believe otherwise. Thus, nonmonotonicity is an equally important issue for numeric approaches.

One might also ask why a nonmonotonic *logic* is desirable. Wouldn't probability, in particular subjective probability, do the trick? There is a long-standing debate between the symbolic and numeric proponents, which we are not going to resolve here. After all, we can do away with logic altogether, by identifying logical truth with the probability value 1, and logical falsity with the probability value 0. However, a logical framework enables us to utilize techniques not available to numeric approaches, in many cases giving us more elegant or efficient solutions. As for modeling uncertainty,

---

[2]The problem of which conclusions we are to retract and how we are going to accomplish it is by no means an easy one. A conclusion can be supported by multiple monotonic or nonmonotonic derivations, and it can also be supporting the derivation of other conclusions.

some argue that notions such as "typicality" are not numeric, and thus cannot be captured by numbers. Subjective probability values can be conjured up out of the blue, but for most purposes we would like the numbers to be somehow *justified*, although it is not clear where such numbers can be obtained. It has also been shown that while people are not good at estimating numeric probabilities or confidence values, varying the numeric values may have no effect on the resulting behavior [Shortliffe, 1976]. Simply accepting nonmonotonic conclusions as opposed to keeping tab of a myriad of probabilities allows us to cut out a lot of the tedious, and arguably mostly nonessential, computation. Additional arguments for nonmonotonic logics and in general accepting uncertain conclusions, symbolic or numeric, can be found in [McCarthy, 1986; Kyburg, 1996].

6.3 Default Logic

Default logic was first introduced by Reiter [Reiter, 1980] as a way to handle non-monotonic reasoning. Ordinary first order logic is augmented with additional inference rules called *default rules*. The successful application of a default rule depends not only on what can be inferred from the theory (body of knowledge), but also what cannot be inferred. Thus, if new information is obtained at a later stage, a conclusion drawn via a default rule might have to be retracted, if the application of that default rule depends on the absence of some piece of information that can now be inferred with the addition of the new information.

6.3.1 Preliminaries

Let us first introduce the preliminary terminology and machinery of default logic. Let $\mathcal{L}$ be a standard first order language. A well-formed formula $\phi \in \mathcal{L}$ is *open* iff it contains free variables; otherwise it is *closed*. We denote the first order provability operator by $\vdash$. For any set of well-formed formulas $S \subseteq \mathcal{L}$, we denote by $\mathbf{Th}(S)$ the set of well-formed formulas provable from S by first order logic; that is, $\mathbf{Th}(S) = \{\phi \mid S \vdash \phi\}$.

The main novel construct in default logic is the default rule, which specifies under what conditions we are allowed to draw a nonmonotonic conclusion.

Definition 6.2. *A default rule is an expression of the form*

$$\frac{\alpha(\bar{x}) : \beta_1(\bar{x}), \ldots, \beta_n(\bar{x})}{\gamma(\bar{x})}$$

where $\bar{x} = x_1, \ldots, x_m$ is a list of variables, and $\alpha(\bar{x}), \beta_1(\bar{x}), \ldots, \beta_n(\bar{x}), \gamma(\bar{x})$ are well-formed formulas of $\mathcal{L}$ whose free variables are among those in $\bar{x}$. The well-formed formula $\alpha(\bar{x})$ is the prerequisite, $\beta_1(\bar{x}), \ldots, \beta_n(\bar{x})$ *are the* justifications, *and $\gamma(\bar{x})$ is the* consequent *of the default rule.*

Intuitively, a default rule of the form

$$\frac{\alpha(\bar{x}) : \beta_1(\bar{x}), \ldots, \beta_n(\bar{x})}{\gamma(\bar{x})}$$

says that if the prerequisite $\alpha(\bar{x})$ is provable, and each of the negations of the justifications $\beta_1(\bar{x}), \ldots, \beta_n(\bar{x})$ is not provable, then by default we assert that the consequent

$\gamma(\bar{x})$ is true in our knowledge base. For example, "birds typically fly" can be formalized by the default rule

$$\frac{\text{bird}(x):\text{fly}(x)}{\text{fly}(x)}. \tag{6.6}$$

This rule indicates that if we can prove that something is a bird, and we cannot prove that it cannot fly, then we may conclude (nonmonotonically) that it can fly.

Definition 6.3. *A default rule is closed iff* $\alpha(\bar{x})$, $\beta_1(\bar{x})$, ..., $\beta_n(\bar{x})$, *and* $\gamma(\bar{x})$ *are all closed formulas; otherwise it is* open. *We write a closed default rule as*

$$\frac{\alpha:\beta_1,\ldots,\beta_n}{\gamma}.$$

The "birds fly" default rule (6.6) is an open default. It can be instantiated to a closed default rule concerning a particular bird Tweety as follows:

$$\frac{\text{bird}(\text{tweety}):\text{fly}(\text{tweety})}{\text{fly}(\text{tweety})},$$

which says: if Tweety is a bird, and it is consistent that it flies, then by default conclude that Tweety flies.

A set of first order formulas augmented by a set of default rules is called a default theory. The first order formulas represent the "facts" in the knowledge base—information that is taken to be certain. The default rules specify under what conditions we may make a nonmonotonic inference.

Definition 6.4. *A default theory* Δ *is an ordered pair* $\langle D, F \rangle$ *where D is a set of default rules and F is a set of closed well-formed formulas (facts). The default theory* $\Delta = \langle D, F \rangle$ *is closed iff all the default rules in D are closed; otherwise it is open.*

Example 6.1. *Tweety is a bird, and by default we conclude that it flies. This scenario can be formulated as the closed default theory* $\Delta = \langle D, F \rangle$, *where*

$$D = \left\{ \frac{\text{bird}(\text{tweety}):\text{fly}(\text{tweety})}{\text{fly}(\text{tweety})} \right\}$$

$$F = \{\text{bird}(\text{tweety})\}.$$

We can conclude $\text{fly}(\text{tweety})$ *from* Δ *by applying its single default rule. This conclusion, however, is not certain. Suppose we learn in addition that Tweety is a penguin, and penguins do not fly. This scenario can be represented by the expanded default theory* $\Delta' = \langle D, F' \rangle$, *where*

$$D = \left\{ \frac{\text{bird}(\text{tweety}):\text{fly}(\text{tweety})}{\text{fly}(\text{tweety})} \right\} \tag{6.7}$$

$$F' = \left\{ \begin{array}{l} \text{bird}(\text{tweety}), \\ \text{penguin}(\text{tweety}), \\ (\forall x)\,(\text{penguin}(x) \rightarrow \neg\text{fly}(x)) \end{array} \right\}. \tag{6.8}$$

We then have to retract $\text{fly}(\text{tweety})$, *as the default rule in (6.7) that was used to derive* $\text{fly}(\text{tweety})$ *is now inapplicable: although the prerequisite* $\text{bird}(\text{tweety})$ *still holds, the negation of the justification* fly(tweety) *is now provable from F'. The justification is no longer consistent with the theory, and thus the default rule is blocked.*

This example shows how the set of conclusions sanctioned by a default theory can shrink as more information is added. Thus, inference in default logic is nonmonotonic.

6.3.2 Transformation of Open Default Theories

Most of the results developed for default logic are for *closed* default theories. However, many default rules of interest contain free variables. For example, it is common to have default rules of the form

$$\frac{\alpha(\bar{x}):\gamma(\bar{x})}{\gamma(\bar{x})}$$

such as

$$D = \left\{ \frac{\texttt{bird}(x):\texttt{fly}(x)}{\texttt{fly}(x)}, \frac{\texttt{tallBuilding}(x):\texttt{hasElevator}(x)}{\texttt{hasElevator}(x)} \right\}. \tag{6.9}$$

These default rules are intended to be general statements about birds and tall buildings: "birds typically fly", and "tall buildings typically have elevators". The rules cover any individual in the language.

In order to make use of the results developed for closed default theories, an open default theory can be first transformed into a closed default theory, and all computations and default inferences are then carried out with the transformed theory. The idea of the transformation is to instantiate each open default rule by all the individuals in the language, generating a set of closed default rules, one for each individual for each open default rule. In each of the above cases, we take the free variable x to be universally quantified over the entire scope of the default rule. Suppose the open default theory $\Delta = \langle D, F \rangle$ in question consists of the above two rules in (6.9) together with the set of facts

$$F = \{\texttt{bird}(\texttt{tweety}), \texttt{tallBuilding}(\texttt{empireState})\}.$$

There are two individuals mentioned in the theory, tweety and empireState. A corresponding closed default theory $\Delta' = \langle D', F \rangle$ would consist of the same set of facts F and the following set of four closed default rules:

$$D' = \left\{ \begin{array}{l} \frac{\texttt{tallBuilding}(\texttt{tweety}):\texttt{hasElevator}(\texttt{tweety})}{\texttt{hasElevator}(\texttt{tweety})}, \\[2mm] \frac{\texttt{tallBuilding}(\texttt{empireState}):\texttt{hasElevator}(\texttt{empireState})}{\texttt{hasElevator}(\texttt{empireState})}, \\[2mm] \frac{\texttt{bird}(\texttt{tweety}):\texttt{fly}(\texttt{tweety})}{\texttt{fly}(\texttt{tweety})}, \\[2mm] \frac{\texttt{bird}(\texttt{empireState}):\texttt{fly}(\texttt{empireState})}{\texttt{fly}(\texttt{empireState})} \end{array} \right\}.$$

Each of the two default rules is instantiated with each of the two individuals occurring in the language. One might think some of the resulting rules do not make much sense: when is the Empire State Building going to fly? However, such a rule would just never be invoked, since its prerequisite `bird(empireState)` would not be true.[3]

Thus, we transform an open default rule by interpreting the free variables as universally quantified over the whole default rule, and instantiating the open rule with all individuals in the language. The case we have discussed above is a simplified version. The reader is referred to [Reiter, 1980] for an extended treatment, including how to deal with existentially quantified variables by Skolemization. Below we will focus on closed default theories, and open defaults will be taken as shorthand for collections of closed default rules.

6.3.3 Extensions

We have discussed informally what a default rule is supposed to represent. Now we need to determine how, given a default theory, the default rules are allowed to interact with each other and with the facts, to give rise to an extended theory containing nonmonotonic conclusions.

Intuitively, in a default theory $\Delta = \langle D, F \rangle$, the known facts about the world constitute F. The theory $\mathbf{Th}(F)$, the deductively closed set of formulas derivable from F by classical logic, is typically an incomplete theory, since we do not know everything about the world. We can complete the theory in many ways, but some are preferable to others. The default rules in D are (nonmonotonic) inference rules that direct the way we extend the incomplete theory in some desirable way. Such a resulting extended theory[4] is called an *extension* of the default theory. Informally, an extension should satisfy the following three criteria:

(1) It includes the set of given facts.
(2) It is deductively closed.
(3) All applicable default rules have been applied.

In addition, an extension should be minimal in the sense that it should not contain spurious formulas that are not supported by the given default rules and facts.

The formal definition of an extension makes use of a fixed point operator Γ as follows. The reason for using a fixed point formulation will become clear in the next section.

Definition 6.5. *Let $\Delta = \langle D, F \rangle$ be a closed default theory over the language $\mathcal{L}$, and E be a subset of $\mathcal{L}$. $\Gamma(E)$ is the smallest set satisfying the following three properties:*

(1) $F \subseteq \Gamma(E)$.
(2) $\Gamma(E) = \mathbf{Th}(\Gamma(E))$.
(3) For every default rule $\frac{\alpha : \beta_1, ..., \beta_n}{\gamma} \in D$, if

    (a) $\alpha \in \Gamma(E)$,
    (b) $\neg\beta_1, ..., \neg\beta_n \notin E$,

    then $\gamma \in \Gamma(E)$.

---

[3] Unless, of course, if `empireState` is actually a bird of yours.
[4] The extension of an incomplete theory $\mathbf{Th}(F)$ may still be incomplete, but it usually contains more information

The set of formulas E is an extension *of the default theory* Δ *iff E is a fixed point of the operator* Γ, *that is,* $E = \Gamma(E)$.

The first two properties of the definition correspond to the two criteria that the extension contains the set of facts and is deductively closed. The third property specifies the applicability condition of a default rule. We would like to say: if the prerequisite α is in the extension, and each of the negations of the justifications β_i is not contained in the extension, then by default conclude the consequent γ. In an extension, all applicable default rules have been used. In other words, no default rule whose applicability condition is satisfied can be used to obtain further information not already contained in the extension.

The reason why the formalization of an extension, in particular the applicability condition of a default rule, needs to make use of a fixed point will be clear in a moment. But first let us look at some examples to see how an extension is selected.

Example 6.2. *Consider the default theory* $\Delta = \langle D, F \rangle$, *where*

$$D = \left\{ \frac{\texttt{bird(tweety)}:\texttt{fly(tweety)}}{\texttt{fly(tweety)}} \right\}$$

$$F = \emptyset.$$

The default theory Δ *contains one default rule and an empty set of facts. This amounts to saying we do not know anything about Tweety, but if Tweety is a bird, then by default we may conclude that it flies. Now consider the set of formulas* $E = \textbf{Th}(\emptyset)$. *We can construct at least three sets of formulas that are candidates for* $\Gamma(E)$:

(1) $S_1 = \textbf{Th}(\emptyset)$,
(2) $S_2 = \textbf{Th}(\{\texttt{bird(tweety)}, \texttt{fly(tweety)}\})$,
(3) $S_3 = \textbf{Th}(\{\neg\texttt{bird(tweety)}\})$.

Each of the above sets satisfies the three conditions in Definition 6.5. Intuitively only S_1 *makes sense. We have no information on what Tweety is, and the assertion in* S_2 *that Tweety is a bird is not justified. Similarly,* S_3 *asserts that Tweety is not a bird without any basis for doing so.*

This consideration is reflected in Definition 6.5 by requiring that $\Gamma(E)$ *be the smallest set satisfying the three conditions. This requirement ensures that everything that is included in* $\Gamma(E)$ *has a reason to be there—either it is a given fact, or there is a chain of default inferences leading to it, or it can be derived (deductively) from a combination of the two. For* $E = \textbf{Th}(\emptyset)$, *the set* S_1 *is the smallest among all candidates satisfying the three conditions.*

We have verified that $\Gamma(E) = \textbf{Th}(\emptyset) = E$, *that is,* E *is a fixed point of* Γ. *The theory* $\textbf{Th}(\emptyset)$ *is thus an extension (in fact, the only extension) of the default theory* Δ.

In the above example, the extension E contains nothing more than what can be derived deductively from the given set of facts F, since the only default rule in Δ is not applicable. Now let us look at a slightly larger example.

Example 6.3. *Now suppose we learnt that Tweety is a bird and this was added to the set of facts. The revised default theory* $\Delta' = \langle D, F' \rangle$ *is as follows:*

$$D = \left\{ \frac{\texttt{bird(tweety)}:\texttt{fly(tweety)}}{\texttt{fly(tweety)}} \right\}$$

The extension $E = \mathbf{Th}(\emptyset)$ we had before in Example 6.2 is no longer an extension of Δ'. To see why, we just have to note that according to condition 1 of Definition 6.5, the theory $\Gamma(E)$, with respect to Δ', must contain the set of facts F'. Therefore any candidate for $\Gamma(E)$ contains at least bird(tweety). *This is not the case for $E = \mathbf{Th}(\emptyset)$. The theory E cannot be a fixed point of Γ and thus cannot be an extension of Δ'.*

Now consider $E_1 = \mathbf{Th}(\{$bird(tweety), fly(tweety)$\})$. We can verify that $\Gamma(E_1)$ is exactly E_1 itself, and thus E_1 is an extension of Δ'.

6.3.4 Need for a Fixed Point

One might wonder why we need a fixed point operator rather than simply substituting E for $\Gamma(E)$ in Definition 6.5. After all, $\Gamma(E)$ is equal to E in the case when E is an extension. The fixed point operator is used to rule out ungrounded formulas in an extension. To illustrate the distinction between E and $\Gamma(E)$, let us for the moment substitute E for $\Gamma(E)$ throughout Definition 6.5, and see what results we obtain. The revised definition of a *pseudoextension* is as follows.

Definition 6.6. *Let $\Delta = \langle D, F \rangle$ be a closed default theory over the language $\mathcal{L}$, and E be a subset of $\mathcal{L}$. Then E is a* pseudoextension *of the default theory Δ iff E is a minimal set satisfying the following three properties:*

(1) $F \subseteq E$.

(2) $E = \mathbf{Th}(E)$.

(3) For every default rule $\frac{\alpha : \beta_1,\dots,\beta_n}{\gamma} \in D$, if

    (a) $\alpha \in E$,

    (b) $\neg\beta_1, \dots, \neg\beta_n \notin E$,

    then $\gamma \in E$.

This definition of a pseudoextension is obtained from Definition 6.5 by uniformly replacing all occurrences of $\Gamma(E)$ by E. Let us compare the extensions with the pseudoextensions obtained in the following example.

Example 6.4. *Given a knowledge base, it is frequently the case that there are many more negative than positive instances. It would be useful to adopt a convention that only positive instances are given, either in the form of ground instances, or as instances derivable from a set of given rules. Negative instances on the other hand can be omitted from the specification of the knowledge base—if a certain instance is not derivable, it is assumed to be negative. This is known as the* closed world assumption, *and is a rather useful shorthand. For example, a newspaper does not need to print articles about events that did not occur: "There is no plane crash at Peaceville today at 3:08 p.m.," "aliens have not invaded Earth," etc.*

One way to think about the closed world assumption[5] is as a set of default rules, each of the form

$$\frac{: \neg\alpha}{\neg\alpha}$$

---

[5]The assumption as usually formulated has a number of undesirable features. It may give rise to inconsistent theories, and the results are dependent on the syntax of the knowledge base. The variant we consider here is harder to compute but avoids some of these problems.

We assert, for each ground instance α, *that if it is consistent for* α *to be false, then by default we can conclude* ¬α. *(An empty prerequisite in a default rule is always true.) Let us consider a default theory with one such default rule. Suppose* Δ = ⟨D, F⟩, *where*

$$D = \left\{ \frac{: \neg \texttt{alienAttack}}{\neg \texttt{alienAttack}} \right\}$$

$$F = \emptyset.$$

There are two pseudoextensions:

$$E_1 = \textbf{Th}(\{\neg \texttt{alienAttack}\})$$
$$E_2 = \textbf{Th}(\{\texttt{alienAttack}\}).$$

We would expect only E_1 *to be an acceptable extension, as there seems to be no reason to include* alienAttack *in* E_2. *However, both* E_1 *and* E_2 *are minimal sets satisfying the conditions for a pseudoextension as specified in Definition 6.6.* E_2 *contains* alienAttack, *which blocks the application of the default rule. Because nothing in Definition 6.6 prevents the inclusion of* alienAttack, *and* E_2 *is minimal (there is no proper subset of* E_2 *that also satisfies the three conditions), we have to accept* E_2 *as a pseudoextension of the default theory* Δ.

On the other hand, only E_1 *satisfies the conditions for being a default extension, as specified in Definition 6.5. For* E_2, *we have* Γ(E_2) = **Th**(∅), *and therefore* E_2 *is not a proper extension of* Δ.

Thus, we have to employ the less intuitive formulation via a fixed point operator, so that arbitrary statements cannot make their way into the extensions. The set of formulas E serves as the proposed extension, and Γ(E) is used to verify that E is indeed an extension. The three conditions in Definition 6.5 ensure that Γ(E) contains every formula that should be in an extension. On the other hand, by requiring Γ(E) to be the smallest set satisfying these conditions, we also make sure that no unjustified formulas are included. When such a Γ(E) is identical to the set of formulas E in question, we know that we have found an extension.

6.3.5   Number of Extensions

So far we have only seen examples where a default theory gives rise to a single extension. This is not always the case, however. A default theory may have multiple extensions, or no extensions at all. Below we give an example of each case.

Example 6.5. *Soft tomatoes are typically ripe, and green ones are typically not ripe. This tomato b, which we got from the grocery store, is soft and green. Is it ripe? This can be formulated as a closed default theory* Δ = ⟨D, F⟩, *where*

$$D = \left\{ \frac{\texttt{soft}(b) : \texttt{ripe}(b)}{\texttt{ripe}(b)}, \frac{\texttt{green}(b) : \neg\texttt{ripe}(b)}{\neg\texttt{ripe}(b)} \right\}$$

$$F = \{\texttt{soft}(b), \texttt{green}(b)\}.$$

Whether we conclude the tomato is ripe or not depends on which default rule we choose to apply. There are two extensions:

$$E_1 = \textbf{Th}(\{\texttt{soft}(b), \texttt{green}(b), \texttt{ripe}(b)\})$$
$$E_2 = \textbf{Th}(\{\texttt{soft}(b), \texttt{green}(b), \neg\texttt{ripe}(b)\}).$$

We can apply one but not both default rules in each case. The extension E_1 is obtained by applying the first default rule. Because ripe(b) *is then contained in this extension, the second default rule is blocked as its justification,* ¬ripe(b), *is not consistent with E_1. Similarly, E_2 is obtained by applying the second default rule, whose consequent* ¬ripe(b) *blocks the first default rule by contradicting its justification* ripe(b).

Multiple extensions pose a problem for default logic (and for other nonmonotonic reasoning formalisms as well), and there is no consensus as to how we should deal with them. We will examine this problem again in Section 6.6.3.

Now let us look at a default theory with no extension.

Example 6.6. *We have another default theory concerning the tomato b. Let $\Delta = \langle D, F \rangle$, where*

$$D = \left\{ \frac{\text{soft}(b) : \text{ripe}(b)}{\neg\text{green}(b)} \right\}$$

$$F = \{\text{soft}(b), \text{green}(b)\}.$$

If it is consistent that a soft tomato is ripe, then we conclude by default that it is not green. The tomato b, as we recall, is soft and green. This default theory has no extension. To see why, first note that any extension E has to contain green(b), *part of the set of facts F. If the default rule were applied, E would be inconsistent (it contains both* green(b) *and* ¬green(b)). *The negation of the default justification,* ¬ripe(b), *along with everything else in the language, would then be derivable from E. Thus, the default rule could not have been applied in the first place.*

On the other hand, if the default rule were not applied, its justification ripe(b) *would be consistent with E and thus we would be forced to invoke the default rule, resulting in a similar situation.*

In fixed point terms, these two scenarios correspond to the following two candidates for an extension:

$$E_1 = \mathcal{L} \qquad \Gamma(E_1) = \textbf{Th}(F)$$
$$E_2 = \textbf{Th}(F) \qquad \Gamma(E_2) = \mathcal{L}.$$

Both E_1 and E_2 fail to be a fixed point of Γ, and therefore neither is an extension of Δ.

The reason why sometimes a default theory has no extension is that the application of a default rule creates an environment that contradicts the applicability condition of some applied default rule (this may be the same or another default rule). One might argue that the default theory is not formulated properly (what happens if we add ¬green(b) to the justification?); however, it remains that the machinery of default logic does not guard against such pathological cases. In Section 6.3.7, we will look at one way of modifying Reiter's formulation to alleviate this problem.

6.3.6 Representation

There are some classes of default theories that have been given special attention. We will look at *normal* default theories and *seminormal* default theories.

Definition 6.7. *A default rule of the form*

$$\frac{\alpha : \gamma}{\gamma}$$

is a (closed) normal default rule. A default theory $\Delta = \langle D, F \rangle$ *is a normal default theory iff every default rule in D is normal.*

The justification and consequent of a normal default rule are identical. Most of the default rules we have considered so far are normal. In fact, it has been argued that all naturally occurring default rules are normal. It certainly seems to be the case intuitively. A normal default rule conveys the idea that if the prerequisite α is true, then one may conclude its consequent γ if it is consistent to do so. To represent "birds typically fly" we have

$$\frac{\texttt{bird}(x) : \texttt{fly}(x)}{\texttt{fly}(x)}.$$

To represent "adults are normally employed", we can have

$$\frac{\texttt{adult}(x) : \texttt{employed}(x)}{\texttt{employed}(x)}.$$

It is thus good news that normal default theories enjoy some nice properties. First of all, an extension always exists.

Theorem 6.1 ([Reiter, 1980]). *Every closed normal default theory has an extension.*

This is not always the case for nonnormal default theories, as we have already seen in Example 6.6. Normality ensures that the justifications and consequents of the applied default rules do not contradict, since the two are identical. We can therefore avoid the situation where no extension can be constructed.

Besides the guaranteed existence of extensions, another property of normal default theories is that they are *semimonotonic*.

Theorem 6.2 (Semimonotonicity [Reiter, 1980]). *Let D and D' be sets of closed normal default rules. If E is an extension of the default theory* $\Delta = \langle D, F \rangle$*, then the default theory* $\Delta' = \langle D \cup D', F \rangle$ *has an extension E' such that* $E \subseteq E'$*.*

Semimonotonicity means that for closed normal default theories, adding more default rules to the theory cannot invalidate the previously drawn conclusions—only adding more "facts" to the theory can make the set of conclusions shrink. Thus, normal theories are monotonic with respect to the default rules.

This is again not true of nonnormal theories. Consider the default theory $\Delta = \langle D, F \rangle$ in Example 6.6 once more. A smaller default theory $\Delta' = \langle \emptyset, F \rangle$ obviously has an extension $\textbf{Th}(F)$, as it contains no default rules. However, by adding the default rule

$$\frac{\texttt{soft}(b) : \texttt{ripe}(b)}{\neg\texttt{green}(b)}$$

to Δ', we obtain Δ, which has no extension.

Thus, we seem to have identified a subclass of default theories that is both interesting and easy to work with. An extension always exists, and we can add more default rules without having to worry about whether we need to retract some of the conclusions or even the whole extension.

However, in many cases, although it seems natural to represent the default reasoning as a normal default rule when we consider the situation in isolation, when viewed in a larger context, normal default rules can interact with each other and produce unintuitive conclusions [Reiter & Criscuolo, 1981]. Consider the following example.

Example 6.7 ([Reiter & Criscuolo, 1981]). *Adults are typically employed. Students are typically unemployed. Sue is an adult university student. This scenario can be represented by the normal default theory* $\Delta = \langle D, F \rangle$, *where*

$$D = \left\{ \frac{\text{adult(sue)}:\text{employed(sue)}}{\text{employed(sue)}}, \frac{\text{student(sue)}:\neg\text{employed(sue)}}{\neg\text{employed(sue)}} \right\},$$

$F = \{\text{adult(sue)}, \text{student(sue)}\}$,

Δ *has two extensions, one containing* employed(sue) *and the other containing* ¬employed(sue). *Intuitively we prefer the second extension, as adult students, for example university students over* 18, *are typically full time and not employed.*

Thus, although each of the two normal defaults in the above example seems quite reasonable when considered independently, when combined in a default theory they can produce dubious extensions. In order to curb unwanted interactions in normal default theories, *seminormal* default theories were proposed.

Definition 6.8. *A default rule of the form*

$$\frac{\alpha : \beta \wedge \gamma}{\gamma}$$

is a (closed) seminormal default rule. A default theory $\Delta = \langle D, F \rangle$ *is a* seminormal default theory *iff every default rule in D is seminormal.*

In a normal default rule, the justification and consequent are the same. Seminormal default rules generalize normal default rules by relaxing the constraint on the justification. The consequent has to be derivable from the justification, but the two do not necessarily have to be identical. Note that a normal default rule is also a seminormal default rule with an empty β.

Example 6.8. *The first default rule in Example 6.7 can be replaced by the seminormal default*

$$\frac{\text{adult(sue)}:\neg\text{student(sue)} \wedge \text{employed(sue)}}{\text{employed(sue)}}.$$

This rule is then applicable only when (among other conditions) Sue is not known to be a student. When student(sue) *is derivable, it is inconsistent with* ¬student(sue) *in the justification of the default rule, thus blocking its application. Therefore, in the revised default theory, even though Sue is known to be both an adult and a student, satisfying the prerequisites of both default rules, only the second rule can be applied. This gives rise to a single extension containing* ¬employed(sue). *The dubious extension containing* employed(sue) *is eliminated, as the first default rule cannot be applied.*

Seminormal rules may be more suited than normal rules for formalizing default reasoning, as they provide more control over the conditions under which a default rule can be applied. However, they do not share the nice properties of normal defaults. For example, seminormal default theories do not always have an extension.

Example 6.9. *Consider the seminormal default theory* $\Delta = \langle D, F \rangle$, *where*

$$D = \left\{ \frac{:p \wedge \neg q}{\neg q}, \frac{:q \wedge \neg r}{\neg r}, \frac{:r \wedge \neg p}{\neg p} \right\}$$

$$F = \emptyset$$

This default theory has no extension. Applying any two of the three default rules would result in a situation where the consequent of one rule undermines the justification of the other. For example, if we apply the first rule

$$\frac{:p \wedge \neg q}{\neg q}$$

the second rule is blocked, as its justification $(q \wedge \neg r)$ *is not consistent with the consequent* $(\neg q)$ *of the applied rule. However, there is nothing to stop us from applying—in fact, we are required to apply—the third default rule*

$$\frac{:r \wedge \neg p}{\neg p}$$

in this case, according to the conditions set out in Definition 6.5 for an extension. But, the consequent $(\neg p)$ *of the third rule is inconsistent with the justification* $(p \wedge \neg q)$ *of the first applied rule, and thus an extension cannot be constructed this way.*

Now if we apply the third rule first, the first rule is blocked, as its justification is not consistent with the consequent of the third rule, but in this case we have to apply the second rule, whose consequent then contradicts the justification of the third rule. (Got it?) A similar scenario can be constructed for any other combination of the default rules. Thus, there is no extension for this default theory.

Moreover, seminormal default theories are not semimonotonic either, as adding more seminormal default rules to a theory that originally has extensions can result in an expanded theory that has no extension. For example, a default theory $\Delta' = \langle D', \emptyset \rangle$ where D' contains any two of the three default rules in Example 6.9 has one extension. However, if the left-out default rule is added back in, we do not get any extension.

Thus, seminormal default theories are not as attractive as normal default theories in ease of manipulation, even though they seem to better capture our intuition about default reasoning.

Apart from identifying interesting subclasses of default theories, another line of work is to formalize alternative interpretations of a default rule and default logic, to overcome some of the difficulties encountered by Reiter's formulation. These variants of default logic have different properties, as we shall see in the next subsection.

6.3.7   Variants of Default Logic

Default logic, as it is formulated in [Reiter, 1980], has a number of unsatisfactory be-haviors. As mentioned earlier, a default theory may not always have an extension. The default consequent of one default rule may contradict the justification or consequent of another applied rule. This motivated the development of a number of variants, starting with Łukaszewicz's *modified default logic* [Łukaszewicz, 1985], in which the existence of extensions is guaranteed for any default theory. In Łukaszewicz's formulation, the justifications of the applied default rules are taken into account when we consider additional rules. A default rule can be applied only if it does not contradict any of these justifications. This extra constraint eliminates the pathological cases where no extension can be constructed.

There has also been much work on modifying Reiter's default logic to reflect different intuitions about what conclusions we should draw by default. One of the controversial issues concerns the treatment of the justifications in a default rule. So far in the examples we have only dealt with single justification default rules of the form

$$\frac{\alpha : \beta}{\gamma}$$

where, although the justification β can be arbitrarily complex, it is a single formula. However, in general the formulation also allows for multiple justifications in a single rule:

$$\frac{\alpha : \beta_1, \ldots, \beta_n}{\gamma}$$

where each justification β_i is a separate formula.

In the original version, the consistency of each justification β_i of a default rule is checked individually, independent of the other justifications of the same rule and those of other applied rules (see item 3(b) of Definition 6.5). Thus, mutually inconsistent justifications can coexist to generate an extension, as long as each justification is consistent with the extension when considered on its own. A rule such as

$$d = \frac{\alpha : \beta, \neg\beta}{\gamma}$$

is thus perfectly acceptable and would generate an extension $\mathbf{Th}(\{\alpha, \gamma\})$ given the default theory $\Delta = \langle\{d\}, \{\alpha\}\rangle$, even though the two justifications β and $\neg\beta$ are inconsistent.

Regardless of whether the justifications are from the same or different default rules, checking the consistency of each individually can lead to some counterintuitive results in some cases. This was illustrated in the broken-arm example, due to Poole [Poole, 1989].

Example 6.10 ([Poole, 1989]). *We know that one of Tom's arms is broken, but we do not remember which one. For each arm, if it is consistent that it is usable and not broken, then by default we conclude that it is usable. A straightforward formulation of this scenario can be represented by the default theory $\Delta = \langle D, F \rangle$, where*

$$D = \left\{ \frac{: \mathtt{usable}(x) \wedge \neg\mathtt{broken}(x)}{\mathtt{usable}(x)} \right\}$$

$$F = \{\mathtt{broken}(\mathtt{leftarm}) \vee \mathtt{broken}(\mathtt{rightarm})\}.$$

There are two (closed) default rules, one for the left arm and one for the right arm. This default theory has a single extension containing the consequents of both rules, usable (leftarm) *and* usable(rightarm). *Although*

$$\mathtt{broken}(\mathtt{leftarm}) \vee \mathtt{broken}(\mathtt{rightarm})$$

is given in F, neither of the disjuncts can be deduced from the disjunction. As a result both rules in D are applicable, as each of their justifications,

$$\text{usable(leftarm)} \land \neg\text{broken(leftarm)}$$
$$\text{usable(rightarm)} \land \neg\text{broken(rightarm)}$$

when considered independently, is consistent with the extension.

This extension amounts to concluding that both arms are functional, which is a bit strange, given that we know one of the arms is broken, even though we do not know which one it is.

One way to provide a more satisfactory treatment of the broken-arm scenario is to evaluate the justifications of the two (closed) default rules jointly rather than separately. Because the two justifications when taken together cannot be consistent with the given fact

$$\text{broken(leftarm)} \lor \text{broken(rightarm)},$$

we can only apply either one but not both of the rules at any given time. This gives rise to two extensions, one containing usable(leftarm) and the other containing usable(rightarm), instead of a single extension containing both assertions.

Before we dismiss individual justification consistency as a complete mistake, let us point out that although joint justification consistency is a desirable feature in the broken-arm example above, there are also situations in which individual justification consistency is more appropriate, as in the following example.

Example 6.11. *Mary and John are the only two suspects of a murder. However, for each suspect, if it is consistent that this person is innocent and not the murderer, then by default we assume this person is innocent. This scenario can be represented by the default theory* $\Delta = \langle D, F \rangle$, *where*

$$D = \left\{ \frac{: \text{innocent}(x) \land \neg\text{murderer}(x)}{\text{innocent}(x)} \right\}$$

$$F = \{\text{murderer(mary)} \lor \text{murderer(john)}\}.$$

Note that this default theory is syntactically identical to the broken-arm default theory in Example 6.10. If we were to evaluate the justifications of the two default rules jointly, we would have two separate extensions, one containing innocent(mary) *and the other containing* innocent(john). *However, to convey the idea that both suspects can be presumed innocent at the same time, it seems more appropriate here to evaluate the justifications of the two default rules separately, and generate a single extension containing the consequents* innocent(mary) *and* innocent(john) *of both rules.*[6]

Thus, the lack of joint justification consistency in default logic is not necessarily a shortcoming; different situations seem to appeal to different intuitions and call for different methods of handling. In many cases people cannot even agree on *what* it is that we should conclude from a given situation, let alone *how* we should arrive at such conclusions. This discussion highlights the lack of a clear-cut directive of what

---

[6]One may argue that it is acceptable to have the two separate extensions representing the two possible scenarios, as according to the premises either one or the other person must be the murderer. Nonetheless it is also desirable to be able to express their presumed innocence in the same context—in other words, in the same extension.

we should expect from a nonmonotonic reasoning system. We will revisit this issue in Section 6.6. Now let us look at some other nonmonotonic reasoning formalisms.

6.4   Autoepistemic Logic

Autoepistemic logic is a modal based formalism. In modal logic, classical logic is augmented with two modal (dual) operators **L** and **M**, through which we are able to express concepts such as necessity, contingency, and possibility. Modal logic is a natural candidate for formalizing nonmonotonic reasoning, since we frequently speak of the modality of an event when we reason under uncertainty.

Below let us first give a brief overview of standard modal logic. For a more complete treatment, refer to [Hughes & Cresswell, 1996; Chellas, 1980].

6.4.1   Modal Logic

Modal logic is based on classical logic, with the addition of two dual modal operators, namely, the necessity operator **L** and the possibility operator **M**. The two operators are interdefinable: **M** is equivalent to $\neg \mathbf{L} \neg$, and **L** is equivalent to $\neg \mathbf{M} \neg$. With the help of these operators, we can express more refined distinctions pertaining to the modality of an event. There can be, for example, the following different states for a statement ϕ:

$\mathbf{L}\phi$:	ϕ is necessarily true
$\phi \wedge \neg \mathbf{L}\phi$:	ϕ is contingently true (ϕ is true but not necessarily so)
$\mathbf{M}\phi$:	ϕ is possibly true
$\mathbf{M}\neg\phi$:	ϕ is possibly false
$\neg\phi \wedge \mathbf{M}\phi$:	ϕ is contingently false (ϕ is false but not necessarily so)
$\mathbf{L}\neg\phi$:	ϕ is necessarily false.

Note the difference between $\mathbf{L}\neg\phi$ and $\neg\mathbf{L}\phi$. In the first case, ϕ is necessarily false; in the second case, we merely deny that ϕ is necessarily true. The second case is weaker: it corresponds to $\neg\neg\mathbf{M}\neg\phi$, which is $\mathbf{M}\neg\phi$. In other words, ϕ is possibly but not necessarily false.

The two modal operators can be given nonmonotonic interpretations relating to belief and consistency. $\mathbf{L}\phi$ is interpreted as "ϕ is believed", and $\mathbf{M}\phi$, which is equivalent to $\neg\mathbf{L}\neg\phi$, is taken to mean $\neg\phi$ is not believed, or simply that ϕ can be consistently believed. We can then express statements such as "it is believed that ϕ is not believed": $\mathbf{L}\neg\mathbf{L}\phi$; or "it is not believed that ϕ is believed": $\neg\mathbf{L}\mathbf{L}\phi$. (What is the difference?)

The semantics of modal logic can be given in terms of possible worlds with different accessibility relations. Informally a world represents a possible state of affairs, and the worlds accessible from a particular world constitute the scenarios considered possible at that world. A relation R over a set of worlds W can have a number of properties:

- *Serial:* $\forall w \in W$, there is some $w' \in W$ such that $R(w, w')$. [Each world can access at least one world (not necessarily itself).]
- *Reflexive:* $\forall w \in W$, we have $R(w, w)$. [Each world is accessible from itself.]

- *Transitive:* $\forall w_1, w_2, w_3 \in W$, if $R(w_1, w_2)$ and $R(w_2, w_3)$, then $R(w_1, w_3)$. [Every world that is accessible from a world is also accessible from its "ancestors".]
- *Euclidean:* $\forall w_1, w_2, w_3 \in W$, if $R(w_1, w_2)$ and $R(w_1, w_3)$, then $R(w_2, w_3)$. [All the worlds that are accessible from the same world are also accessible from each other.]
- *Symmetric:* $\forall w_1, w_2 \in W$, if $R(w_1, w_2)$, then $R(w_2, w_1)$. [A world is accessible from all the worlds it has access to.]
- *Equivalence*: Accessibility is reflexive, transitive, and symmetric. [Every world is accessible from every world in its class.]

Some of the well-known axioms and their corresponding constraints on the accessibility relation of the possible world structures are given below [Kripke, 1963]:

Symbol	Axiom	Accessibility relation
K	$L(\phi \rightarrow \psi) \rightarrow (L\phi \rightarrow L\psi)$	No constraints
D	$L\phi \rightarrow M\phi$	Serial
T	$L\phi \rightarrow \phi$	Reflexive
4	$L\phi \rightarrow LL\phi$	Transitive
5	$ML\phi \rightarrow L\phi$	Euclidean

These axioms can be interpreted in terms of knowledge or belief. The **D**-axiom corresponds to saying that the knowledge or belief is consistent. The **T**-axiom corresponds to stating that one cannot know anything false. The **4**-axiom corresponds to positive introspection, that is, the agent knows what is known. The **5**-axiom corresponds to negative introspection, that is, the agent knows what is not known.

Many modal systems are defined using a combination of the above axioms. Some of the standard modal systems are given below[7]:

Modal system	Axioms	Accessibility relation
K	K	No constraints
T	K, T	Reflexive
S4	K, T, 4	Reflexive and transitive
S5	K, T, 4, 5	Equivalence

The "correct" mechanism for reasoning can be modeled by incorporating various axioms, or equivalently, by imposing various constraints on the structure of the possible worlds. The **T**-axiom is usually adopted for systems modeling perfect knowledge, where anything an agent knows has to be true. The systems **T**, **S4**, and **S5** all have been considered as the basis for a family of nonmonotonic logics developed by McDermott and Doyle [McDermott & Doyle, 1980; McDermott, 1982], which are closely related to autoepistemic logic.

One way to characterize the difference between knowledge and belief is to consider the truth status of the statements in question. If we *know* something, it is true in the real world. In contrast, if we *believe* something, we think it is true but in reality it could be false. This distinction is captured by the **T**-axiom, which requires that whatever is known has to be true. Formal systems modeling belief typically drop the **T**-axiom, allowing an agent to hold false beliefs.

In everyday usage, however, we do not make a hard distinction between "knowledge" and "belief". When we say we know something, sometimes we just mean

---

[7]The 4-axiom in the system **S5** may be dropped, since it is deducible from the **5**-axiom.

that we think it is true, or that we believe it to be true. We will use these two terms interchangeably sometimes when the meaning is clear from the context.

There are various arguments, which we will not go into here, for each of the following *intermediate* modal systems to be a suitable basis for representing belief:

Modal system	Axioms	Accessibility relation
Weak **S4**	**K, 4**	Transitive
K45 (weak **S5**)	**K, 4, 5**	Transitive and euclidean
KD45	**K, D, 4, 5**	Serial, transitive, and euclidean

Note the absence of the **T**-axiom in each system. Moore's autoepistemic logic [Moore, 1984; Moore, 1985] is based on the modal system **K45**.

6.4.2 Autoepistemic Reasoning vs Default Reasoning

Early attempts to formalize nonmonotonic reasoning in a modal logic framework include McDermott and Doyle's *nonmonotonic logic I* [McDermott & Doyle, 1980] and a subsequent version *nonmonotonic logic II* [McDermott, 1982]. Autoepistemic logic [Moore, 1984; Moore, 1985] models the set of total beliefs possessed by an ideally rational agent, who is capable of perfect introspection. The agent is able to examine and reason about its own beliefs, it knows all the consequences of its beliefs, and it knows what it knows and what it does not know.

This is obviously an idealized situation. It is regarded as a "competence model", specifying a normative set of beliefs an agent should possess given unbounded time and resources for reasoning. The belief set of any finite agent should be expected to converge to the autoepistemic belief set as more resources are available.

Note, however, that although an ideal agent has access to unlimited resources, the quality of the initial premise set given to the agent is not guaranteed. Thus, even ideal agents may possess irrational beliefs if the premises given are unreasonable. There are two ways to approach this problem. One is to insist on ideally rational premises to begin with. The other approach is to equip the agent with ways to recognize and discard irrational beliefs. We will come back to this issue in Section 6.4.5.

Moore argues that autoepistemic reasoning and default reasoning are two distinct notions. Default reasoning concerns the inference of plausible conclusions from incomplete evidence based on typicality. For example, birds typically fly, and in the absence of any evidence that Tweety is atypical, such as that Tweety is a penguin, we can conclude by default reasoning that Tweety flies.

Autoepistemic reasoning, on the other hand, concerns inferences about introspective beliefs. Suppose $\phi_1, \phi_2, \ldots$ are instances of ϕ. We have the following reasoning principle:

If ϕ_i were true, I would know ϕ_i; since I do not know ϕ_i, it cannot be true.

This kind of reasoning assumes that all occurrences of ϕ's are known, and thus if a particular instance ϕ_i is not known, one can infer that ϕ_i is not true. It can be taken as a succinct means of representation, where all positive instances are represented. We can therefore infer a negative instance from the absence of its positive counterpart in our belief set. This is also the basic idea behind the closed world assumption discussed in Example 6.4.

An example Moore gave was that he believes he is the eldest child of his parents, not because he has been told about this explicitly, nor because "typically" he is the eldest (which would then be an instance of default reasoning), but simply because if he indeed had an elder sibling, he would have known it. Because he does not know about an elder sibling, he must be the eldest. In autoepistemic logic, this line of reasoning can be formulated as follows:

$$(\forall x)(\mathbf{M}\mathrm{eldest}(x) \rightarrow \mathrm{eldest}(x))$$

This conveys the idea that if it is consistent for an individual (in this case Moore) to believe that he himself is the eldest child, then we can conclude autoepistemically that he is the eldest.

Moore continues to argue that, unlike default reasoning, whose conclusions may have to be retracted as we learn more about an incomplete static situation, autoepistemic reasoning is not defeasible, because one has to believe that all positive instances of ϕ's are already known in order to infer $\neg\phi_i$ when a particular instance ϕ_i is not known. Thus, autoepistemic reasoning is nonmonotonic not because of its defeasibility, but because of its indexical nature: whether a statement is believed or not is determined with respect to the entire context of the theory.

Although default reasoning and autoepistemic reasoning seem to differ in motivation and underlying principles, operationally both of them are concerned with drawing conclusions that are not directly derivable from the premises. To add to the confusion, Moore also formulated the classical default reasoning example of Tweety the bird in autoepistemic terms [Moore, 1985].

Example 6.12. *Consider the following autoepistemic premise set:*

$$\left\{ \begin{array}{l} (\forall x)\,(\mathrm{bird}(x) \wedge \mathbf{M}\mathrm{fly}(x) \rightarrow \mathrm{fly}(x)), \\ \mathrm{bird}(\mathrm{tweety}) \end{array} \right\} \qquad (6.10)$$

For all individuals x, if x is a bird, and it is consistent to believe that it flies, then we are to conclude that it flies. With respect to this premise set, we can conclude that Tweety flies. However, suppose in addition we have the following:

$$\left\{ \begin{array}{l} (\forall x)\,(\mathrm{penguin}(x) \rightarrow \neg\mathrm{fly}(x)), \\ \mathrm{penguin}(\mathrm{tweety}) \end{array} \right\} \qquad (6.11)$$

That is, penguins do not fly, and Tweety is a penguin. We then have to retract the conclusion fly(tweety), *as it is no longer justified with respect to the expanded premise set. The rule*

$$(\forall x)\,(\mathrm{bird}(x) \wedge \mathbf{M}\mathrm{fly}(x) \rightarrow \mathrm{fly}(x))$$

is not applicable in the expanded context. $\mathbf{M}\mathrm{fly}(\mathrm{tweety})$ *in the antecedent of the rule is false, as* fly(tweety) *is not possible given that we can infer* $\neg$fly(tweety) *from (6.11).*

Note the parallel between this example and Example 6.1. Autoepistemic reasoning and default reasoning share a lot of common structure. They can be utilized to formulate the same types of situations. Let us introduce below the notion of an autoepistemic *stable expansion*, the equivalent of an extension in default logic.

6.4.3   Stable Expansions

We restrict the discussion of autoepistemic logic to the propositional level. Let $\mathcal{ML}$ be a propositional modal language. An *autoepistemic theory* $T \subseteq \mathcal{ML}$ is a set of modal formulas representing a belief set of an agent reflecting upon its own belief. For example, the autoepistemic theory

$$T = \{\text{sunny}, \mathbf{L}\text{hot}\}$$

represents the following beliefs of an agent: it is sunny, and the agent believes it is hot.

An autoepistemic theory can be any subset of the language; there is no constraint on internal consistency or the relationship between the various pieces of belief in the same set. In particular, there is no prescribed connection between a belief ϕ and its modal projections $\mathbf{L}\phi$ and $\mathbf{M}\phi$. For example, we may have the belief set

$$T = \{\text{sunny}, \neg\mathbf{L}\text{sunny}\}.$$

This amounts to saying that according to T, the agent believes that it is sunny, but it also believes simultaneously that it does not believe it is sunny. This does not seem quite rational, but nonetheless it is a legitimate theory in autoepistemic logic.

To represent a belief set of an *ideal* agent, we need to impose additional constraints on an autoepistemic theory. Let $A \subseteq \mathcal{ML}$ be an initial set of premises an agent is given to reason about. An autoepistemic theory T, based on a set of premises A, of an ideal rational agent is called a *stable expansion*. It possesses two properties, *stability* and *groundedness*.

Definition 6.9. *An autoepistemic theory $T \subseteq \mathcal{ML}$ is stable ([Stalnaker, 1980]) iff it satisfies the following conditions.*

(1) If $\alpha_1, \ldots, \alpha_n \in T$ and $\alpha_1, \ldots, \alpha_n \vdash \beta$, then $\beta \in T$.
(2) If $\alpha \in T$, then $\mathbf{L}\alpha \in T$.
(3) If $\alpha \notin T$, then $\neg\mathbf{L}\alpha \in T$.

Stability characterizes the "rational" relationship between the components of a belief set. The first condition requires a stable theory to be deductively closed. The second condition specifies that the agent believes that it believes what it believes: if the agent believes α, then it believes that it believes α, and it believes that it believes that it believes α, and.... This corresponds to positive introspection. The third condition corresponds to negative introspection: if the agent does not believe α, then it believes that α is not believed.

For example, suppose the agent believes that it is hot, but does not know whether it is sunny. Its belief set T would contain hot but neither sunny nor $\neg$sunny. If the belief set T is stable, it would also contain, among other formulas, the following:

$$T = \{\text{hot}, \mathbf{L}\text{hot}, \neg\mathbf{L}\text{sunny}, \neg\mathbf{L}\neg\text{sunny}, \mathbf{L}\mathbf{L}\text{hot}, \mathbf{L}\neg\mathbf{L}\text{sunny}, \ldots\} \qquad (6.12)$$

These formulas are derived according to the three rules in Definition 6.9, as follows:

$$
\begin{array}{ll}
\text{hot}: & \text{rule 1} \\
\text{L\,hot}: & \text{rule 2 applied to hot} \\
\neg\text{L\,sunny}: & \text{rule 3 applied to sunny} \\
\neg\text{L}\,\neg\text{sunny}: & \text{rule 3 applied to } \neg\text{sunny} \\
\text{LL\,hot}: & \text{rule 2 applied to L\,hot} \\
\text{L}\,\neg\text{L\,sunny}: & \text{rule 2 applied to } \neg\text{L\,sunny}.
\end{array}
\qquad (6.13)
$$

The set of nonmodal formulas in an autoepistemic theory is referred to as its *kernel*. Let us denote the kernel of a theory T by T^*. One useful property of stability is that each stable autoepistemic theory is uniquely determined by its kernel. The membership of all the modal formulas can be determined from the nonmodal base, and thus we only need to specify the kernel when we refer to a stable theory. We start with the kernel and successively add into the theory modal formulas sanctioned by the three rules for stability in Definition 6.9. For example, given the kernel $T^* = \mathbf{Th}(\{\text{hot}\})$, we can reconstruct the whole theory T in (6.12) by following the line of derivation in (6.13).

Stability specifies which formulas need to co-occur in a belief set, but it does not specify which formulas should *not* be there. We may insert as many reasonable or unreasonable formulas as we like into a stable belief set. For example, the set of all formulas in the language $\mathcal{ML}$ is stable. The fact that it is also inconsistent and contains a lot of contradictory information is not a concern for stability.

To characterize an ideal belief set, we in addition need to restrict the formulas that are allowed into the set. Whether a belief is considered reasonable is relative to the premises the agent is given, as different premises would lead to different conclusions. This is captured by the notion of *groundedness*, which is defined with respect to a premise set.

Definition 6.10. *An autoepistemic theory $T \subseteq \mathcal{ML}$ is* grounded *in a set of premises $A \subseteq \mathcal{ML}$ iff every formula in T is included in the tautological consequences of $A \cup \{\text{L}\alpha : \alpha \in T\} \cup \{\neg\text{L}\alpha : \alpha \notin T\}$.*

For an autoepistemic theory to be grounded in a set of given premises, only formulas that are deductively derivable from the premises together with positive and negative introspection are allowed. Consider the autoepistemic theory T in (6.12) again. T is a stable theory with the kernel $\mathbf{Th}(\{\text{hot}\})$. Let the premise set A contain a single formula hot. The theory T is grounded in A, as it contains only those formulas that are derivable from the premise set $\{\text{hot}\}$ and the stability rules. On the other hand, although (as we have observed) the set of the whole language $\mathcal{ML}$ is stable, $\mathcal{ML}$ is not grounded in A, as it contains many extra formulas, for example $\neg\text{hot}$ and $\text{L}\,\neg\text{hot}$, that are neither part of the premises nor essential for stability.

Now we are ready to give a formal definition of a stable expansion.

Definition 6.11. *An autoepistemic theory $T \subseteq \mathcal{ML}$ is a* stable expansion *of a set of premises $A \subseteq \mathcal{ML}$ iff it satisfies the following three conditions:*

(1) $A \subseteq T$.
(2) T is stable.
(3) T is grounded in A.

A stable expansion characterizes the belief set of an ideal rational agent with respect to a given premise set. It includes the premises, and is stable and grounded in the premises. Thus, an ideal belief set contains all and only those formulas deemed "rational" given the premises. Stability determines what formulas need to be included, and groundedness ensures that the extension to the premise set in order to achieve stability is minimal.

Example 6.13. *Consider the autoepistemic premise set*

$$A = \{\mathbf{L}\text{lose} \rightarrow \text{lose}, \neg\mathbf{L}\text{lose} \rightarrow \text{win}\}.$$

It embodies a self-fulfilling prophecy: if we believe we shall lose, then we shall lose; if we do not believe we shall lose, then we shall win.

We can construct two stable expansions with respect to this premise set:

$$T_1^* = \mathbf{Th}(\{\text{lose}\}); \quad \text{i.e.,} \quad T_1 = \{\text{lose}, \mathbf{L}\text{lose}, \neg\mathbf{L}\text{win}, \ldots\}$$
$$T_2^* = \mathbf{Th}(\{\text{win}\}); \quad \text{i.e.,} \quad T_2 = \{\text{win}, \neg\mathbf{L}\text{lose}, \mathbf{L}\text{win}, \ldots\}.$$

Depending on which antecedent of the two premises we believe, we can obtain a stable expansion T_1 in which we believe we shall lose, or T_2 in which we believe we shall win. For T_1, the first premise $\mathbf{L}\text{lose} \rightarrow \text{lose}$ is invoked to conclude lose and therefore $\mathbf{L}\text{lose}$, which in turn makes the antecedent of the other premise $\neg\mathbf{L}\text{lose} \rightarrow \text{win}$ false. For T_2, the second premise is invoked to conclude win. The antecedent of the first premise cannot be true in this stable expansion, as otherwise the antecedent of the second premise would be falsified. One may check that both theories satisfy the three conditions required of a stable expansion in Definition 6.11.

6.4.4 Alternative Fixed–Point Formulation

Both default logic and autoepistemic logic make use of fixed points. A default extension (Definition 6.5) is specified via an operator Γ, whose definition makes use of Γ itself. The specification of an autoepistemic stable expansion T (Definition 6.11) makes reference to the theory T itself. These two fixed-point definitions may seem quite different on the surface, but we can put them into analogous forms so that they can be more easily compared.

Before we give the alternative fixed-point formulation for a stable expansion, let us first introduce a normal form for autoepistemic formulas. Konolige [Konolige, 1988; Konolige, 1989] showed that in the modal system **K45**, every well-formed formula of $\mathcal{ML}$ is equivalent to a formula of the normal form

$$\neg\mathbf{L}\alpha \vee \mathbf{L}\beta_1 \vee \cdots \vee \mathbf{L}\beta_n \vee \gamma$$

where $\alpha, \beta_1, \ldots, \beta_n, \gamma$ are all nonmodal formulas, and any of the formulas $\alpha, \beta_1, \ldots, \beta_n$ may be absent.[8] Formulas in normal form do not contain any nested modal operators. In our discussion here, we assume that all autoepistemic formulas are given in a slightly different normal form

$$\mathbf{L}\alpha \wedge \neg\mathbf{L}\beta_1 \wedge \cdots \wedge \neg\mathbf{L}\beta_n \rightarrow \gamma.$$

---

[8]Note that the nonmodal disjunct γ has to be present, and thus the normal form of the formula $\mathbf{L}\alpha$ is $\mathbf{L}\alpha \vee \bot$, where $\bot$ is the contradiction symbol.

Note the parallel between the structure of this normal form and that of a default rule. Informally, the positive and negative modal components in the antecedent correspond respectively to the prerequisite and justifications of a default rule, while the nonmodal part corresponds to the consequent of the default rule. This similarity has been observed by many ([Konolige, 1988; Marek & Truszczyński, 1993] for example), and it will become even more apparent when we look at the alternative fixed-point formulation presented below.

We can characterize stable expansions using a fixed-point operator Ω similar to the fixed point operator Γ used to define default extensions.

Theorem 6.3. *Let $A \subseteq \mathcal{ML}$ be a set of premises in normal form, and T be a consistent subset of $\mathcal{ML}$. Then $\Omega(T) \subseteq \mathcal{ML}$ is the set with the smallest kernel satisfying the following two properties:*

(1) $\Omega(T)$ is stable.
(2) For every formula $\mathbf{L}\alpha \wedge \neg\mathbf{L}\beta_1 \wedge \cdots \wedge \neg\mathbf{L}\beta_n \to \gamma \in A$, if

> *(a) $\alpha \in T$,*
> *(b) $\beta_1, \ldots, \beta_n \notin T$,*
> *then $\gamma \in \Omega(T)$.*

T is a stable expansion of A iff T is a fixed point of the operator Ω, that is, $T = \Omega(T)$.

This fixed-point formulation is laid out in a way very similar to that for a default extension (Definition 6.5). The "smallest" requirement is imposed on the kernel of $\Omega(T)$ rather than on $\Omega(T)$ itself, in contrast to the case for $\Gamma(E)$ of default logic, since a consistent stable set cannot have a proper stable subset and therefore we cannot talk about the smallest stable set here. Note again that knowing the kernel is sufficient to recover the modal components of a stable set, and thus the whole theory $\Omega(T)$ is fully determined.

The first property of Theorem 6.3 is similar to the second property of Definition 6.5. The stability condition in autoepistemic logic includes the deductive closure condition required of a default extension.

The first property of Definition 6.5 does not have a counterpart in Theorem 6.3, since in autoepistemic logic "facts" (formulas with no modal components) are not formally distinguished. A more significant difference between the two formulations is that in Theorem 6.3, we have $\alpha \in T$ instead of $\alpha \in \Omega(T)$ in condition 2(a), whereas the corresponding specification (condition 3(a)) for a default extension calls for $\Gamma(E)$ and not E. This is the one main difference between the two formalisms.

Recall that in default logic, the set E is the proposed extension, and $\Gamma(E)$ is the set in which we accumulate formulas that are considered justified according to the rules in Definition 6.5. When the two sets match up, we have verified that E is an extension. The same relationship holds between T and $\Omega(T)$ for autoepistemic stable expansions. The formula α corresponds to the prerequisite of a default rule in default logic, and it corresponds to the positive modal component in a formula put in normal form in autoepistemic logic.

The difference between the two formalisms can then be characterized as the *groundedness* requirement of this α component. In default logic, the prerequisite needs to have a justifiable derivation, just like every formula in $\Gamma(E)$. In autoepistemic logic, on the other hand, the positive modal term only needs to be "hypothesized"—it needs only to be present in the proposed stable expansion E, which does not require any justification at the start. The formula α does need to be justified in the end, of course, but this justification can be grounded in a line of reasoning that is circular: α is used to derive another formula, which in turn is used to justify α itself.

Thus, it appears that autoepistemic logic allows for conclusions that are not as firmly grounded as those obtained in a corresponding default logic setting. This is the topic of discussion of the next subsection.

6.4.5 Groundedness

The formulation of autoepistemic logic gives rise to some unintuitive results. In particular, it allows beliefs to be self-grounded, as shown in the following example.

Example 6.14. *Consider the premise set* $A = \{\mathbf{L}\,\text{sunny} \to \text{sunny}\}$. *This amounts to saying that if the agent believes that it is sunny today, then it is in fact sunny today. Intuitively, one would not expect that the weather can be changed by our will, and thus a belief in* sunny *should not be justified given this premise set.*

However, there is a stable expansion with the kernel $\mathbf{Th}(\{\text{sunny}\})$. *The reasoning involved is circular: We start with the assumption that we believe it is sunny, that is,* $\mathbf{L}\,\text{sunny}$, *and from that and the premise* $\mathbf{L}\,\text{sunny} \to \text{sunny}$ *we can deduce the consequent* sunny. *This in turn justifies, by virtue of stability, the inclusion of* $\mathbf{L}\,\text{sunny}$ *in the belief set in the first place, and thus we have a stable expansion containing* sunny *and* $\mathbf{L}\,\text{sunny}$.

Thus, even an ideally rational agent can be justified in believing some statement on the grounds that it believes that it believes it. This example indicates that the notion of groundedness in Moore's formulation is not strong enough to prevent circular beliefs, and thus does not fully capture the notion of ideally rational belief. One may argue that formulas such as

$$\mathbf{L}\,\text{sunny} \to \text{sunny}$$

should never appear in the premise set to begin with, since they are not "ideally rational". However, it is also desirable to be able to guard against the proliferation of such unwanted beliefs given any premises. We would like an ideally rational agent to reject any belief that appeals to such "irrational" premises.

Note that the modal operator $\mathbf{L}$ in classical modal logic is interpreted as *necessity*. The schema

$$\mathbf{L}\phi \to \phi$$

is the axiom $\mathbf{T}$: if ϕ is necessarily true, then it is true in reality. This makes perfect sense. In autoepistemic logic, however, the operator $\mathbf{L}$ is interpreted as *belief*, and it is usually not the case that we can will something to become true by simply choosing to believe in it. This circularity makes the $\mathbf{T}$-axiom difficult to justify in systems modeling belief, and as we have already noted in Section 6.4.1, the $\mathbf{T}$-axiom is indeed dropped in most such systems, including autoepistemic logic, which is based on $\mathbf{K45}$. The $\mathbf{T}$-axiom however has been reincarnated in the form of a premise in Example 6.14.

Attempts have been made to strengthen the groundedness criterion of autoepistemic logic. *Moderately grounded* stable expansions [Konolige, 1988] are stable expansions that are minimal in the sense that the premise set is not true in any stable theory (not necessarily a stable expansion of the premise set) with a smaller kernel. This avoids some of the circularity, such as that exhibited in Example 6.14. The undesirable stable expansion with the kernel **Th**({sunny}) we obtained in the example is not moderately grounded, as the single premise **L**sunny $\rightarrow$ sunny is also true in the stable set with the smaller kernel **Th**($\emptyset$).

Moderate groundedness rules out some but not all occurrences of circular derivation. The groundedness criterion can be further strengthened. The idea of *strongly grounded* stable expansions [Konolige, 1988; Konolige, 1989], also known as *supergrounded* stable expansions [Marek & Truszczyński, 1989], is to ensure that every nonmodal sentence ϕ has a derivation that is independent of $\mathbf{L}\phi$, by disallowing the derivation of $\mathbf{L}\phi$ except as a consequence of the derivation of ϕ. However, strong groundedness is sensitive to the syntactic form of the theory: logically equivalent theories can differ in the strongly grounded stable expansions sanctioned.[9] *Robust expansions* [Marek & Truszczyński, 1989], another attempt to strengthen the groundedness, also suffer from the same problem. It is not clear how we can strengthen the groundedness condition sufficiently and elegantly to ward off all circular derivations.

6.5 Circumscription

In this section we will discuss very briefly circumscription [McCarthy, 1980]. Unlike default logic and autoepistemic logic, where an incomplete theory is extended under some consistency constraints, circumscription is an approach to nonmonotonic reasoning based on minimization. Some particular aspects of the language, such as the domain of individuals or the extension of a predicate, are restricted so that their "coverage" is as small as possible.

There are many forms of circumscription, each of which minimizes a different construct in a different way. One of the earliest formalisms is *predicate circumscription*, where, as one might have guessed, the target of minimization is one or more predicates.

Informally, given a belief set A and a target predicate relation $P(\bar{x})$, we can minimize the positive instances of this predicate so that the only tuples of individuals satisfying P are those that can be shown to do so in the belief set. In other words, any tuple that does not provably satisfy the predicate is assumed *not* to be a positive instance of the predicate.

Example 6.15. *Suppose our belief set A concerning two individuals Tweety and Polly is as follows:*

$$A = \{\text{penguin(tweety)}, \text{polly} \neq \text{tweety}\}.$$

Tweety is a penguin, and Polly is not the same individual as Tweety. From this belief set A we can conclude (monotonically) that Tweety is a penguin. But what about Polly? We do not have information to either confirm or deny that Polly is a penguin.

Operating under the minimization principle, we want to assert that Tweety is the only penguin there is, because no other individual can be proved to be a penguin. In particular, because Polly is not the same individual as Tweety, we would like to conclude that Polly is

---

[9]Konolige suggested that this property can be a desirable feature in syntax-directed procedural systems.

not a penguin. This idea can be captured in the formula

$$(\forall x)\,(\mathtt{penguin}(x) \to x = \mathtt{tweety}). \tag{6.14}$$

That is, for every object x, if x is a penguin, then x is Tweety. This corresponds to minimizing the predicate penguin, *so that it holds for as few individuals as possible. If we incorporate the formula (6.14) into our belief set A, we would then be able to conclude* ¬penguin(polly), *since we also know that* polly ≠ tweety.

Generally speaking, we can devise a set of formulas $C_{A,P}$ based on a given theory A and a target predicate P, such that when $C_{A,P}$ is added to the original theory A, the predicate P is minimized in $A \cup C_{A,P}$. The formulas in $C_{A,P}$ are called *circumscription formulas*, and the augmented theory $A \cup C_{A,P}$ is the *circumscribed theory*. The general form of a circumscription formula in $C_{A,P}$ is

$$[A(\Phi) \wedge ((\forall \bar{x})\,(\Phi(\bar{x}) \to P(\bar{x})))] \to (\forall \bar{x})\,(P(\bar{x}) \to \Phi(\bar{x})) \tag{6.15}$$

where $A(\Phi)$ denotes the set of formulas that is identical to A except that every occurrence of the predicate symbol P is replaced by the predicate symbol Φ. This formula can be interpreted as follows:

- The antecedent of (6.15):

    (1) $A(\Phi)$ says that the predicate Φ satisfies all the conditions set out in A for the target predicate P.
    (2) $(\forall \bar{x})\,(\Phi(\bar{x}) \to P(\bar{x}))$ says that the set of objects satisfying Φ is a subset of that satisfying P, since every object x that satisfies Φ also satisfies P.

- The consequent of (6.15): $(\forall \bar{x})\,(P(\bar{x}) \to \Phi(\bar{x}))$ says the converse of item 2 above: the set of objects satisfying P is a subset of that satisfying Φ.

Taken together, these statements say that if the antecedent of (6.15) is true (that is, Φ satisfies the constraints on P and Φ's interpretation is a subset of P's), then the consequent is true, that is, P's interpretation is a subset of Φ's. In other words, the extension of P is restricted to the extension of Φ.

In general, (6.15) is taken as a schema. We circumscribe a theory with a set of formulas of the form (6.15), varying the predicate Φ exhaustively. Formal expression of this schema requires second order logic—a logic where we can quantify over relations. The *circumscription schema* can then be written as a second order formula

$$(\forall \Phi)\,[\,[A(\Phi) \wedge ((\forall \bar{x})\,(\Phi(\bar{x}) \to P(\bar{x})))\,] \to (\forall \bar{x})\,(P(\bar{x}) \to \Phi(\bar{x}))\,], \tag{6.16}$$

which quantifies over the predicate Φ.

Example 6.16. *Let us continue with Example 6.15. The predicate P to be circumscribed is* penguin. *We have*

$$A = \{\mathtt{penguin(tweety)}, \mathtt{polly} \neq \mathtt{tweety}\}$$

$$A(\Phi) = \{\Phi(\mathtt{tweety}), \mathtt{polly} \neq \mathtt{tweety}\}$$

and the circumscription schema

$$(\forall \Phi)[[\Phi(\mathtt{tweety}) \wedge \mathtt{polly} \neq \mathtt{tweety} \wedge ((\forall x)\,(\Phi(x) \to \mathtt{penguin}(x)))]$$
$$\longrightarrow (\forall x)\,(\mathtt{penguin}(x) \to \Phi(x))]. \tag{6.17}$$

The above schema includes many formulas corresponding to various instantiations of Φ. *Some of them are trivially true, for example in the case of* $\Phi(x) \equiv \text{penguin}(x)$. *The instantiation of particular interest is*

$$\Phi(x) \equiv (x = \text{tweety}).$$

By applying this instantiation to (6.17), we obtain the formula

$$[\text{tweety} = \text{tweety} \wedge \text{polly} \neq \text{tweety} \wedge ((\forall x)\,(x = \text{tweety} \rightarrow \text{penguin}(x)))]$$
$$\longrightarrow (\forall x)\,(\text{penguin}(x) \rightarrow x = \text{tweety}).$$

The antecedent of this circumscription formula consists of three conjuncts, which are all true in the given theory A. Thus, the circumscribed theory contains the consequent

$$(\forall x)\,(\text{penguin}(x) \rightarrow x = \text{tweety}).$$

This is just what we set out to get in (6.14). It specifies that the only penguin that can exist is Tweety. Together with the original premises A, we can now conclude $\neg\text{penguin}(\text{polly})$.

Circumscription allows us to draw conclusions that are not derivable classically from a theory by augmenting the theory with additional circumscription formulas. This process is nonmonotonic, as a conclusion that can be derived from a circumscribed theory is not necessarily retained in the theory obtained by circumscribing an expanded theory. This is illustrated in the following example.

Example 6.17. *Reconsider the theory A in Example 6.16. Suppose now we know in addition that Polly is a penguin. We then have the following scenario:*

$$A' = \{\text{penguin}(\text{tweety}), \text{penguin}(\text{polly}), \text{polly} \neq \text{tweety}\}.$$

The circumscription schema is then amended to

$$(\forall\Phi)\,[[\Phi(\text{tweety}) \wedge \Phi(\text{polly}) \wedge \text{polly} \neq \text{tweety} \wedge ((\forall x)\,(\Phi(x) \rightarrow \text{penguin}(x)))]$$
$$\longrightarrow (\forall x)\,(\text{penguin}(x) \rightarrow \Phi(x))].$$

Using the same instantiation $\Phi(x) \equiv (x = \text{tweety})$ *as in Example 6.16, the circumscription formula obtained from the above schema is trivially true, as one of the conjuncts in the antecedent* $\Phi(\text{polly})$ *translates to* $\text{polly} = \text{tweety}$, *which is false.*

We cannot derive the previous result $\neg\text{penguin}(\text{polly})$ *from circumscribing the amended theory A'. Instead, one may check that with the instantiation*

$$\Phi(x) \equiv (x = \text{tweety} \vee x = \text{polly})$$

we obtain the formula

$$(\forall x)\,(\text{penguin}(x) \rightarrow (x = \text{tweety} \vee x = \text{polly})),$$

that is, any individual that is neither Tweety nor Polly is deemed to be not *a penguin.*

This illustrates that circumscription is a nonmonotonic reasoning formalism. By circumscribing the theory A, we can conclude $\neg\text{penguin}(\text{polly})$, *but by circumscribing the extended theory A', this conclusion is no longer justified.*

Circumscription enjoys an advanced level of technical development, but its formulation is one of the hardest to visualize. One of the main difficulties with circumscription is that it involves second order formulas, and in many cases it is not easy to determine which formulas from the circumscription schema (6.16) are of interest. The process seems to be backward: we need to know the circumscription formula in advance, and then we can pull it out of the schema with the appropriate instantiation, and show that it is indeed the formula we need.

Circumscription is different from both default logic and autoepistemic logic in a number of ways. We have already mentioned that default and autoepistemic logics are consistency based, whereas circumscription is minimization based. In addition, the two logics may produce multiple extensions from a given theory, while circumscription gives rise to a single extended theory at a time. In default logic, classical first order logic is augmented with default rules. In autoepistemic logic, a modal system is used. In circumscription, the reasoning is formalized within a second order framework, although the circumscription formulas in many cases may be reduced to first order, via some clever instantiations.

6.6   Unresolved Issues

A number of issues are common to many formalisms of nonmonotonic reasoning. These problems concern the general task of extending classical logics to allow for nonmonotonicity, independent of the specific formalism one has in mind. We will highlight a number of issues in this section.

6.6.1   "Intuition": Basis of Defaults

Nonmonotonic reasoning is modeled after the "jumping to conclusions" behavior of human reasoning. People arrive at nonmonotonic conclusions all the time, and they justify their conclusions with expressions such as "typically," "usually," "probably," "intuitively," "by common sense." The nonmonotonic logic community has largely relied on "intuition" as the basis for evaluating the behaviors of various nonmonotonic reasoning systems. When a default theory gives rise to an extension that does not conform to our expectation, for example, either the specific theory given is revised, or at a higher level, the formalism itself is amended to reflect a different intuition about what constitutes an intuitive leap of reasoning. Many variants of nonmonotonic formalisms have been proposed, each motivated to circumvent some counterintuitive results, some of which have been discussed in this chapter.

Although people do seem to agree on what counts as intuitive or unintuitive in most everyday situations, our intuitions do differ in many ambiguous cases. This is particularly true when we have to reason about carefully engineered theories designed to represent curious or even pathological scenarios. What does our intuition tell us for example about the autoepistemic premise $\neg L$ sunny $\rightarrow$ sunny? Does it make sense to say that if we do not believe it is sunny, then really it is sunny?

There are two issues here. First, what constitutes a good nonmonotonic rule? And second, given a set of rules and base facts, what conclusions *should* we draw? We need to have a solid understanding of nonmonotonic inference and a formal basis other than whimsical intuition before we can make meaningful evaluations of the various nonmonotonic reasoning systems.

6.6.2 Computational Complexity

One appeal of nonmonotonic reasoning is that it can be faster than its classical logic counterpart. We can cut out the lengthy formalization and proof procedures we have to go through in order to use a deductive logic to arrive at the same conclusions given by a nonmonotonic logic that can "jump to conclusions." This computational advantage has mostly not been realized, however, and in many cases the problem is even more serious than that.

Provability in classical first order logic is *semidecidable*. To determine whether $T \vdash \phi$, there are algorithms that always terminate when the answer is "yes," but may not terminate when the answer is "no." There is no algorithm that can be guaranteed to terminate when $T \nvdash \phi$. This poses a problem for the nonmonotonic logics that are consistency based. For example, in default logic, part of the applicability condition of a default rule

$$\frac{\alpha : \beta_1, \ldots, \beta_n}{\gamma}$$

requires that each of the justifications β be consistent with the default extension E. In other words, the negation of each β has to be *not* provable from the extension, that is, $E \nvdash \neg\beta$. Such nonmonotonic logics involving a consistency check are harder than classical first order logic. These logics are not even semidecidable, and there is no effective existing computational mechanism for these formalisms in general.

Minimization approaches such as circumscription do not employ consistency checks, but circumscription involves second order logic in the general case, which is also messy to compute.

6.6.3 Multiple Extensions

A nonmonotonic theory may give rise to more than one extension, to use the terminology of default logic. We have seen such cases in Example 6.5 and Example 6.13. Each extended theory represents a scenario obtained from one possible way of augmenting the premises using conclusions drawn via the given nonmonotonic rules. The number of extensions can grow exponentially with the number of rules, and we can easily end up with a lot of scenarios to consider. In addition, not all of these extensions are equally desirable, as illustrated in the adult students example (Example 6.7).

Sometimes there is a good reason to have multiple extensions to represent alternative scenarios of what is possible, but at other times it is desirable to be able to generate one or only a few models of the world, especially when not all the extensions are of equal standing. One way of handling the proliferation of extensions is to impose a prioritized order on the default rules. When multiple rules are applicable at the same time, the rule(s) with the highest priority are applied first.

A natural ordering of rules can be extracted from an inheritance hierarchy [Touretzky, 1984; Horty et al., 1987]. For example, given a taxonomy of animals, a rule about flying that applies to a more specific set of animals (for example penguins) takes precedence over one that applies to a more general set (for example birds). A default logic allowing for reasoning about rule priorities within the system itself has been proposed by Brewka [Brewka, 1994].

Instead of considering individual extensions separately, another way of dealing with the problem of multiple extensions is to take the intersection of the extensions as the set of conclusions sanctioned. This approach gives us only one theory, which consists of only those conclusions that are true in all possible scenarios. We have cut down the number of alternatives we need to consider to only one, but the sanctioned set of conclusions may not be as informative, as the resulting theory is essentially a disjunction of all the extensions. Note, however, that this intersection may still contain statements that are nonmonotonically derived, if there exist rules that have been applied in every possible extension.

Using the terminology in [McDermott, 1982], retaining multiple extensions is *brave*. There is more information contained in each extension, even though it may be overcommitting, as the possibly contradictory information in other extensions is not taken into account. On the other hand, taking the intersection of all the extensions is *cautious*. Although it contains less information, we are less likely to be misled after considering the information contained in all possible extensions. Which approach suits our purposes better depends on, among other considerations, how safe we need to be in a given situation.

6.7 Summary

Deductive inference, as exemplified by classical logic, is monotonic. A conclusion deduced from a set of premises can never be falsified when we incorporate more premises. The set of deductive conclusions can only increase monotonically with the expansion of the premise set. This, however, does not reflect our experience dealing with uncertain information. We draw tentative conclusions which later may have to be retracted. When reasoning under uncertainty, the set of sanctioned conclusions may shrink, and thus the inference process is nonmonotonic.

We discussed several formalisms of nonmonotonic reasoning: default logic, autoepistemic logic, and circumscription. All three are based on classical logic. Default logic augments classical logic with extralogical rules specifying the conditions under which we may make a default conclusion. Autoepistemic logic models the belief set of an ideal rational agent within a modal logic framework. Both logics are consistency based and have very similar characterizations. The third approach, circumscription, is based on second order logic. A theory is extended by minimizing certain constructs in the language.

These nonmonotonic reasoning formalisms suffer from a number of difficulties, ranging from technical problems such as *how* we can deal with the high computational complexity, to conceptual problems such as a lack of consensus on *what* conclusions we should draw nonmonotonically in a given situation. Nevertheless, nonmonotonic reasoning is a pervasive form of reasoning, and these and other approaches provide a useful framework for formalizing the problems and solutions.

6.8 Bibliographical Notes

A collection of foundational papers can be found in [Ginsberg, 1987]. A comprehensive text on nonmonotonic reasoning, in particular default logic, is [Antoniou,

1997]. For an introduction to modal logic, see [Hughes & Cresswell, 1996; Chellas, 1980]. A number of variants to Reiter's default logic have been proposed, including [Łukaszewicz, 1985; Brewka, 1991; Delgrande et al., 1994]. The relationship between default logic and autoepistemic logic has been studied extensively ([Lin & Shoham, 1990; Marek & Truszczyński, 1993; Gottlob, 1993], for example). There are many forms of circumscription, such as domain circumscription [McCarthy, 1977], formula circumscription [McCarthy, 1986], prioritized circumscription [Lifschitz, 1985], and pointwise circumscription [Lifschitz, 1987].

6.9 Exercises

(1) Is the probability calculus nonmonotonic? Why?

(2) Write out the closed default theory corresponding to the default theory in Example 6.11.

(3) Definition 6.5 implies that Γ is a function. Prove that this is the case.

(4) For each of the following default theories $\Delta = \langle D, F \rangle$, enumerate the extensions, or explain why there is no extension.

(a) $D = \left\{ \dfrac{p:q}{r}, \dfrac{p:\neg r}{\neg q} \right\}$ and $F = \{p\}$.

(b) $D = \left\{ \dfrac{p:q}{r}, \dfrac{p:\neg q}{\neg q} \right\}$ and $F = \{p\}$.

(c) $D = \left\{ \dfrac{p:q}{r}, \dfrac{r:\neg q}{\neg q} \right\}$ and $F = \{p\}$.

(d) $D = \left\{ \dfrac{p:q}{r}, \dfrac{r:q}{q} \right\}$ and $F = \{p\}$.

(5) Consider the default theory $\Delta = \langle D, F \rangle$, where

$$D = \left\{ \frac{p:r}{r}, \frac{q:r}{r} \right\} \quad \text{and} \quad F = \{p \vee q\}$$

What are the extensions? Does the result correspond to what you expect? Why?

(6) It has been suggested that every default rule of the form

$$\frac{\alpha : \beta}{\gamma}$$

can be better specified as a seminormal default rule

$$\frac{\alpha : \beta \wedge \gamma}{\gamma}$$

What are the advantages of adopting such a scheme? Give an example.

(7) What is the difference between the following two default rules?

$$d_1 = \frac{\alpha : \beta_1, \beta_2}{\gamma}$$

$$d_2 = \frac{\alpha : \beta_1 \wedge \beta_2}{\gamma}$$

Illustrate with an example.

(8)   Are the following formulas contained in the stable set whose kernel is $\mathbf{Th}(\{p\})$?

    (a) $\mathbf{L}\,p$ (b) $\mathbf{L}\neg p$ (c) $\neg\mathbf{L}\,p$ (d) $\neg\mathbf{L}\neg p$ (e) $\mathbf{L}\neg\mathbf{L}\,p$

    (f) $\mathbf{L}\,q$ (g) $\mathbf{L}\neg q$ (h) $\neg\mathbf{L}\,q$ (i) $\neg\mathbf{L}\neg q$ (j) $\mathbf{L}\neg\mathbf{L}\,q$

    (k) $\mathbf{L}(\neg\mathbf{L}\,p \vee \neg\mathbf{L}\,q)$.

(9)   Show that no stable autoepistemic theory can be a proper subset of another consistent stable autoepistemic theory.

(10)   What happens if we add to a stable autoepistemic theory T a modal formula that is not already contained in T? Is the resulting theory stable?

(11)   Give the stable expansions of the following autoepistemic sets.

    (a) $A = \{p, r, \mathbf{L}\,p \wedge \neg\mathbf{L}\neg q \rightarrow q, \mathbf{L}\,r \wedge \neg\mathbf{L}\,q \rightarrow \neg q\}$.

    (b) $A = \{\neg\mathbf{L}\,p \rightarrow q, \neg q\}$.

(12)   What is the circumscription to minimize `bird` in

$$A = \{\texttt{bird(tweety)} \vee \texttt{bird(polly)}\}?$$

Another approach is to "complete" a predicate, such that in the augmented theory A', for any instantiation $\bar{a}$ of $\bar{x}$,

    if $P(\bar{a}) \in A$, then $P(\bar{a}) \in A'$,

    if $P(\bar{a}) \notin A$, then $\neg P(\bar{a}) \in A'$.

How does this approach compare to predicate circumscription as applied to the above theory A?

(13)   How would you formulate the "birds fly, but penguins don't" scenario in circumscription?

(14)   Discuss possible ways of dealing with the unresolved issues in Section 6.6.

(15)   What are some of the complications that arise when a statement in a set of nonmonotonic conclusions turns out to be false?

(16)   How may we extend the nonmonotonic formalisms to capture the notion of varying degrees of typicality?

Bibliography

[Antoniou, 1997] Grigoris Antoniou. *Nonmonotonic Reasoning*. MIT Press, 1997.

[Brewka, 1991] Gerhard Brewka. Cumulative default logic: In defense of nonmonotonic inference rules. *Artificial Intelligence*, 50:183–205, 1991.

[Brewka, 1994] Gerhard Brewka. Reasoning about priorities in default logic. In *Proceedings of the Twelfth National Conference on Artificial Intelligence*, pp. 940–945, 1994.

[Chellas, 1980] B. F. Chellas. *Modal Logic: An Introduction*. Cambridge University Press, 1980.

[Delgrande et al., 1994] J. P. Delgrande, T. Schaub, and W. K. Jackson. Alternative approaches to default logic. *Artificial Intelligence*, 70 (1-2), pp. 167–237, 1994.

[Ginsberg, 1987] Matthew Ginsberg, editor. *Readings in Nonmonotonic Reasoning*. Morgan Kaufman, 1987.

[Gottlob, 1993] Georg Gottlob. The power of beliefs or translating default logic into standard autoepistemic logic. In *Proceedings of the Thirteenth International Joint Conference on Artificial Intelligence*, pp. 570–575, 1993.

[Horty et al., 1987] J. F. Horty, D. S. Touretzky, and R. H. Thomason. A clash of intuitions: The current state of nonmonotonic multiple inheritance systems. In *Proceedings of the Tenth International Joint Conference on Artificial Intelligence*, pp. 476–482, 1987.

[Hughes & Cresswell, 1996] G. E. Hughes and M. J. Cresswell. *A New Introduction to Modal Logic*. Routledge, London, 1996.

[Konolige, 1988] Kurt Konolige. On the relation between default and autoepistemic logic. *Artificial Intelligence*, 35:343–382, 1988.

[Konolige, 1989] Kurt Konolige. On the relation between default and autoepistemic logic. Errata. *Artificial Intelligence*, 41:115, 1989.

[Kripke, 1963] S. Kripke. Semantical considerations of modal logic. *Zeitschrift fur Mathematische Logik und Grundlagen der Mathematik*, 9:67–96, 1963.

[Kyburg, 1996] Henry E. Kyburg, Jr. Uncertain inferences and uncertain conclusions. In *Proceedings of the Twelfth Conference of Uncertainty in Artificial Intelligence*, pp. 365–372, 1996.

[Lifschitz, 1985] Vladimir Lifschitz. Computing circumscription. In *Proceedings of the Ninth International Joint Conference on Artificial Intelligence*, pp. 121–127, 1985.

[Lifschitz, 1987] Vladimir Lifschitz. Pointwise circumscription. In Matthew Ginsberg, editor, *Readings in Nonmonotonic Reasoning*, pp. 179–193. Morgan Kaufman, 1987.

[Lin & Shoham, 1990] Fangzhen Lin and Yoav Shoham. Epistemic semantics for fixed-point non-monotonic logics. In *Proceedings of the Third Conference on Theoretical Aspects of Reasoning about Knowledge*, pp. 111–120, 1990.

[Łukaszewicz, 1985] Witold Łukaszewicz. Two results on default logic. In *Proceedings of the Ninth International Joint Conference on Artificial Intelligence*, pp. 459–461, 1985.

[Marek & Truszczyński, 1989] Wiktor Marek and Miroslaw Truszczyński. Relating autoepistemic and default logics. In *Proceedings of the First International Conference on Principles of Knowledge Representation and Reasoning*, pp. 276–288, 1989.

[Marek & Truszczyński, 1993] V. W. Marek and M. Truszczyński. *Nonmonotonic Logic: Context-Dependent Reasoning*. Springer–Verlag, 1993.

[McCarthy, 1977] John McCarthy. Epistemological problems of artificial intelligence. In *Proceedings Artificial Intelligence*, pp. 1038–1044, 1977.

[McCarthy, 1980] John McCarthy. Circumscription—a form of nonmonotonic reasoning. *Artificial Intelligence*, 13:27–39, 1980.

[McCarthy, 1986] John McCarthy. Applications of circumscription to formalizing common sense. *Artificial Intelligence*, pp. 89–116, 1986.

[McDermott, 1982] Drew McDermott. Non-monotonic logic II: Non-monotonic modal theories. *Journal of the ACM*, 29:32–57, 1982.

[McDermott & Doyle, 1980] Drew McDermott and Jon Doyle. Non-monotonic logic I. *Artificial Intelligence*, 13:41–72, 1980.

[Moore, 1984] Robert C. Moore. Possible-world semantics for autoepistemic logic. In *Proceedings of AAAI Non-monotonic Reasoning Workshop*, pp. 344–354, 1984.

[Moore, 1985] Robert C. Moore. Semantical considerations on nonmonotonic logic. *Artificial Intelligence*, 25:75–94, 1985.

[Poole, 1989] David Poole. What the lottery paradox tells us about default reasoning. In *Proceedings of the First International Conference on Principles of Knowledge Representation and Reasoning*, pp. 333–340, 1989.

[Reiter, 1980] R. Reiter. A logic for default reasoning. *Artificial Intelligence*, 13:81–132, 1980.

[Reiter & Criscuolo, 1981] Raymond Reiter and Giovanni Criscuolo. On interacting defaults. In *Proceedings of the Seventh International Joint Conference on Artificial Intelligence*, pp. 270–276, 1981.

[Shortliffe, 1976] Edward Hance Shortliffe. *Computer-Based Medical Consultations: MYCIN*. Elsevier, 1976.

[Stalnaker, 1980] R. Stalnaker. A note on non-monotonic modal logic. Department of Philosophy, Cornell University, Ithaca, NY, 1980. Unpublished manuscript.

[Touretzky, 1984] David S. Touretzky. Implicit ordering of defaults in inheritance systems. In *Proceedings of the Fifth National Conference on Artificial Intelligence*, pp. 322–325, 1984.

7

Theory Replacement

7.1 Introduction

We form beliefs about the world, from evidence and inferences made from the evidence. Belief, as opposed to knowledge, consists of defeasible information. Belief is what we *think* is true, and it may or may not be true in the world. On the other hand, knowledge is what we are aware of as true, and it is always true in the world.[1]

We make decisions and act according to our beliefs, yet they are not infallible. The inferences we base our beliefs on can be deductive or uncertain, employing any number of inference mechanisms to arrive at our conclusions, for instance, statistical, nonmonotonic, or analogical. We constantly have to modify our set of beliefs as we encounter new information. A new piece of evidence may complement our current beliefs, in which case we can hold on to our original beliefs in addition to this new evidence. However, because some of our beliefs can be derived from uncertain inference mechanisms, it is inevitable that we will at some point encounter some evidence that contradicts what we currently believe. We need a systematic way of reorganizing our beliefs, to deal with the dynamics of maintaining a reasonable belief set in the face of such changes.

The state of our beliefs can be modeled by a logical theory K, a deductively closed set of formulas. If a formula ϕ is considered accepted in a belief set, it is included in the corresponding theory K; if it is rejected, its negation $\neg\phi$ is in K. In general the theory is incomplete. There are formulas that we neither accept nor reject, and thus it may be the case that neither ϕ nor $\neg\phi$ is contained in K.

By representing the set of beliefs as a logical theory, we have a well-defined mechanism (that of classical logic, for example) for reasoning about our beliefs. For example, ducklings quack. There is WaWa here, but we do not know whether it is a duckling. This can be represented by the propositional theory

$$K = \mathbf{Th}(\{\texttt{duckling\_WaWa} \rightarrow \texttt{quack\_WaWa}\}) \tag{7.1}$$

where $\mathbf{Th}(S)$ is the deductive closure of the formulas in S. There are not many interesting things we can say at this point. We may deduce that if WaWa does not quack

---

[1] This distinction is captured in modal logic by the axiom $\mathbf{T}$: $\mathbf{L}\phi \rightarrow \phi$, which reads "if ϕ is necessarily true, then it is true in *this* world". Axiom $\mathbf{T}$ is rejected by the logics of belief, but is acceptable for the logics of knowledge.

then it is not a duckling:

$$\neg\texttt{quack\_WaWa} \rightarrow \neg\texttt{duckling\_WaWa}$$

because this is one of the deductive consequences contained in K. However, we do not know whether WaWa does or does not quack, as the truth value of quack_WaWa is unspecified in K. The same is true for duckling_WaWa. Thus, according to the belief set K, we know nothing about WaWa.

This is all very well when we are talking about a static situation. A theory represents a snapshot of the state of affairs. The status of each formula is fixed—accepted, rejected, or unspecified—within a given theory. The theory is deductively closed, and no amount of computation is going to modify the truth value of a formula. To effect a change in the status of a formula would require a change in the theory itself, by replacing the original theory K with one that contains the desired formulas. These changes may be due to the addition of a newly discovered piece of information, a correction of an existing "fact", or a retraction of some previously accepted information. We may come about these changes by way of some factual evidence, or via some means of uncertain inference, in which case we are less sure of but nonetheless accept the new information.

Regardless of the source and nature of the change, we should not construct the new theory from the original theory by arbitrarily adding or removing formulas. Furthermore, since a belief set is modeled as a deductively closed theory, any change to the theory would involve an infinite number of formulas. Theory replacement is concerned with formalizing the relationship between a theory and its replacement when we need to incorporate or retract information, and with the constraints governing what constitutes a rational replacement.

7.2 Theory Change

As we mentioned before, a formula may be accepted, rejected, or unspecified with respect to a belief set. This corresponds to the formula having a truth value *true* (t), *false* (f), or neither in the belief set.[2]

Theory change can be classified into three categories: expansion, contraction, and revision, according to the kind of change in the status of a formula we may encounter in a theory.

7.2.1 Expansion

An expansion occurs when a formula that was previously unknown enters into the belief set. The status of the formula in the belief set changes from unspecified to definite—either true or false.

$$? \rightsquigarrow t$$
$$? \rightsquigarrow f.$$

---

[2]Note that in a classical logic framework, there are only two truth values: true and false. However, the status of a formula in a belief set can be categorized into three classes according to its derivability: true, false, and *neither*, the last class due to the incompleteness of the belief set. This idea is the basis of three-valued logic, which we will not go into here.

This situation may arise when we learn of some new information that was previously not available, and the theory is expanded to include the additional information. For example, we may learn that WaWa is indeed a duckling. This requires us to expand the theory K in (7.1) by the additional formula

duckling_WaWa.

We may simply append this formula to the theory K, arriving at an expanded theory $\overset{\circ}{K}$:

$$\overset{\circ}{K} = \textbf{Th}\left(\left\{\begin{array}{l} \text{duckling\_WaWa} \rightarrow \text{quack\_WaWa} \\ \text{duckling\_WaWa} \end{array}\right\}\right). \tag{7.2}$$

This in turn brings in new consequences that can now be inferred from $\overset{\circ}{K}$, such as

quack_WaWa.

With the expansion, we can now say that WaWa is a quacky duckling.

Note that the new formula duckling_WaWa cannot contradict any of the existing formulas in the original theory K, as neither duckling_WaWa nor ¬duckling_WaWa belongs to the theory before the expansion. This is what we meant when we said an expansion occurs when we learn of new, *previously unknown* information. If the addition of the new formula would result in a contradiction, then the expansion operation is not applicable; rather, the theory would undergo a *revision*, which we will discuss in Section 7.2.3.

7.2.2 Contraction

A contraction occurs when a formula of known truth value is retracted without any addition of new formulas. The status of the contracted formula in the belief set is left unspecified in the contracted theory:

$$t \rightsquigarrow ?$$
$$f \rightsquigarrow ?.$$

Consider the theory $\overset{\circ}{K}$ in (7.2) again. In $\overset{\circ}{K}$ we believe that WaWa is a duckling, and that it quacks. Suppose now we are not so sure whether WaWa quacks after all. Then we need to contract $\overset{\circ}{K}$ by the formula

quack_WaWa.

Note that removing this formula does not mean that its negation is accepted. All that is required is that after the contraction quack_WaWa be not contained in the theory.

Because a theory is a deductively closed set of formulas, in many cases removing the target formula (and its deductive consequences) alone is not enough. We need to consider the effect of inference chaining, as the target formula may still be derivable from the remaining formulas in the theory. We therefore need to make sure all the chains of inference that lead to the target formula are broken. In our case, the two formulas

$$\text{duckling\_WaWa} \rightarrow \text{quack\_WaWa}$$
$$\text{duckling\_WaWa} \tag{7.3}$$

in $\overset{\circ}{K}$ conspire to give rise to quack_WaWa. We need to remove or modify at least one of these two formulas, in order to avoid deriving the target formula quack_WaWa all over again in the contracted theory.

There are in general many ways to accomplish a contraction. For example, we could remove either one of the two formulas in (7.3), or we could even remove both. We could also relax one of the formulas. Instead of requiring WaWa to quack always if it is a duckling:

$$\text{duckling\_WaWa} \rightarrow \text{quack\_WaWa}, \tag{7.4}$$

we could specify that it only needs to quack when it is not sick (and if it is a duckling):

$$\text{duckling\_WaWa} \land \neg\text{sick\_WaWa} \rightarrow \text{quack\_WaWa}. \tag{7.5}$$

Formula (7.4) is modified to incorporate an additional condition in the antecedent, to arrive at formula (7.5). This corresponds to removing, among other formulas,

$$\text{duckling\_WaWa} \land \text{sick\_WaWa} \rightarrow \text{quack\_WaWa} \tag{7.6}$$

from the original theory $\overset{\circ}{K}$. The target formula quack_WaWa is not derivable from this modified and smaller theory.

In any case, a contraction operation determines which formulas to give up in a theory. There are many possibilities, each involving an infinite set of formulas. A large part of our discussion in this chapter concerns the rationale for picking out the appropriate replacement theories, and their desirable properties.

7.2.3 Revision

A revision occurs when a formula of specified truth value in a belief set is contradicted:

$$t \rightsquigarrow f$$
$$f \rightsquigarrow t.$$

Perhaps we have accepted in the original theory some formula ϕ that is later found to be false. To revise the theory, the formula ϕ has to be replaced by its negation $\neg\phi$. For example, given the theory $\overset{\circ}{K}$ in (7.2), we believe that WaWa quacks. If we later realize that WaWa really does not quack, we need to revise $\overset{\circ}{K}$ by the formula

$$\neg\text{quack\_WaWa}.$$

Revision is a stronger change than contraction. In Section 7.2.2, we are ambivalent about the truth value of quack_WaWa; it is neither true nor false in the contracted theory. Here, we not only remove quack_WaWa, but we also assert its negation in the revised theory.

Again it is not enough to simply reverse the sign of the offending formula in the original theory. In order to preserve consistency, the effect of inference chaining should be taken into account in a way similar to that for contraction. In our case, having all three formulas

$$\text{duckling\_WaWa} \rightarrow \text{quack\_WaWa}$$
$$\text{duckling\_WaWa}$$
$$\neg\text{quack\_WaWa}$$

in the same theory would lead to a contradiction. One way to avoid the inconsistency is to remove the quacking rule (and associated formulas). This amounts to concluding that WaWa is a nonquacking duckling, and we can have a consistent revised theory

$$K' = \mathbf{Th}\left(\left\{\begin{array}{l} \text{duckling\_WaWa} \\ \neg\text{quack\_WaWa} \end{array}\right\}\right). \tag{7.7}$$

Note that this is by no means the only way to revise $\overset{\circ}{K}$. For example, we may also decide that WaWa is not a duckling. (*When* have you last seen one that doesn't quack?) Then, the theory

$$K'' = \mathbf{Th}\left(\left\{\begin{array}{l} \text{duckling\_WaWa} \rightarrow \text{quack\_WaWa} \\ \neg\text{duckling\_WaWa} \\ \neg\text{quack\_WaWa} \end{array}\right\}\right) \tag{7.8}$$

would also work. Or we may modify some of the formulas in the theory to avoid deriving quack_WaWa. We shall have more to say about this in Section 7.3. But first let us reiterate how the replacement theories differ in the three change categories.

When we *revise* the theory $\overset{\circ}{K}$ in (7.2) by the formula ¬quack_WaWa, we obtain a theory that includes ¬quack_WaWa but not quack_WaWa. When we *contract* $\overset{\circ}{K}$ by quack_WaWa, the resulting theory contains neither quack_WaWa nor ¬quack_WaWa. If we were to *expand* the theory $\overset{\circ}{K}$ by ¬quack_WaWa, by simply appending the formula to $\overset{\circ}{K}$, we would get an inconsistent theory containing both quack_WaWa and ¬quack_WaWa (and everything else, for that matter). However, this is not a situation such as expansion is supposed to handle; for expansion assumes that the truth value of quack_WaWa in the original theory is unspecified before the operation. Revision is the more suitable way of replacing the theory in this case.

7.3   Rationality Considerations

We have chosen to represent a belief set as a *deductively closed* logical theory. There are a few basic considerations concerning such a rational theory and a rational theory change. First of all, we do *not add arbitrary formulas* into the theory. Except for the target formula by which a theory is expanded or revised, and those formulas deemed necessary for deductive closure, no other formula is added to the theory. This ensures that every formula in the replacement theory is rationally accepted, provided the original theory and the target formula are both rationally accepted.

Second, we would like to preserve *consistency*. In the framework of classical logic, an inconsistent theory includes every formula (and its negation) in the whole language, and it is hardly rational to believe contradictory information. Thus, one of the criteria of rational theory replacement is that if the original theory and the target formula are each individually consistent, the replacement theory should be consistent as well. An exception to this rule is when a consistent theory is expanded by a consistent formula but the two when taken together are inconsistent. However, as we have noted earlier, this is not a proper use of expansion, and the intuitive treatment leads to an inconsistent replacement in this case.

Third, apart from adding or removing the target formula, we may need to modify parts of the original theory as well. For example, we may revise a theory with a formula which, although consistent by itself, contradicts what we have in the theory. As the theory is deductively closed, in addition to removing the offending formula and all associated consequences from the theory, any chain of inference that would lead to inconsistency also needs to be broken. We may achieve this in any number of ways, including such drastic measures as removing all the formulas from the theory. This is obviously not a very good solution. We would like to lose as little of the original information as possible. Thus, the changes we induce to restore consistency (and to maintain deductive closure) should be *minimal* in some sense. Roughly speaking, we would like the replacement theory to resemble the original theory as much as possible, and retract only those formulas that we must retract.

How do we measure "minimal change", or "closeness" to the original theory? Consider the two revised theories (7.7) and (7.8). In essence, in (7.7) we gave up the formula

$$\text{duckling\_WaWa} \rightarrow \text{quack\_WaWa},$$

while in (7.8) we gave up the formula

$$\text{duckling\_WaWa}.$$

Which theory is closer to the original theory $\overset{\circ}{K}$ in (7.2)? It seems that the answer is extralogical: if we suspect that WaWa is *the* ugly duckling we heard about, we might opt to keep the quacking rule and retract the assertion that WaWa is a duckling, arriving at the revised theory (7.8). On the other hand, if we strongly believe WaWa is a duckling, then we may retract the quacking rule and go with the revised theory (7.7). In addition, there is a plethora of other possible ways to revise the theory, involving adding and subtracting various sets of formulas.

Similar alternatives are available for contraction in the scenario in Section 7.2.2. We can choose to eliminate either the quacking rule or the assertion that WaWa is a duckling. Both alternatives seem to conform to our intuitive notion of a rational contraction, but as we shall see in Section 7.4.2, removing the quacking rule creates fewer complications. (Besides, we really do not believe WaWa is an ugly duckling.) Again, we may also eliminate or modify some other formulas in the theory, as in the case of (7.5), to achieve a contraction.

Logical cues alone are not enough to provide a basis for choosing between the various alternatives. Each replacement theory can be justified under different circumstances. However, even though we may not be able to identify a unique theory, we can formulate logical constraints that should be satisfied by all rational replacements. The AGM postulates we shall discuss in the next section were motivated by such considerations.

One should be cautioned against indiscriminately using the extent of modification (in the set-theoretic sense)[3] to determine how good a replacement theory is. In some cases more changes may be preferable. Consider, for example, the theory K' further expanded from $\overset{\circ}{K}$ in (7.2) by the introduction of another duckling, Cookie:

$$K' = \mathbf{Th} \left(\left\{ \begin{array}{l} \text{duckling\_WaWa} \rightarrow \text{quack\_WaWa} \\ \text{duckling\_Cookie} \rightarrow \text{quack\_Cookie} \\ \text{duckling\_WaWa} \\ \text{duckling\_Cookie} \end{array} \right\} \right).$$

Both WaWa and Cookie are ducklings, and both quack. Suppose we are subsequently told that WaWa and Cookie are not both ducklings, but we do not know exactly which

---

[3] So far belief sets are modeled as deductively closed theories, and thus any modification involves an infinite

one is a duckling and which one is not. We then need to contract the theory by the conjunction

$$\text{duckling\_WaWa} \wedge \text{duckling\_Cookie}.$$

To accomplish this contraction with the removal of a minimal set of formulas, we can either retract duckling_WaWa or retract duckling_Cookie. It is not necessary to remove both formulas to remove their conjunction. However, in the absence of any grounds for choosing one formula over the other, it may be argued that retracting *both* duckling assertions is a more rational and unbiased treatment than arbitrarily choosing between the two set-minimal alternatives.

We will revisit these topics later in the chapter. But first let us introduce the AGM framework, in which we can formalize some of the ideas of theory replacement we have discussed.

7.4   The AGM Postulates

Alchourrón, Gärdenfors, and Makinson [Alchourrón et al., 1985] proposed a set of *rationality postulates* (the AGM postulates) that formalize the considerations in Section 7.3, in particular the notion of *minimal change* between a theory and its replacement. Formally, we have a propositional language $\mathcal{L}$. A theory K is a subset of the language: $K \subseteq \mathcal{L}$. A theory change operator is a function from a theory and a target formula to a (replacement) theory: $2^{\mathcal{L}} \times \mathcal{L} \rightarrow 2^{\mathcal{L}}$. Although each theory change mapping is functional (one to one), the AGM postulates in general are not specific enough to uniquely determine a single mapping (except in the case of expansion). Instead the postulates pick out a set of candidate mappings that are *not irrational*. The AGM postulates can be regarded as the minimal constraints that should be satisfied by any rational theory change.

Below we will give the rationality postulates for each of the three types of theory change operations.

7.4.1   Expansion

Consider a theory K and a formula ϕ that is to be incorporated. Let K_{ϕ}^{+} denote the replacement theory resulting from an expansion of K by ϕ. The expansion operator should satisfy the following postulates.

($^{+}1$) **K_{ϕ}^{+} is a theory.**
The replacement is a theory itself.

($^{+}2$) **$\phi \in K_{\phi}^{+}$.**
Expansion always succeeds: the target formula is always included in the expanded theory.

($^{+}3$) **$K \subseteq K_{\phi}^{+}$.**
The expanded theory includes the original theory.

($^{+}4$) **If $\phi \in K$, then $K_{\phi}^{+} = K$.**
Expanding with a formula that is already in the original theory does not change the theory.

($^{+}5$) **If $K \subseteq H$, then $K_{\phi}^{+} \subseteq H_{\phi}^{+}$.**
The set inclusion relation between a theory and its subset remains the same after

a subset of another theory H, and we expand both by the same formula ϕ, the expanded theory K_ϕ^+ obtained from K will still be a subset of the expanded theory H_ϕ^+ obtained from H.

($^+$6) **K_ϕ^+ is the smallest theory satisfying ($^+$1)–($^+$5).**

The expanded theory is the smallest possible—it does not include any formulas not required to satisfy the above five postulates. Note that the set of theories satisfying ($^+$1)–($^+$5) is closed under intersection. If two distinct theories both satisfy the five postulates, we can generate a smaller theory that also satisfies the same postulates by taking their intersection.

As we have noted, one way to expand a theory K is to simply add the target formula ϕ to K and form the deductive closure, that is, we can replace K with $K' = \mathbf{Th}(K \cup \{\phi\})$. This is a fairly straightforward operation: recall that in theory expansion the target formula ϕ is not supposed to contradict any of the formulas in K, and so K' should be consistent as long as K is consistent (and ϕ does not contradict K). Otherwise—if ϕ contradicts K—then K' is inconsistent. It turns out that the expansion postulates ($^+$1)–($^+$6) have constrained the expansion mapping to exactly this one operation.

Theorem 7.1. $K_\phi^+ = \mathbf{Th}(K \cup \{\phi\})$.

Thus, there is only one way to expand a theory by a target formula according to the AGM postulates: add the target formula and the associated deductive consequences to the theory. This operation conforms to our intuitive notion of consistency and minimal change. In the scenario discussed in Section 7.2.1, the expansion is carried out as specified in Theorem 7.1:

$$K_{\texttt{duckling\_WaWa}}^+ = \mathbf{Th}(K \cup \{\texttt{duckling\_WaWa}\}) = \mathring{K}$$

where K and $\mathring{K}$ are the theories (7.1) and (7.2). It is easy to check that the theory $\mathring{K}$ satisfies the postulates ($^+$1)–($^+$6) for being $K_{\texttt{duckling\_WaWa}}^+$.

Note again that syntactically nothing in the postulates prevents us from applying an expansion operator to a theory that is not consistent with the formula to be added, although that is not the *intended* usage. In such cases, a revision operation is more appropriate than an expansion.

7.4.2 Contraction

The expansion postulates and especially Theorem 7.1 seem fairly intuitive. Now let us turn to the more controversial contraction postulates. Let K_ϕ^- denote the replacement theory resulting from a contraction of a theory K by a formula ϕ. The contraction operator should satisfy the following postulates.

($^-$1) **K_ϕ^- is a theory.**

The replacement is a theory itself.

($^-$2) **$K_\phi^- \subseteq K$.**

The contracted theory is a subset of the original theory; that is, the contracted theory does not contain any new formulas.

($^-$3) **If $\phi \notin K$, then $K_\phi^- = K$.**

If the formula to be contracted is not contained in the original theory, then

($^-$4) **If $\not\vdash \phi$, then $\phi \notin K_\phi^-$.**
Unless the target formula is a tautology, contraction always succeeds, that is, the target formula is absent from the contracted theory.

($^-$5) **If $\phi \in K$, then $K \subseteq (K_\phi^-)_\phi^+$.**   **[recovery postulate]**
If the formula to be contracted is in the original theory, then a contraction followed by an expansion by this same formula gives rise to a replacement theory in which all the formulas in the original theory are restored. In other words, all the formulas in the original theory that have been removed because of the contraction can be recovered by adding the target formula and the associated deductive consequences back into the contracted theory.

($^-$6) **If $\vdash \phi \leftrightarrow \psi$, then $K_\phi^- = K_\psi^-$.**
Logically equivalent formulas give rise to identical contractions.

($^-$7) **$K_\phi^- \cap K_\psi^- \subseteq K_{\phi \wedge \psi}^-$.**
The formulas that are retained in both the theory contracted by ϕ and the one contracted by ψ are also retained in the theory contracted by their conjunction $\phi \wedge \psi$. Note that contracting a theory by a conjunction $\phi \wedge \psi$ is not the same as contracting by ϕ *and* contracting by ψ. What we are removing in the first case is the joint truth of the two formulas, and we can accomplish this by removing either one but not necessarily both of the conjuncts. Postulate ($^-$7) requires that all the formulas that are not involved in the contraction of either one of the conjuncts ϕ and ψ be retained in the contraction of their conjunction $\phi \wedge \psi$.

($^-$8) **If $\phi \notin K_{\phi \wedge \psi}^-$, then $K_{\phi \wedge \psi}^- \subseteq K_\phi^-$.**
If a contraction by a conjunction is successful in removing one of the conjuncts, then every formula that is removed by a contraction with that conjunct alone is also removed by the contraction with the conjunction. One might think that because removing ϕ would falsify the conjunction $\phi \wedge \psi$ as well, the two theories $K_{\phi \wedge \psi}^-$ and K_ϕ^- would be identical. This is sometimes the case, but we also need to allow for the situation where *both* conjuncts ϕ and ψ could have been removed in $K_{\phi \wedge \psi}^-$. The latter is an arguably more rational contraction when we have no reason to prefer removing one conjunct over another.

Reconsider the scenario in Section 7.2.2. The theory $\overset{\circ}{K}$ in (7.2) is to be contracted by the formula quack_WaWa. As we have noted, we may try to retract either one of the two formulas in (7.3). Now consider one way to retract duckling_WaWa $\rightarrow$ quack_WaWa,

$$K' = \mathbf{Th}(\{\text{duckling\_WaWa}\}).$$

Let us check if K' satisfies the contraction postulates for being $\overset{\circ}{K}{}_{\text{quack\_WaWa}}^-$.
Postulates ($^-$1) to ($^-$4) and ($^-$6) are trivially satisfied. For ($^-$5), we have

$$(K')_{\text{quack\_WaWa}}^+ = \mathbf{Th}(K' \cup \{\text{quack\_WaWa}\}) = \overset{\circ}{K},$$

and therefore the recovery postulate is also satisfied. ($^-$7) and ($^-$8) deal with composite changes. Thus, there can be a contraction function satisfying all the AGM contraction postulates, such that

$$\overset{\circ}{K}{}_{\text{quack\_WaWa}}^- = K'.$$

Postulate ($^-$5) is known as the *recovery postulate*, and it turns out to be very controversial [Hansson, 1991; Fuhrmann, 1991; Levi, 1991; Niederee, 1991; Tennant, 1994]. The postulate was motivated by the minimal change principle: we should remove from a theory as little as possible when we perform a contraction. The idea behind the recovery postulate is that we should be able to recover from a contraction all the information removed from the original theory, by simply reintroducing the target formula into the contracted theory through an expansion.

The recovery postulate does not always hold for some otherwise intuitively satisfactory ways of contracting a theory. For example, consider again the contraction of the theory $\overset{\circ}{K}$ by quack_WaWa. Instead of retracting from $\overset{\circ}{K}$ the rule duckling_WaWa $\rightarrow$ quack_WaWa, we can also remove the formula duckling_WaWa. One possible candidate for such a contracted theory $\overset{\circ}{K}^-_{\text{quack\_WaWa}}$ is

$$K'' = \mathbf{Th}(\{\text{duckling\_WaWa} \rightarrow \text{quack\_WaWa}\}).$$

This seems like an equally reasonable treatment of the contraction. However, K'' cannot be constructed from a function that satisfies the recovery postulate. The reexpanded theory

$$(K'')^+_{\text{quack\_WaWa}} = \mathbf{Th}(K'' \cup \{\text{quack\_WaWa}\})$$

does not give us all the formulas in the original theory $\overset{\circ}{K}$. In particular, it does not contain the formula duckling_WaWa that was removed from $\overset{\circ}{K}$. Thus, K'' is not acceptable as $\overset{\circ}{K}^-_{\text{quack\_WaWa}}$ according to the AGM contraction postulates.

There is, however, a way to accomplish a contraction along the same lines while satisfying the AGM constraints. Instead of K'', consider a theory H, where $K'' \subset H \subset \overset{\circ}{K}$. This theory H does not contain either duckling_WaWa or quack_WaWa, just like K'', but it does contain the formula

$$\psi = \text{duckling\_WaWa} \vee \neg\text{quack\_WaWa},$$

which is included in $\overset{\circ}{K}$ (as a deductive consequence of duckling_WaWa) but not in K''.

By retaining ψ in H, the recovery postulate is satisfied in addition to the other AGM contraction postulates. When we add the target formula quack_WaWa back into H, we get duckling_WaWa back as well, as a deductive consequence of ψ and quack_WaWa. Thus, the whole original theory $\overset{\circ}{K}$ can be recovered from H in accordance to the specification of ($^-$5).

Whereas the theory H does conform to the contraction postulates, it may seem odd that the formula ψ is retained in H even though the reason for its existence, namely, duckling_WaWa, has been retracted. However, it may also be argued that since ψ does not lead to quack_WaWa in H, it should be retained after the contraction to minimize information loss.[4] This is one of the topics of debate between the *foundations* and *coherence* schools of thought, which will be discussed again in Section 7.8.

7.4.3 Revision

Let K^*_ϕ denote the replacement theory resulting from a revision of a theory K by a formula ϕ. The revision operator should satisfy the following postulates.

---

[4]Recall that a belief state is represented only as a deductively closed theory, and there is no provision for recording the justification or dependency information of the constituent formulas. All formulas in a theory are therefore accorded equal status.

(*1) K_ϕ^* **is a theory.**
The replacement is a theory itself.

(*2) $\phi \in K_\phi^*$**.**
Revision always succeeds: the target formula is always contained in the revised theory.

(*3) $K_\phi^* \subseteq K_\phi^+$**.**
A revision never incorporates formulas that are not in the theory expanded by the same target formula; that is, a revised theory does not contain any formula that is not a deductive consequence of the target formula together with the original theory.

(*4) **If** $\neg\phi \notin K$**, then** $K_\phi^+ \subseteq K_\phi^*$**.**
Together with (*3), this implies that if the target formula ϕ is consistent with the original theory K, then a revision is identical with an expansion, that is, $K_\phi^* = K_\phi^+$.

(*5) $K_\phi^* = \perp$ **if and only if** $\vdash \neg\phi$**.**
If we attempt to revise a theory with a contradiction, we end up with an inconsistent theory ($\perp$). This is the only case where a revision results in a contradiction.

(*6) **If** $\vdash \phi \leftrightarrow \psi$**, then** $K_\phi^* = K_\psi^*$**.**
Logically equivalent formulas give rise to identical revisions.

(*7) $K_{\phi \wedge \psi}^* \subseteq (K_\phi^*)_\psi^+$**.**
Every formula in the theory revised by a conjunction is contained in the theory constructed by the following two-step process: first revise the original theory by one conjunct, and then expand the revised theory by the other. From (*3) we have $K_{\phi \wedge \psi}^* \subseteq K_{\phi \wedge \psi}^+ = (K_\phi^+)_\psi^+$. Because $K_\phi^* \subseteq K_\phi^+$ (postulate (*3) again), postulate (*7) provides a tighter upper bound on $K_{\phi \wedge \psi}^*$ than (*3).

(*8) **If** $\neg\psi \notin K_\phi^*$**, then** $(K_\phi^*)_\psi^+ \subseteq K_{\phi \wedge \psi}^*$**.**
Together with (*7), this entails that the two-step process described in (*7) is identical to the revision by the conjunction as a whole, that is, $(K_\phi^*)_\psi^+ = K_{\phi \wedge \psi}^*$, when the conjunct used in the second step (the expansion) is consistent with the initial revision in the first step. Otherwise we have $\neg\psi \in K_\phi^*$, and $(K_\phi^*)_\psi^+ = \perp$, which is identical to $K_{\phi \wedge \psi}^*$ if and only if $\phi \wedge \psi$ is a contradiction.

There is a close correspondence between the revision and contraction postulates. A noted exception is the recovery postulate for contraction, which has no counterpart in revision.

Let us see if the example in Section 7.2.3 satisfies the AGM revision postulates. The theory $\overset{\circ}{K}$ in (7.2) is to be revised by $\neg$quack_WaWa. We considered two candidates for a revised theory:

$$K' = \mathbf{Th}\left(\left\{ \begin{array}{l} \text{duckling\_WaWa} \\ \neg\text{quack\_WaWa} \end{array} \right\}\right)$$

$$K'' = \mathbf{Th}\left(\left\{ \begin{array}{l} \text{duckling\_WaWa} \rightarrow \text{quack\_WaWa} \\ \neg\text{duckling\_WaWa} \\ \neg\text{quack\_WaWa} \end{array} \right\}\right)$$

in (7.7) and (7.8) respectively. It is easy to check that both theories satisfy the revision postulates for $\overset{\circ}{K}{}^*_{\neg\text{quack\_WaWa}}$, and thus both are possible rational replacements.

Note that the AGM expansion postulates uniquely identify a mapping function from a theory and a target formula to an expanded theory, whereas this is not the case for contraction and revision. There are in general multiple mappings that satisfy the contraction and revision postulates, as there is usually more than one way to accomplish these changes. We have already seen some examples of such multiple alternatives for contraction and revision in this and the previous sections.

The AGM postulates are the *logical* constraints placed on a rational theory change. They are best thought of as a characterization of the minimal requirements for any theory change function to be considered rational. The postulates do not dictate which function we should adopt among the various alternatives that satisfy these postulates. To further differentiate between these mappings, we need to appeal to information that is not available from the logical structure alone. This *nonlogical* information will let us identify the appropriate revision or contraction function to use in a particular situation. We will discuss one such nonlogical measure, called *epistemic entrenchment*, in Section 7.7. But let us first look at the connections between the three theory change operators.

7.5   Connections

The AGM contraction and revision postulates in the previous sections have been given independently of each other. However, we can also define the two functions in terms of one another, via the *Levi identity* and the *Harper identity*.

Theorem 7.2 ([Gärdenfors, 1988]). *Given a contraction function $^-$ that satisfies $(^-1)$– $(^-4)$ and $(^-6)$, and an expansion function $^+$ that satisfies $(^+1)$–$(^+6)$, the revision function * as defined by the* Levi identity

$$K^*_\phi = (K^-_{\neg\phi})^+_\phi \qquad \text{[Levi Identity]}$$

satisfies $(^*1)$–$(^*6)$. Furthermore, if $(^-7)$ is satisfied, then $(^*7)$ is satisfied, and if $(^-8)$ is satisfied, then $(^*8)$ is satisfied.

The Levi identity [Levi, 1977] specifies that a revision can be achieved by first contracting by the negation of the target formula and then expanding by that formula. In other words, by the contraction we first remove from the theory any potential inconsistency with the target formula, and then the formula is added to the theory using an expansion. This procedure ensures that the resulting revised theory contains the target formula and is consistent (unless the target formula is itself a contradiction).

The Levi identity is only one of many ways of defining a revision function in terms of contraction and expansion, but it provides a handy and intuitive mapping. Note that the controversial recovery postulate $(^-5)$ was not used in this result.

We can also define a contraction function in terms of a revision function, as follows.

Theorem 7.3 ([Gärdenfors, 1988]). *Given a revision function * that satisfies $(^*1)$–$(^*6)$, the contraction function $^-$ as defined by the* Harper identity

$$K^-_\phi = K \cap K^*_{\neg\phi} \qquad \text{[Harper Identity]}$$

satisfies $(^-1)$–$(^-6)$. Furthermore, if $(^*7)$ is satisfied, then $(^-7)$ is satisfied, and if $(^*8)$ is satisfied, then $(^-8)$ is satisfied.

The Harper identity [Harper, 1977] specifies how a contraction function can be constructed from a revision function. The theory $K^*_{\neg\phi}$ is obtained by revising the original theory K with the negation of the target formula $\neg\phi$. Because revision preserves consistency in most cases, $K^*_{\neg\phi}$ is a theory that is minimally changed from K to include $\neg\phi$ but exclude ϕ (unless ϕ is a tautology). $K^*_{\neg\phi}$ approximates the contracted theory K^-_ϕ, in the sense that K^-_ϕ is also a theory that is minimally changed from K to exclude the target formula ϕ, except that it does not necessarily also include $\neg\phi$. ($\neg\phi$ should be contained in K^-_ϕ only if it is contained in the original theory K, since a contraction should not invent any new formulas.)

Thus, $K^*_{\neg\phi}$ may contain additional formulas that should not be found in K^-_ϕ. We can obtain the contracted theory K^-_ϕ from the revised theory $K^*_{\neg\phi}$ by removing those formulas that are not originally found in K.

Note that the recovery postulate ($^-5$) holds for the contraction function defined by the Harper identity. We need to show that

$$\text{if} \quad \phi \in K, \quad \text{then} \quad K \subseteq (K \cap K^*_{\neg\phi})^+_\phi. \tag{7.9}$$

Note that

$$(K \cap K^*_{\neg\phi})^+_\phi = K^+_\phi \cap (K^*_{\neg\phi})^+_\phi = K^+_\phi \cap \bot = K^+_\phi$$

because $(K^*_{\neg\phi})^+_\phi$ is inconsistent ($\neg\phi$ is contained in $K^*_{\neg\phi}$, and adding ϕ to it makes it inconsistent). Thus, (7.9) amounts to saying that if ϕ is contained in K, then K is a subset of K^+_ϕ, which is always the case (even if ϕ is not contained in K).

Again, the Harper identity is only one way of defining a contraction function in terms of a revision function. Although there are other theory change operators satisfying the AGM postulates, nevertheless these two theorems have established a strong connection between the operators. These operators can in some sense be defined interchangeably.

7.6   Selecting a Contraction Function

The AGM rationality postulates pose logical constraints that any rational theory replacement operator should satisfy. Now let us briefly discuss another line of approach for characterizing the process of theory change. We shall see that the two approaches are closely related.

We will examine only the contraction function, as the revision function can be determined from the contraction function via the Levi identity. Let K be a deductively closed theory, and ϕ be a formula. We would like to determine a replacement theory K^-_ϕ obtained from contracting K by the formula ϕ. Based on the rationality considerations, in particular the minimal change criterion, it is reasonable to focus on a special set of candidate theories for K^-_ϕ, namely the set of theories that are maximal subsets of K not containing ϕ. Let us denote this set by $K_{\perp\phi}$. This is the set of theories obtained from K by removing as little as possible, while making sure that ϕ is absent. Note that $K_{\perp\phi}$ is empty if ϕ is a tautology, since every (deductively closed) theory contains the tautologies.

We can formulate a number of ways to construct a contraction function from $K_{\perp\phi}$. One idea is to select a theory from this set using a selection function δ which returns a single element of the set.

$$K_\phi^- = \begin{cases} \delta(K_{\perp\phi}) & \text{if } \not\vdash \phi \\ K & \text{otherwise.} \end{cases} \qquad \textbf{[Maxichoice]}$$

The *maxichoice* contraction function specifies that the contracted theory K_ϕ^- can be identified with the theory selected from $K_{\perp\phi}$ by the function δ. If the formula to be contracted is a tautology, then the contracted theory is just the original theory K.

Another idea for constructing a contracted theory is to take the intersection of all the theories in the set $K_{\perp\phi}$:

$$K_\phi^- = \begin{cases} \cap K_{\perp\phi} & \text{if } \not\vdash \phi \\ K & \text{otherwise.} \end{cases} \qquad \textbf{[Full Meet]}$$

The *full meet* contraction function specifies that a formula ψ is in the contracted theory K_ϕ^- if it is present in every theory that is a maximal subset of K not containing ϕ.

A third formulation is to take the intersection of a subset of theories picked out from $K_{\perp\phi}$ by a selection function Δ which returns a nonempty subset of $K_{\perp\phi}$:

$$K_\phi^- = \begin{cases} \cap \Delta(K_{\perp\phi}) & \text{if } \not\vdash \phi \\ K & \text{otherwise.} \end{cases} \qquad \textbf{[Partial Meet]}$$

The *partial meet* contraction function specifies that a formula ψ is in the contracted theory K_ϕ^- if it is present in every theory selected from $K_{\perp\phi}$ by the function Δ.

It is clear that both maxichoice and full meet contractions are special cases of partial meet contraction. In all three formulations, K_ϕ^- is just K itself if we attempt to contract the theory by a tautology. The selection function Δ can be tailored to select a single element in the case of maxichoice contraction, or the whole set $K_{\perp\phi}$ in the case of full meet contraction.

In the very simple case where the formula ϕ to be contracted is not present in the original theory K, there is only one maximal subset of K not containing ϕ, namely, the theory K itself. Thus, the set $K_{\perp\phi}$ contains only one element: $K_{\perp\phi} = \{K\}$. The three alternatives presented above for selecting a contraction function coincide in this case. Maxichoice contraction would return K as the contracted theory, as this is the only choice the selection function δ can make in $K_{\perp\phi}$. Likewise, the intersection of all (in the case of full meet contraction) or some (in the case of partial meet contraction) elements in the singleton set $K_{\perp\phi}$ would also give rise to K. Thus, $K_\phi^- = K$ in all three cases.

Now let us look at a slightly less simple example. Consider a theory

$$K = \textbf{Th}(\{\text{quack\_WaWa}\}).$$

Because K is deductively closed, it contains an infinite number of formulas, including

$$\begin{aligned} &\text{quack\_WaWa} \lor \text{quack\_WaWa} \\ &\text{quack\_WaWa} \lor \neg\text{quack\_WaWa} \end{aligned} \qquad (7.10)$$

$$\begin{aligned} &\text{quack\_WaWa} \lor \text{sick\_WaWa} \\ &\text{quack\_WaWa} \lor \neg\text{sick\_WaWa} \end{aligned} \qquad (7.11)$$

$$\vdots$$

K contains both (quack_WaWa $\vee$ ϕ) and (quack_WaWa $\vee$ $\neg\phi$) for any formula ϕ, since we can count on the disjunct quack_WaWa to hold. Now, let us contract K by quack_WaWa. We must remove quack_WaWa from every set in $K_{\perp\text{quack\_WaWa}}$, but in order for these sets to be maximal, we should remove as few additional formulas as possible (even if some of these formulas were derived from quack_WaWa). For any pair of formulas

$$\begin{aligned} &\text{quack\_WaWa} \vee \phi \\ &\text{quack\_WaWa} \vee \neg\phi, \end{aligned} \tag{7.12}$$

we only need to remove one of the two. Sometimes it is clear which one it should be, as in the case when $\phi \equiv$ quack_WaWa (7.10). Sometimes we have a choice, as in the case when $\phi \equiv$ sick_WaWa (7.11). We can obtain different maximal subsets depending on whether we give up (quack_WaWa $\vee$ sick_WaWa) or (quack_WaWa $\vee$ $\neg$sick_WaWa).

Thus, $K_{\perp\text{quack\_WaWa}}$ will contain many maximal subsets, each corresponding to giving up a different set of formulas, for example of the sort (7.12). A maxichoice contraction corresponds to committing to a particular series of such choices, by selecting via the function δ a single element out of $K_{\perp\text{quack\_WaWa}}$. A full meet contraction corresponds to *not* committing to any of the particular sets, but admitting into the contracted theory only those formulas that are common in all the maximal subsets. It is clear that the theory constructed from full meet contraction is a subset of any constructed from maxichoice. Intuitively, maxichoice is not restrictive enough, whereas full meet is too restrictive in terms of the formulas retained in the theory. By varying the selection function Δ in partial meet contraction, we attempt to strike a balance between the two extremes.

Without going into the details here, let us just note that all three kinds of contraction functions satisfy the AGM contraction postulates ($^-1$)–($^-6$). The reverse is also true with some additional constraints characterizing special subclasses of contraction functions.

For the above results, the mechanism by which the selection functions δ (which returns one theory) and Δ (which returns a set of theories) operate has been left open. We can incorporate postulates ($^-7$) and ($^-8$) in addition to ($^-1$)–($^-6$) by imposing some restrictions on the selection functions. These special selection functions are known as *transitively relational* selection functions.

Informally, a transitively relational selection function picks out those theories from $K_{\perp\phi}$ that are deemed most desirable, and a formula is in the contracted theory K_{ϕ}^- if it is present in all such most preferred theories. We can establish a correspondence between contraction functions that satisfy the AGM postulates ($^-1$)–($^-8$) and those that are partial meet and transitively relational.

The relative desirability can also be expressed by an ordering of the formulas within a theory, independent of the formula to be contracted. This is the basis for the notion of epistemic entrenchment, which we will discuss in the next section.

7.7 Epistemic Entrenchment

We have seen that a theory can be changed in many possible ways, all of which satisfy the AGM rationality postulates for contraction and revision. Logical constraints are the minimal requirements for a rational theory change, but they alone are not strong enough to uniquely identify a desirable change operation. Because belief states are modeled as sets of formulas in classical logic, relative ordering is not expressible in the language. For this we can turn to an extralogical structure called *epistemic entrenchment* [Gärdenfors & Makinson, 1988; Grove, 1988; Spohn, 1988].

Some information is considered more desirable or valuable than other information. Faced with several possible ways of changing a theory, we are more inclined to give up or change the less important or less reliable information, and retain the more essential information. Let us assume that each formula has associated with it a degree of epistemic entrenchment. The more deeply entrenched formulas are more resistant to change than the less entrenched formulas. The formulas then form an ordering based on their degrees of entrenchment, and this ranking provides a basis for determining the "right" contraction and revision function to use. The less entrenched formulas are given up in a contraction before the more entrenched formulas if possible.

More formally, an epistemic entrenchment relation can be denoted by $\leq_e$. For two formulas ϕ and ψ, let

$$\phi \leq_e \psi$$

stand for "ψ is at least as epistemically entrenched as ϕ". Let $\phi <_e \psi$ stand for $\phi \leq_e \psi$ and $\psi \not\leq_e \phi$, that is, ψ is strictly more entrenched than ϕ. Let $\phi =_e \psi$ stand for $\phi \leq_e \psi$ and $\psi \leq_e \phi$, that is, ϕ and ψ are equally entrenched. There are a number of postulates that the relation $\leq_e$ should satisfy.

(EE1) If $\phi \leq_e \psi$ and $\psi \leq_e \chi$, then $\phi \leq_e \chi$.
The epistemic entrenchment relation is transitive.

(EE2) If $\phi \vdash \psi$, then $\phi \leq_e \psi$.
A formula is at most as entrenched as the formulas it logically entails. When we retract the entailed formula ψ, we need to retract ϕ as well to avoid rederiving ψ in the new theory. Thus, ϕ should not be more entrenched than ψ, so that when we give up ψ, we can give up ϕ as well. On the other hand, ϕ may be given up alone without retracting ψ at the same time.

(EE3) For all ϕ and ψ, either $\phi \leq_e \phi \wedge \psi$ or $\psi \leq_e \phi \wedge \psi$.
A conjunction is at least as entrenched as one of its conjuncts. From (EE2), we have the opposite relations $\phi \wedge \psi \leq_e \phi$ and $\phi \wedge \psi \leq_e \psi$. Together with (EE3) here, we have either $\phi \wedge \psi =_e \phi$ or $\phi \wedge \psi =_e \psi$. In other words, a conjunction is as entrenched as its least entrenched conjunct. By retracting the least entrenched conjunct, we retract the conjunction as well.

(EE4) When $K \neq \bot$, then $\phi \notin K$ if and only if $\phi \leq_e \psi$ for all ψ.
Formulas that are not in the theory are the least entrenched, and (if the theory is consistent) vice versa.

(EE5) If $\psi \leq_e \phi$ for all ψ, then $\vdash \phi$.
The most entrenched formulas are the tautologies.

The epistemic entrenchment ordering of formulas provides a constructive way to choose a unique contraction operator from the set of all possible operators satisfying the AGM contraction rationality postulates. A revision function can then be obtained from the contraction function via the Levi identity. We can also reverse the process: an epistemic entrenchment ordering can be uniquely determined from a given contraction function. There are two conditions we will make use of in the construction.

(C⁻) $\psi \in K_\phi^-$ **if and only if** $\psi \in K$ **and either** $\phi <_e \phi \vee \psi$ **or** $\vdash \phi$.

(C⁻) specifies what formulas are retained in a contraction given an epistemic entrenchment relation. Only formulas that are in the original theory K can be included in the contracted theory. In addition, if the target formula is a tautology, then all formulas are retained. Otherwise, if the target formula ϕ is less entrenched than the disjunction $\phi \vee \psi$, then ψ is retained.[5]

(C≤ᵉ) $\phi \leq_e \psi$ **if and only if** $\phi \notin K_{\phi \wedge \psi}^-$ **or** $\vdash \phi \wedge \psi$.

An epistemic entrenchment relation can be constructed from a contraction function by (C≤ᵉ). If one of the conjuncts ϕ is not retained in a theory contracted by the conjunction $\phi \wedge \psi$, then ϕ cannot be more entrenched than the other conjunct ψ. Note that both conjuncts can be absent from the contracted theory, in which case the two conjuncts are equally entrenched, that is, $\phi =_e \psi$. Moreover, if the conjunction is a tautology, then both conjuncts are also equally entrenched.

These two conditions give rise to the following correspondence results.

Theorem 7.4 ([Gärdenfors & Makinson, 1988]). *Given an epistemic entrenchment ordering $\leq_e$ that satisfies (EE1)–(EE5), condition (C⁻) uniquely determines a contraction function that satisfies the AGM contraction postulates (⁻1)–(⁻8) and condition (C≤ᵉ).*

Theorem 7.5 ([Gärdenfors & Makinson, 1988]). *Given a contraction function that satisfies the AGM contraction postulates (⁻1)–(⁻8), condition (C≤ᵉ) uniquely determines an epistemic entrenchment ordering $\leq_e$ that satisfies (EE1)–(EE5) and condition (C⁻).*

Thus, whereas the rationality postulates put up a set of candidate theory change functions, an epistemic entrenchment relation allows us to pick out a specific function among this class. However, note that the epistemic entrenchment ordering is relative to the theory which it describes, and two different theories may have different entrenchment orderings even if their repertoires of formulas are identical. The ordering does not carry over from a theory to its replacement, and thus we need to reestablish the relation after each change. We will examine this again in Section 7.8.4.

7.8 Must It Be?

The presentation of this chapter has followed closely the AGM framework of theory change [Alchourrón et al., 1985]. Let us briefly examine some of the basic assumptions that underly this framework, and a number of viable alternatives.

7.8.1 Belief Bases

In the AGM framework, a belief state is modeled as a deductively closed theory. There is no distinction between "basic facts" and "derived facts", and there is no bookkeeping of the justifications involved in deriving a particular formula. Thus, when we retract a formula, we do not necessarily retract all the formulas derived from

---

[5]The use of $\phi \leq_e \phi \vee \psi$ instead of the more intuitive $\phi \leq_e \psi$ here is due to technicalities involving the recovery postulate. Rott showed that with $\phi \leq_e \psi$, results similar to those presented here can be obtained by leaving out the recovery postulate [Rott, 1991].

it at the same time. This is known as the *coherence* approach to theory change, where as much information is preserved as possible, regardless of how the information was originally derived, and whether that derivation still holds in the replacement theory.

The opposing view is the *foundations* approach, where a belief base consisting of only a finite set of non-deductively-closed formulas is kept. Theory changes are always made by changes to the formulas in the belief base. In this approach, we are always dealing with a finite set of formulas, as opposed to the coherence approach, where any change involves an infinite number of formulas. Base revisions and base contractions have properties that are different from the AGM style revisions and contractions. Different bases might represent the same deductively closed theory, yet react differently to changes depending on the particular base formulas involved. For example, consider the following two bases:

$$B_1 = \{\phi, \phi \rightarrow \psi\}$$
$$B_2 = \{\phi, \psi\}.$$

Both bases have the same deductive closure, which contains, among other formulas, ϕ, ψ, and $\phi \rightarrow \psi$. Let us now remove from the bases the formula ϕ. We then have the following contracted bases:

$$B_1' = \{\phi \rightarrow \psi\}$$
$$B_2' = \{\psi\}.$$

The formula ψ remains in the deductive closure of the contracted base B_2', but it is removed alongside ϕ from the deductive closure of the contracted base B_1'.

It has been argued [Gärdenfors, 1990] that the coherence approach is psychologically more realistic and computationally less expensive, because we do not need to keep track of and hunt down all the implicit changes caused by a change in a base formula. It is also more economical, as it conserves more of the original information. A critique of these arguments can be found in [Doyle, 1992]. For example, it is infeasible for practical systems to maintain a deductively closed and thus infinite set of sentences. Furthermore, a belief revision system does not necessarily have to be descriptive of human behavior. It would be desirable to have a normative system, which we can utilize to help us reason "rationally", rather than one which mimics our own fallible reasoning patterns.

Work involving belief bases includes [Makinson, 1985; Fuhrmann, 1991; Hansson, 1992]. Truth maintenance systems [Doyle, 1979] provide structures to keep track of the support for various formulas explicitly, and changes are propagated through backtracking mechanisms. A formula is no longer believed when all of the reasons for believing it have been removed.

7.8.2 Updates

Revision concerns changes to a theory about a static, unchanging world. The changes are brought about by the discovery or correction of information, but the underlying condition of the world being described remains the same. This is in contrast to *updates*, which are changes brought about by actions in a dynamic world [Katsuno & Mendelzon, 1989]. The revision and update processes behave differently. For example, suppose we know that one but not both ducklings, Cookie and WaWa, has hatched (that was before they learnt to quack), but we do not know which one. Formally, this can be represented by the theory

$$K = \mathbf{Th}(\{(\texttt{hatched\_WaWa} \wedge \neg\texttt{hatched\_Cookie})$$
$$\vee (\neg\texttt{hatched\_WaWa} \wedge \texttt{hatched\_Cookie})\}).$$

Now, at a later time, we learnt that WaWa has just hatched: `hatched_WaWa`. If we revise the theory K by the formula `hatched_WaWa`, we can infer that Cookie has not hatched[6]:

$$\neg\texttt{hatched\_Cookie} \in K^{*}_{\texttt{hatched\_WaWa}}.$$

This follows from postulate (*4): if $\neg\phi \notin K$, then $K^{+}_{\phi} \subseteq K^{*}_{\phi}$.

But why should Cookie be returned to her egg just because WaWa has hatched? In revision, it is assumed that the state of the world has not changed; only the information we have access to has changed. It has been specified statically that only one of the ducklings has hatched, and the revision operation preserves this piece of information. Thus, when we learnt that WaWa has hatched, it was concluded at the same time that Cookie has not hatched, since only one of them could have hatched.

A more appropriate way to handle this change is to recognize that the new information is brought about by a change in the state of the world itself. An *update*, instead of a *revision*, allows us to infer that both ducklings have hatched. Updating is characterized by its own set of postulates, which are similar to those for revision, but nonetheless with some interesting differences. In particular, updating rejects the revision postulate (*4). An analogous counterpart for contraction, called *erasure* [Katsuno & Mendelzon, 1989], has also been formulated for situations where the target formula is no longer true due to a change in the state of the world, rather than a change in the information we have about the (unchanged) world.

7.8.3   Rationality Revisited

There are a number of AGM postulates that have attracted some scrutiny. The most controversial of all is the recovery postulate (⁻5). At times it dictates the inclusion of information that we would rather not possess, as we have seen in Section 7.4.2. Makinson studied a *withdrawal* function which satisfies only the postulates (⁻1)–(⁻4) and (⁻6), dropping the recovery postulate and thus avoiding some of its problems [Makinson, 1987].

Another complaint is that there appears to be no formal restriction on the nature of the target formulas and belief sets involved. The theory change operators can be applied with arbitrary target formulas to arbitrary theories. For example, there is nothing to prevent us from expanding a theory by an inconsistent formula. Along the same lines, it is not obvious how we should interpret operations such as contracting a theory by a tautology or revising it by a contradiction.

One of the principal considerations for the postulates is that the changes have to be minimal; that is, the new belief set should resemble the original one as much as possible. However, there are different ways of defining minimality, as the change can

---

[6]The same would be true if we performed an expansion instead of a revision.

be characterized in several ways, for example in terms of the set of formulas changed, the number of base formulas or justifications changed (in the case of belief bases), or the epistemic entrenchment of the formulas changed. There are also cases in which minimality gives rise to dubious results. For example, given the theory

$$K = \mathbf{Th}(\{\phi, \psi\}),$$

if we were to contract K by $\phi \wedge \psi$, the minimal change criterion would direct us to give up only one of the two formulas ϕ and ψ, but not both. However, it may also be argued that we should consider giving up both formulas, as without any information for choosing between the two formulas, giving up both is a fairer practice than giving up one of them arbitrarily. Thus, the notion of minimality needs to be examined more closely to arrive at a rational guideline for theory change.

7.8.4   Iterated Change

Iterated change concerns the cumulative effect of executing a series of theory change operations. Recall that the epistemic entrenchment relation is relative to the particular theory it is associated with. The ordering is not passed down from the original theory to the changed theory. For example, if a formula ϕ is not in a theory K, then by (EE4) we have $\phi \leq_e \psi$ for all ψ. Now suppose we revise K by ϕ, so that ϕ is in K_ϕ^*. By (EE4) again, in the entrenchment ordering associated with K_ϕ^*, it is no longer the case that $\phi \leq_e \psi$ for all ψ, since now $\phi \in K_\phi^*$. This poses a problem for iterated theory change, since we do not know how to update the epistemic entrenchment relation for use with the new theory.

Another concern involves situations such as performing a revision with a formula ϕ already in a belief set K. According to the postulates (*3) and (*4), such a revision will not change the belief set (if the target formula is consistent): $K_\phi^* = K$. However, it seems intuitively more reasonable to expect that the repeated encounter of the target formula will boost its strength or epistemic entrenchment, which will affect future theory change operations.

The AGM postulates are focused on noniterated theory change, where we do not need to be concerned about updating the epistemic entrenchment ordering while we make changes to the belief set. Not only is the ordering not preserved from the original to the changed theories, but the two orderings need not bear any relationship to each other, even though the theories they are associated with are closely related. Work toward specifying ways to rationally update the epistemic entrenchment relation includes [Rott, 1992; Boutilier & Goldszmidt, 1993; Nayak, 1994; Williams, 1994; Darwiche & Pearl, 1997].

7.9   Summary

A large number of our beliefs about the world are derived from uncertain inference mechanisms. These beliefs are defeasible, and we need to adjust them from time to time, based on the new and possibly contradictory information we encounter. The relationship between a belief set and its replacement as a result of an adjustment should follow a set of rational principles. The most extensively studied framework for formalizing such principles is perhaps that proposed by Alchourrón, Gärdenfors, and Makinson (AGM). Three types of theory change operators have been identified: expansion, contraction, and revision, corresponding to three different ways the status of a belief can be adjusted. The AGM rationality postulates are construed as the logical constraints, which should be satisfied by all rational theory change operators. In general the AGM postulates do not uniquely identify a most desirable way to perform a belief change; for this we need to appeal to extralogical apparatus such as an epistemic entrenchment ordering of the beliefs. Whereas the AGM framework provides an elegant basis for formalizing theory change, some of its assumptions and postulates are questionable. Alternative views on the nature of belief change and how best to capture our notion of rationality have led to different characterizations.

7.10 Bibliographical Notes

Much of this chapter follows the presentation in [Gärdenfors, 1988]. Other writings from the authors Alchourrón, Gärdenfors, and Makinson on the topic of theory change include [Alchourrón & Makinson, 1982; Alchourrón et al., 1985; Gärdenfors & Makinson, 1988; Gärdenfors, 1992]. The relationship between theory replacement and nonmonotonic reasoning has been studied in [Makinson & Gärdenfors, 1991].

7.11 Exercises

(1) Show that the set of theories satisfying $(^+1)$–$(^+5)$ is closed under intersection.

(2) Prove or disprove the "reverse" of $(^-5)$:

$$\text{If} \quad \phi \in K \quad \text{then} \quad (K_\phi^-)_\phi^+ \subseteq K$$

Give an example where $\phi \in K$ but $(K_\phi^-)_\phi^+ \neq K$.

(3) Consider $(^-7)$. Give an example where $K_\phi^- \cap K_\psi^- \neq K_{\phi \wedge \psi}^-$.

(4) Prove or disprove the following:

    (a) $K_{\phi \wedge \psi}^+ \subseteq (K_\phi^+)_\psi^+$.

    (b) $K_{\phi \wedge \psi}^- \subseteq (K_\phi^-)_\psi^-$.

    (c) $K_{\phi \wedge \psi}^* \subseteq (K_\phi^*)_\psi^*$.

(5) Some of the AGM postulates can be simplified. For example,

    $(^-3)$ If $\phi \notin K$, then $K_\phi^- = K$

    can be rewritten as

    $(^-3')$ If $\phi \notin K$, then $K \subseteq K_\phi^-$.

This is because the original $(^-3)$ follows logically from $(^-2)$ and $(^-3')$. Show also the following simplifications:

    (a) $(^-5)$ If $\phi \in K$, then $K \subseteq (K_\phi^-)_\phi^+$.

           $(^-5')$ $K \subseteq (K_\phi^-)_\phi^+$.

    (b) $(^*5)$ $K_\phi^* = \perp$ if and only if $\vdash \neg\phi$.

           $(^*5')$ $K_\phi^* = \perp$ only if $\vdash \neg\phi$.

(6) Prove or disprove the following:

    (a)   Either $K_{\phi \wedge \psi}^- \subseteq K_\phi^-$ or $K_{\phi \wedge \psi}^- \subseteq K_\psi^-$.

    (b)   Either $K_{\phi \wedge \psi}^- = K_\phi^-$ or $K_{\phi \wedge \psi}^- = K_\psi^-$.

(7)   What happens when we contract a theory by a tautology?

(8)   Prove Theorem 7.2.

(9)   Prove Theorem 7.3.

(10) Prove or disprove the reverse of (EE5):

$$\text{If } \vdash \phi \text{ then } \psi \leq_e \phi \text{ for all } \psi.$$

(11) Show that any epistemic entrenchment relation that satisfies (EE1)–(EE5) is a total ordering.

(12) Consider the two belief bases in Section 7.8.1. What happens when we contract the two bases by ψ?

Bibliography

[Alchourrón & Makinson, 1982] C. E. Alchourrón and D. Makinson. On the logic of theory change: Contraction functions and their associated revision functions. *Theoria*, 48:14–37, 1982.

[Alchourrón et al., 1985] C. E. Alchourrón, P. Gärdenfors, and D. Makinson. On the logic of theory change: Partial meet contraction and revision functions. *Journal of Symbolic Logic*, 50:510–530, 1985.

[Boutilier & Goldszmidt, 1993] C. Boutilier and M. Goldszmidt. Revising by conditional beliefs. In *Proceedings of the Eleventh National Conference on Artificial Intelligence*, pp. 648–654, 1993.

[Darwiche & Pearl, 1997] A. Darwiche and J. Pearl. On the logic of iterated belief revision. *Artificial Intelligence*, 89:1–29, 1997.

[Doyle, 1979] Jon Doyle. A truth maintenance system. *Artificial Intelligence*, 12(3):231–272, 1979.

[Doyle, 1992] Jon Doyle. Reason maintenance and belief revision. In Peter Gärdenfors, editor, *Belief Revision*, pp. 29–51. Cambridge University Press, 1992.

[Fuhrmann, 1991] A. Fuhrmann. Theory contraction through base contraction. *Journal of Philosophical Logic*, 20:175–203, 1991.

[Gärdenfors, 1988] Peter Gärdenfors. *Knowledge in Flux: Modeling the Dynamics of Epistemic States*. MIT Press, 1988.

[Gärdenfors, 1990] Peter Gärdenfors. The dynamics of belief systems: Foundations vs. coherence theories. *Revue Internationale de Philosophie*, 172:24–46, 1990.

[Gärdenfors, 1992] Peter Gärdenfors, editor. *Belief Revision*. Cambridge University Press, 1992.

[Gärdenfors & Makinson, 1988] Peter Gärdenfors and David Makinson. Revisions of knowledge systems using epistemic entrenchment. In *Second Conference on Theoretical Aspects of Reasoning about Knowledge*, pp. 83–96, 1988.

[Grove, 1988] A. Grove. Two modellings for theory change. *Journal of Philosophical Logic*, 17:157–170, 1988.

[Hansson, 1991] Sven Hansson. Belief contraction without recovery. *Studia Logica*, 50:251–260, 1991.

[Hansson, 1992] Sven Hansson. In defense of base contraction. *Synthese*, 91:239–245, 1992.

[Harper, 1977] W. L. Harper. Rational conceptual change. *PSA 1976*, 2:462–494, 1977. Philosophy of Science Association.

[Katsuno & Mendelzon, 1989] H. Katsuno and A. O. Mendelzon. A unified view of propositional knowledge base updates. In *Proceedings of the Eleventh International Joint Conference on Artificial Intelligence*, pp. 269–276, 1989.

[Levi, 1977] Issac Levi. Subjunctives, dispositions and chances. *Synthese*, 34:423–455, 1977.

[Levi, 1991] Issac Levi. *The Fixation of Belief and Its Undoing*. Cambridge University Press, 1991.

[Makinson, 1985] David Makinson. How to give it up: A survey of some formal aspects of the logic of theory change. *Synthese*, 62:347–363, 1985.

[Makinson, 1987] David Makinson. On the status of the postulate of recovery in the logic of theory change. *Journal of Philosophical Logic*, 16:383–394, 1987.

[Makinson & Gärdenfors, 1991] D. Makinson and P. Gärdenfors. Relations between the logic of theory change and nonmonotonic logic. In A. Fuhrmann and M. Morreau, editors, *Logic of Theory Change, Lecture Notes in Artificial Intelligence*. Springer–Verlag, 1991.

[Nayak, 1994] Abhaya C. Nayak. Iterated belief change based on epistemic entrenchment. *Erkenntnis*, 41:353–390, 1994.

[Niederee, 1991] R. Niederee. Multiple contraction: A further case against Gärdenfors' principle of recovery. In A. Fuhrmann and M. Morreau, editors, *Logic of Theory Change, Lecture Notes in Artificial Intelligence*, pp. 322–334. Springer–Verlag, 1991.

[Rott, 1991] Hans Rott. Two methods of constructing contractions and revisions of knowledge systems. *Journal of Philosophical Logic*, 20:149–173, 1991.

[Rott, 1992] Hans Rott. Preferential belief change using generalized epistemic entrenchment. *Journal of Logic, Language and Information*, 1:45–78, 1992.

[Spohn, 1988] W. Spohn. Ordinal conditional functions: A dynamic theory of epistemic states. In W. L. Harper and B. Skyrms, editors, *Causation in Decision, Belief Change and Statistics*, Volume 2, pp. 105–134. Reidel, 1988.

[Tennant, 1994] Neil Tennant. Changing the theory of theory change: Toward a computational approach. *British Journal for Philosophy of Science*, 45:865–897, 1994.

[Williams, 1994] Mary-Anne Williams. Transmutations of knowledge systems. In *Proceedings of the Fourth International Conference on Principles of Knowledge Representation and Reasoning*, pp. 619–629, 1994.

8

Statistical Inference

8.1 Introduction

We consider a group of puppies, take what we know about that group as a premise, and infer, as a conclusion, something about the population of all puppies. Such an inference is clearly risky and *invalid*. It is nevertheless the sort of inference we must make and do make. Some such inferences are more cogent, more rational than others. Our business as logicians is to find standards that will sort them out.

Statistical inference includes inference from a sample to the population from which it comes. The population may be actual, as it is in public opinion polls, or hypothetical, as it is in testing an oddly weighted die (the population is then taken to be the hypothetical, population of possible tosses or possible sequences of tosses of the die). Statistical inference is a paradigm example of uncertain inference.

Statistical inference is also often taken to include the uncertain inference we make from a population to a sample, as when we infer from the fairness of a coin that roughly half of the next thousand coin tosses we make will yield heads—a conclusion that might be false. Note that this is not probabilistic inference: the inference from the same premises to the conclusion that the probability is high that roughly half of the next thousand tosses will yield heads is *deductive* and (given the premises) not uncertain at all.

The inference from a statistical premise about a population to a nonprobabilistic conclusion about part of that population is called *direct inference*. The inference from a premise about part of a population to the properties of the population as a whole is called *inverse inference*. The latter is a fundamental kind of inductive inference. Statistical inference is a paradigm and a well-explored example of uncertain inference.

Consider an urn containing black and white balls. Suppose we know that the proportion of black balls in the urn is 0.3. Suppose that we know that black and white balls are chosen in this experiment in the same ratio as their frequency in the urn, and that the draws are independent—i.e., are made with replacement. If we draw a sample of ten balls from the urn, what is the most likely composition of the sample? The population is known. We may apply the methods of Chapter 3, and calculate that the most likely sample composition, given the proportion of black balls in the urn, is three black balls and seven white balls. Writing $S_{3/10}$ for "the sample composition is three black balls out of ten balls" and $U_{0.3}$ for "the urn contains 30% black balls", we have $P(S_{3/10}|U_{0.3}) = \binom{10}{3}(0.3)^3(0.7)^7 = 0.2668$. The probability of getting a sample of this composition is 0.2668; the probability of getting a sample of any other composition is less. This is not an uncertain inference. The only uncertainty infecting the conclusion is due to the uncertainty of the premises. Were we to conclude that the next sample of ten balls *will* contain three black and seven white balls, this would be an uncertain inference. Obviously it would be a poor one, since the chances are it would be false, despite the truth of its premises, and despite the fact that this is the most probable composition.

Here is another example, which might naively be thought to embody an inverse inference: Suppose that we have three urns, one of which contains 30% black balls and two of which contain 70% black balls. We choose one of the three urns by a method that ensures that each urn has probability $\frac{1}{3}$ of being chosen. We draw a sample of ten balls (with replacement) from that urn. Six of the balls are black. Which urn is most likely to have been chosen?

Again the answer is given by the methods of Chapter 3. The probability of getting six black balls if we have one of the 70% urns is 0.2001. The probability of getting six black balls if we have the 30% urn is 0.0368. The probability that we have a 70% urn, given that we have gotten a sample containing six black balls, is thus $(0.67 \times 0.2001)/(0.67 \times 0.2001 + 0.33 \times 0.0368) = 0.9179$.

Observe that although we are making an inference concerning one population (the population of balls in the selected urn), we are making it on the basis of our knowledge of another population. This becomes clear if we think of the *sample space* in terms of which we present the argument. It must consist of a set of *pairs* consisting of the selection of an urn, and the selection of ten balls from that urn. The first member of the pair may have any one of three values (corresponding to the three urns), and the second member of the pair may have any one of eleven values, corresponding to the possible proportions of white and black balls.

Note that all this is quite independent of how we interpret probability. If we construe probability empirically, this means that we are taking the sample space to have a measure on it that is determined by some kind of frequency or propensity. If we construe probability subjectively, then the measure comes from a different source— presumably the degrees of belief of the agent. If we construe probability logically, the measure is determined on logical grounds alone—that is, as a characteristic of a formal language. *In every case what is involved in this inference is no more than the application of the probability calculus to a completely specified problem.* There is no uncertain inference involved; all the inference is deductive and mathematical.

Contrast a case of inverse inference: In this case we draw ten balls from an urn of *unknown* provenance. We draw ten balls (with replacement) and observe that six are black. We infer that about 60% of the balls in the urn are black. Intuitively this seems reasonable, though we would like to be able to say something about the degree of appropriateness of our inference, as well as something more precise than "about" 60%. This is the inverse inference in pure form: we want to infer the probable composition of a population from the known composition of a sample alone, as opposed to inferring the probable composition of a sample from the known composition of a population.

Much early work in statistics (eighteenth century to early twentieth century) was exploring the possibilities for reducing inverse inference to direct inference. Laplace's principle of indifference [Laplace, 1951] was designed as one way of attacking this problem. The idea was that *if* we could reconstruct inductive inference as direct inference, it would be possible, first, to assign probabilities to inductive conclusions, and second, to accept or believe those conclusions whose probabilities were sufficiently high.

We can follow these ideas more easily if we simplify our example of inverse inference. Suppose that we consider just two kinds of urns—one kind with 30% black balls and one kind with 70% black balls. Note that we are *not* drawing one of two urns of these two kinds with equal probability. Nevertheless, we can attempt to use the sample space we constructed before: It is a set of pairs consisting of the selection of an urn and the selection of ten balls from that urn. Suppose that the sample contains three black and ten white balls. Given our assumptions about the conditions of the drawing, we can confidently assign a conditional probability to this sample composition, *given* the kind of urn from which the sample was drawn. (These conditional probabilities are often referred to as *likelihoods*. They play an important role in applications of Bayes' theorem.)

This allows us to attempt to use Bayes' theorem to compute the probability that the unknown composition of the urn is 30% black balls:

$$P(U_{0.3}|S_{3/10}) = \frac{P(U_{0.3})P(S_{3/10}|U_{0.3})}{P(U_{0.7})P(S_{3/10}) + P(U_{0.3})P(S_{3/10}|U_{0.3})}$$

However, in the previous example we assumed that we had grounds for assigning probabilities to each kind of urn, whereas we have no grounds for making such an assignment in this case: $P(U_{0.3})$ and therefore $P(U_{0.7}) = 1 - P(U_{0.3})$ remain unknown. For all we know, the urn is obtained by a random drawing from a collection of urns, of which 10% are of the kind with 30% black balls. Without this number, we cannot make a direct inference; we do not have the statistical data.

Faced with this problem, there is a simple solution, first explored in mathematical detail by Laplace [Laplace, 1951]: If you don't have the statistics, make them up. Put more generously, Laplace offered a principle of indifference, according to which if your knowledge about two alternatives is symmetrical, you should assign equal probabilities to each alternative.[1] In the case described, Laplace's principle would lead to the assignment of a probability of $\frac{1}{2}$ to each kind of urn, and thus in turn to a probability of 0.8439 that the urn was a 70% urn.

Laplace extended this reasoning to the case where there are $N + 1$ urns, containing each possible ratio of black balls to white balls among N balls. Laplace calculated the probability that the next ball is black, given that we have sampled n balls, of which m have turned out to be black. (It is clear how to do this from our earlier examples.) Laplace then elegantly let N increase without limit, so that, in effect, all possible ratios are represented equally.[2] From these assumptions he derived his famous (or infamous) *rule of succession*: If you consider an experiment, which has two kinds of outcome, success and failure, and observe m outcomes out of n to be success, the

---

[1] This principle has been discussed in Chapter 3.

[2] At this time the theory of limits was not well understood. We have seen that there *is* no uniform σ-additive measure on any denumerably infinite set. To understand Laplace's point, it is sufficient to consider that his rule

probability that the next trial will yield a success is $(m + 1)/(n + 2)$.[3] On this basis, it is alleged that Laplace calculated that the probability that the sun would rise tomorrow is 2114811/2114812, assuming that the world was created in 4004 B.C., and that the sun had risen every day from then to 1790 A.D. If he did so, then surely Laplace, the great astronomer, meant this calculation as no more than a jocund illustration of his rule of succession.

We have already considered the difficulties raised by the principle of indifference. Nevertheless, its use in this context serves to illustrate the most profound difference between inverse inference and direct inference. The examples of direct inference displayed above are relatively uncontroversial, no matter what interpretation of probability is being considered. Whether probabilities represent propensities, or frequencies, or logical measures, or personal subjective opinions embodying the same numerical probabilities, the most likely ratio in the sample of ten balls is 0.3 black, and the probability of that ratio is 0.2668. Similarly, in the second example, everybody from von Mises to Savage would agree that the probability that we have the 70% urn is 0.9179.

The moment we pass to inverse inference, that agreement vanishes. The holders of an empirical interpretation of probability assert that we do not have the data to arrive at any probability in the case of inverse inference: we simply do not have the measure on the sample space that is required. The holders of the subjective or logical interpretations of probability do accept inverse inference, but they can do so just because they convert inverse inference to direct inference by supplying the missing measures on the sample space. The logical interpreters of probability find these measures in the characteristics of a formal language; the subjectivists find them in the opinions of the users of probability.

This difference creates a great divide in the foundations of statistics. *Classical statistics* construes probability as an empirical notion, reflecting, if not defined in terms of, long run frequencies. For classical statistics inverse inference is simply invalid, and of no interest. (But remember that when we have the appropriate statistical knowledge (and this need not be based *directly* on statistical data), what may superficially be thought of as inverse inference can be construed as direct inference.) The alternative is *Bayesian statistics*, in which probability is construed as subjective or logical. For Bayesian statistics, inverse inference simply picks out a particular kind of direct inference: the required measures on the sample space always can be found in the opinions of the agent making the inference, or in the structure of the language being employed. Classical statistics is the most pervasive form of statistics, and is certainly the most common form of statistics for applications in the sciences. We shall review some of the content of classical statistics as it bears most directly on inductive questions in the following section. In Section 8.3 we will return to Bayesian statistics, which is so popular as to warrant the adjective "prevalent" both in symbolic artificial intelligence and in philosophy.

8.2   Classical Statistics

Probability is construed empirically in classical statistics. A probability statement says something about real or hypothetical long run frequencies characteristic of the

---

[3]Notice the similarity to Carnap's $c^\dagger$; the difference is that Carnap takes into account the "logical width" of the predicates involved.

real world. Thus, probabilities of particular singular occurrences—the probability of heads on *the next toss of this coin*—are simply undefined (or are trivially taken as admitting only the two values 0 and 1). The propensity interpretation of probability is an exception to this generalization, because each toss of a coin, including the next toss, but also including the last toss, which we observed to land heads, has a propensity of a half to land heads. So far as inference is concerned, a probability is a general statement; even propensities are, primarily, the propensities of *kinds* of chance setups or trials to yield results of certain sorts.

The outcome of a statistical inference thus cannot yield the assignment of a probability to a hypothesis. It can provide *tests* for hypotheses, or characterize the outcome of experiments in other ways. The following sections will explore *significance testing*, frequently used in the agricultural, psychological, and social sciences; general *hypothesis testing*, also much used in the reporting of statistical data; and *confidence interval methods*, which are most often employed in assessing quantitative data, including measurement. Often included among confidence interval analyses is the conceptually distinct method, due to R. A. Fisher, of *fiducial inference*.

8.2.1   Significance Tests

Significance testing was formalized by the English statistician R. A. Fisher in the 1920s [Fisher, 1922; Fisher, 1930]. The general idea is that, at least in some disciplines (Fisher was particularly concerned with agricultural research), we are concerned with whether an outcome is produced by chance, or as a result of our experimental intervention. Did the increase in yield of beans on our experimental plot reflect our application of potash, or was it due to accidental variation? We test a hypothesis by constructing an experiment that will have a specified outcome by chance (i.e., if the hypothesis is false) only very rarely. If we observe such an outcome, we reject the hypothesis of chance and pursue the contrary hypothesis, the hypothesis under test— in our example, the hypothesis that the application of potash increases productivity. The methodology rests on the *rejection* of chance hypotheses when highly improbable results are observed.

The mechanism of the procedure can be illustrated by a time-honored, oft-cited example due to Fisher himself [Fisher, 1971]:

Example 8.1 (The Lady Tasting Tea). *A certain lady makes an empirical claim: that she can tell by tasting a cup of tea with milk whether the cup was prepared by adding milk to the tea, or by adding tea to the milk. Of course, she is not claiming to be infallible in doing this; just that she can do better than chance.*

Dr. Fisher proposes to test her claim. He will do so by offering her five cups of tea, each prepared in one of the two ways. The experiment must be designed in such a way that we can apply direct inference in the case that the lady's ability is absent. That is: The sample space consists of trials of the experiment. Each trial is characterized by a 5-sequence of random quantities $X_1, \ldots, X_5$, where X_i has the value 1 for a trial if the lady guesses correctly on the i th cup, and the value 0 if the lady guesses incorrectly.

If the lady lacks the ability to discriminate, we want the number of correct guesses out of the five trials to be binomially distributed with the parameter $p = 0.5$. The experiment must be carefully designed to ensure this. The lady receives the cups in a certain order: if *every other cup were prepared differently, the lady could perhaps get all the cups right by just guessing correctly on the first cup. The first step in designing this experiment, therefore, is to randomize the preparation of the cups of tea.*

Suppose that the cups prepared with milk first all had thick rims, and the others thin rims. Or that a different kind of tea were used in the preparation of the milk first cups. Or that the milk in the milk first cups were introduced the night before the experiment. If any of these factors were present, we could try to take them into account, though that might be difficult. Instead we arrange, or try to arrange, that they are absent. We can choose cups that are as much alike as possible, use the same kind of tea and the same kind of milk in each cup, etc. However hard we try, though, there will be accidental differences between the presentations of each kind. The cups will not be really indistinguishable; the tea in the later cups will have steeped a bit longer, etc.

If the ideal conditions were satisfied and the lady lacked the ability to discriminate, then the empirical probability (corresponding to the long run in a hypothetical series of trials) that the lady is correct on a single trial would be 0.5. The probability that she would be correct on four out of five trials is $\binom{4}{5}(0.5)^4(0.5) = \frac{5}{32}$; the probability that she would be correct on all five trials is $\frac{1}{32}$.

Two natural significance tests suggest themselves: one in which we reject the lady's claim unless she identifies four or five cups correctly, and a more rigorous test in which we reject the lady's claim unless she identifies all five cups correctly. The significance level is the probability on the hypothesis that the lady cannot really discriminate that nevertheless we obtain results, purely by chance, suggesting that she does have that ability. It is the probability of mistakenly accepting her claim. The former test has a significance level of $\frac{5}{32}$, or 0.156; the latter a significance level of $\frac{1}{32}$, or 0.0325.

Let us now formulate the structure of significance tests more generally and more formally.

We must first identify a *null hypothesis* to contrast with the hypothesis of interest. The null hypothesis is often the hypothesis of "no effect" or "mere chance variation". Testing a cold remedy, it would be the hypothesis that those who took that remedy got better no faster than those who didn't. Testing an agricultural treatment, the null hypothesis would be that the treatment made no difference in productivity.

Once we have settled on a null hypothesis, we must try to design an experimental situation in such a way that under the null hypothesis a *known probability distribution* holds for the sample space of the experiment. This may not be easy. In the case of the lady, we need to ensure that if she lacks the ability to discriminate, the chance of her identifying *m* out of *n* cups correctly is simply given by the binomial distribution. In the case of testing a cold remedy, one of the constraints on the experiment might be that those who were not getting the remedy under test would get a treatment that was indistinguishable, to them, from the treatment the other patients were getting. Otherwise, those who received the new treatment might simply *think* they were getting better. (In general it is true that those who think they are receiving the new treatment get better faster; this is called the *placebo effect*.) A further step is to randomize doubly, so that not even the physicians involved know which patients are getting the new treatment. This is designed precisely to ensure that the probability distribution over the sample space, on the null hypothesis, is the chance distribution. In the case of testing an agricultural treatment, we need not worry about the placebo effect; the placebo effect does not occur in radishes. But there are other things to worry about. The natural fertility of the soil varies in streaks and pockets; we design our agricultural experiment so that these variations in natural fertility have minimal effect on our results: that is, again, so that if the null hypothesis is true, we will encounter only chance variations whose overall distribution we know.

Once we have designed the experiment to ensure that under the null hypothesis the results of the experiment will really be results due to a known chance distribution over the sample space, then we must decide under what circumstances we will reject the null hypothesis. That is, we must pick out a subset of the sample space such that *if* our experiment results in a point in that subset, *then* we reject the null hypothesis. This subset of the sample space is called the *rejection region*; we must choose the rejection region in such a way as to make it reasonably unlikely that we will reject the null hypothesis (the hypothesis of chance effects) if it is true.

Definition 8.1 (Rejection Region). *The rejection region of an experiment is a subset of the sample space corresponding to the set of possible experimental outcomes that we take to warrant the rejection of the null hypothesis H_0.*

What corresponds to "reasonably unlikely" when we say that the rejection region should make it reasonably unlikely that we will reject the null hypothesis if it is true? There is, of course, a relation between the selection of a rejection region and the risk of error: the probability of rejecting the hypothesis H_0 though in fact it is true. This probability is often set at 0.05 or 0.01. It is called the *significance level* or the *size* of the rejection region.

Definition 8.2 (Significance Level). *The significance level of an experiment is the probability, under the null hypothesis H_0, that the test will lead to the false rejection of H_0.*

Note that the null hypothesis makes many assumptions about the experiment. It is stated as a simple statistical hypothesis, for example, that the distribution of errors by the lady tasting tea is binomial with a parameter of 0.5. But for this hypothesis to hold, as we shall see, requires that the experimental setup satisfy various constraints, which we may call background conditions.

Suppose we regard 5% as a reasonable risk to run of falsely rejecting the null hypothesis. We are still not done, since there are various alternatives that carry the same probability of false rejection. For example, if the lady is wrong five times out of five, that is something that could only happen by chance $\frac{1}{32}$ of the time, but we would not (ordinarily) take that as indicating that the lady had the ability to discriminate. A natural constraint that we might impose on the rejection region is that the unspecified hypothesis we have in mind must yield a result in the rejection region more often than the null hypothesis. To be more specific would lead us to considerations that are more germane to the next section.

If we were content with a probability of false rejection of $\frac{5}{32}$, we could choose as a rejection region the part of the sample space corresponding to exactly four correct identifications. But it would be odd indeed if we were to reject the null hypothesis if the lady classified four of the five cups correctly, but failed to reject it if she identified all five cups correctly.

In a sense, the lady possesses the ability to discriminate if in the long run she can tell which cup is prepared in which way 0.5001 of the time. This would be very difficult to distinguish from a complete lack of ability to discriminate. Surely to have an ability requires her to do better than this. To take account of these natural conditions we should attend to the alternatives to the null hypothesis. We shall do this in the next section, but to do so involves going beyond simple tests of significance.

Suppose we conduct a significance test and the null hypothesis is rejected: an event with probability less than 0.02, given the null hypothesis, is observed. What does this mean epistemically? Many writers argue that to reject the null hypothesis H_0 of no effect is not to accept the "alternative" hypothesis H_1 [Baird, 1992, p. 133], for example that the treatment being considered is effective. Fisher [Fisher, 1956] is very clear that in his opinion passing a significance test is only the *beginning* of research, not its end. But Fisher also writes: "The force with which such a conclusion [the rejection of the null hypothesis] is supported is logically that of the simple disjunction: *Either* an exceptionally rare chance has occurred, or the theory of random distribution is not true" [Fisher, 1956, p. 39]. If we refuse to accept the occurrence of a very rare event, then we are obliged to accept the rejection of the null hypothesis. If we reject the null hypothesis, and the alternative *is* the logical alternative, then we have, ipso facto, accepted the alternative hypothesis. It seems clear that the common claim that to reject the null hypothesis is not to accept the alternative depends for its plausibility on construing the "alternative" in two ways: first as simply the logical denial of the null hypothesis, and second as embodying more content than that logical denial— for example: the lady can dependably discriminate whether milk or tea was added first.

The null hypothesis is still ambiguous, however. What does it come to? If we take it to include merely H_0 as it is often stated (the medicine is ineffective, the fertilizer has no effect, etc.), then it is not the case that the observation of a sample point in the rejection region of the sample space requires the rejection of H_0 itself. What is involved in the computation of the significance level is not merely the null hypothesis H_0, but the auxiliary hypotheses that make up the conditions imposed on the experiment. What we have is a conjunction of conditions that spell out not only the simple hypothesis H_0 of no effect, but also many other conditions (for example, that the lips of the cups do not differ detectably in thickness, that the milk and tea are of the same temperature in each trial, etc.) that are required in order that the hypothesis H_0 correctly characterize the sample space. It is this conjunction of conditions that must be rejected if we reject the occurrence of a rare event. To reject that conjunction of conditions is indeed to accept a general hypothesis, but it is a complex hypothesis. If, as is often the case, we have reason to accept some of those components, the onus of rejection falls on the disjunction of the remainder. (Of course, in this case as in all inductive cases, both rejection and acceptance are tentative and can be undone by further evidence.)

8.2.2 Hypothesis Testing

In performing significance tests, the hypothesis of interest is often allowed to remain quite vague (the lady can discriminate). In the general theory of testing hypotheses [Lehman, 1959], we focus on a uniform class of hypotheses. For example, in the case of the lady tasting tea, it would make sense to limit our concerns to two explicit hypotheses: the probability p with which the lady can correctly identify cups of tea is greater than 0.5 or it is equal to 0.5. In other words, we explicitly ignore the possibility that she has wonderful taste buds, but is confused. We do so, of course, on the grounds that this is an alternative we feel we can reject—that is, its negation follows from things we accept.

The hypothesis that the experiment is correctly described by a binomial distribution with parameter $p = \frac{1}{2}$ is a *simple* hypothesis. The hypothesis that the distribution in question is binomial with a parameter $p < \frac{1}{2}$ is a *compound* hypothesis.

In the most transparent form of hypothesis test, we are testing a simple hypothesis against a simple alternative. Consider the example of two kinds of urns used to introduce the idea of inverse inference. (Do not forget that from a classical point of view, inverse inference is impossible.) Let us consider a variant of that example. There are just two urns; one contains 30% black balls, and one contains 70% black balls. We are presented with one of the urns. We are going to draw ten balls (with replacement, with thorough mixing, etc.), and on the basis of that sample, we are going to decide which urn we have.

There are a number of facts we typically accept if we are to conduct a useful experiment in this context. We believe that the balls are drawn from the urn in such a way that each ball has the same probability of being drawn. This is sometimes stated as "The balls are well mixed, they differ only in color," etc. We assume that the draws are independent: The draws are made with replacement; there is no causal connection between one draw and another. These are sometimes called "assumptions," but to call something an assumption provides no argument for accepting it; and any normative reconstruction of statistical inference requires that these conditions be accepted on reasonable grounds. They are better called "background conditions" or "background facts".

Against the background of these empirical facts, we may design an experiment for testing hypothesis H_1 (the 30% hypothesis, the hypothesis under test) against hypothesis H_2 (the 70% hypothesis). Because both are simple hypotheses, we can (as in the case of testing a null hypothesis) compute the probability, relative to the truth of H_1, of each subset of the sample space. The first step in designing a test is to pick a probability α (we call it the size or the level of the test) that is small enough to be disregarded as characterizing a real possibility. As in the case of significance tests, such numbers as 0.01, 0.02, 0.05 are often chosen. Similarly, the region of the sample space chosen is called, as before, the rejection region: if we obtain a sample falling in this region, we reject the hypothesis H_1. But, as before, there are many such regions carrying a probability less than α.

There are two kinds of error that we may commit in applying a statistical test. The size, α, reflects the *probability of committing an error of the first kind*, or Type I error: that of falsely rejecting the test when the hypothesis is true. This probability is the probability associated with the region of rejection, according to the hypothesis H_1 under test. Error of the second kind, or Type II error, consists in *failing* to reject a hypothesis when it is false. This is usually denoted by $1 - \beta$, where β is the probability of obtaining a point not in the rejection region when the hypothesis under test is false. The *power* of a test is its power to discriminate between the hypothesis being tested and the alternative hypothesis or hypotheses; it is measured by $1 - \beta$, the probability of not rejecting the hypothesis under test when it is true. More formally:

Definition 8.3 (Type I and Type II Error). *Given an experimental design for the test of a simple statistical hypothesis H_1 against a simple alternative H_2:*

- *Type I error, or error of the first kind, is the probability α that the test will reject H_1 when it is true.*
- *Type II error, or error of the second kind, is the probability $1 - \beta$ that the test will fail to reject H_1 when it is false, i.e., when H_2 is true. Thus, β is the probability, given H_2, of obtaining a point in the rejection region.*

In the test of one simple hypothesis against a simple alternative, the case under discussion, we can actually calculate both the size (α) and power ($1 - \beta$) of a test. Suppose we take as our rejection region the set of samples containing eight or more black balls. The sample space reflects the fact that we know that we are drawing balls in such a way that each ball in the urn has the same chance of being selected; that the drawing is done with replacement, so that the draws are independent; etc. Because the distribution under these circumstances is binomial, we may take the partition of the sample space to consist of the eleven possible outcomes of taking a sample of ten: $0, 1, \ldots, 10$ black balls. Let the hypothesis under test be H_1, the 30% hypothesis. If this hypothesis is true, the measure of the subset of the sample space containing trials with i black balls is given by the corresponding distribution:

$$B(10, i, p) = \binom{10}{i} p^i (1 - p)^{10-i}.$$

The probability, under H_1, of getting eight or more black balls is $\binom{10}{10}0.3^{10} + \binom{10}{9}0.3^9 0.7 + \binom{10}{8}0.3^8 0.7^2 = 0.00159$. Thus, the test is of very small size: the probability of falsely rejecting H_1 is about 0.00159 if we adopt this rejection region.

How about the power of this test? Because the alternative to H_1, H_2, is also a simple hypothesis, we can calculate the power of the test to discriminate: to reject H_1 when it is false and H_2 is true. So what is the probability that this test will erroneously fail to detect the truth of H_1? It is the probability of getting eight or more black balls under the alternative hypothesis, i.e. $\binom{10}{10}0.7^{10} + \binom{10}{9}0.7^9 0.3 + \binom{10}{8}0.7^8 0.3^2 = 0.38277$. This means that the chance of an error of the second kind is significant. The power of the test, $1 - \beta$, is 0.61723. Thus, despite the fact that this test is very unlikely to reject H_1 when it is true, it suffers a significant chance of failing to reject H_1 when it is false.

It is perhaps obvious that by altering the test, we can decrease the chance of an error of the second kind at the cost of increasing the chance of a error of the first kind. Suppose we change the rejection region to consist of *seven* or more black balls. The size of the test becomes 0.0106, while the power increases to 0.6496. If we go one more step, we can improve the power yet more. If the rejection region is taken to be *six* or more black balls, then the size of the test becomes 0.04735, and the power becomes 0.84973.

Of course, we could decrease the chances of committing either kind of error by increasing the size of the sample. But it costs time, money, and effort to increase the size of the sample, and statistics has often focused on getting the best information out of a sample of fixed size rather than increased sample sizes.

Suppose that we felt that a size (a chance of false rejection) of 0.0106 was too conservative, but that a size of 0.04735 was too big? Perhaps we are willing to take a chance of 1 in 50 of falsely rejecting the hypothesis under test. Even this can be arranged, according to the Neyman–Pearson approach to hypothesis testing. We arrange it by conducting a *mixed*, or *randomized*, test.

A mixed test is a test in which we choose two (or more) rejection regions and reject the hypothesis under test according to the outcome of an auxiliary experiment as well as according to the outcome of the observation of the sample. For example, in the case at hand, the test would be a mixture of the second and third tests we have described.

We begin by performing an auxiliary experiment, say with a deck of 100 cards, 74 of which are black. We draw a card, and if we get a black card, we perform test 2, with a critical region of seven or more black balls. If we don't get a black card, we perform test 3, in which the critical region is six or more black balls. Because we falsely reject H_1 0.0106 of the time when we get a black card, and 0.04735 when we get a nonblack card, the *overall long run frequency* with which we falsely reject H_1 is $0.74 \times 0.0106 + 0.26 \times 0.04735 = 0.0200$—exactly our target. The power of the mixed test is similarly a mixture of the powers of the component tests.

Of course, it makes no difference whether we perform the auxiliary test before or after we have drawn our sample. An equivalent formulation of the mixed test is therefore: if there are seven or more black balls, reject H_1. If there are five or less black balls, do not reject H_1. If there are exactly six black balls, then perform the auxiliary test, and reject H_1 if and only if the card drawn is a black card.

Intuitively there is something a bit odd about the mixed test. What has the black card to do with the contents of the urn? It seems odd to take the result of a draw from a deck of cards as determining whether or not we should reject a hypothesis about an urn, after we have already obtained and observed the sample from the urn. There is a rationale for doing this, though. According to the classical point of view, statistics can give us no information about a particular case—probabilities do not apply to particular cases. All statistics can do is to characterize the rejection rules we adopt. This characterization is in terms of long run frequencies of error. It is thus only the long run frequencies of error we should take to heart. *In the long run*, the mixed test does indeed lead to error exactly 2% of the time. If that strikes us as an appropriate long run frequency of error of the first kind, then the mixed test is the one to employ.

Our analysis so far has concerned a particularly simple case: the testing of one simple statistical hypothesis against a simple alternative. We may also consider testing a simple alternative against a compound alternative, as in the case of the lady tasting tea, where we are testing the simple hypothesis $p = 0.5$ that the lady has no ability to discriminate against the alternative compound hypothesis $p > 0.5$ that says that she some unspecified ability to discriminate. The significance levels we computed for the tests we devised for this example correspond to sizes: they are the probabilities of falsely rejecting the precise hypothesis $p = 0.5$. What can we say about power? Given any particular hypothesis in the compound alternative, say $p = 0.55$, the test of rejecting the claim unless the lady correctly identifies all five cups is the most powerful test of $p = 0.5$ against this alternative. This is so for *each* alternative among the set of alternatives we are considering. Such a test is called *uniformly most powerful*, because it is most powerful against any alternative in the set of alternatives under consideration.

We have been concerned so far with elementary binomial hypotheses. The same ideas can of course be applied to tests concerning other statistical hypotheses. The fundamental lemma of Neyman and Pearson spells out the very broad conditions under which most powerful tests exist.

They don't always exist. There are many circumstances under which statistical tests are appropriate that are more complicated than those we have been considering, and in which tests satisfying other desiderata must suffice. The basic idea of statistical testing, and the two kinds of error that one may make in testing a statistical hypothesis, are fundamental, however. The quality of a statistical inference is to be measured in terms of long run frequencies of the two kinds of error.

8.2.3 Confidence Intervals

Our concern in statistical inference is often to find the approximate value of a parameter in a statistical distribution. For example, we want to know the approximate relative frequency with which a loaded die will land with the two up, or we may want to know how often a certain machine produces parts that fail to pass inspection, or we may want to know the frequency with which albinism occurs in a certain species. In each case, we may take the distribution of the frequency in a sample to be binomial, and use a sample to make an interval estimate of the parameter. Or, taking for granted that the distribution of mature weights in a certain species is approximately normal, we may want know the approximate mean (average) weight—that is, we seek bounds on the parameter μ of that distribution.

A pervasive example of this form of inference is provided by *measurement*. Every measurement is subject to error. Usually that error is taken to be distributed approximately normally with a mean of 0 and a known standard deviation that is characteristic of the method of measurement. It follows that the true value of a quantity is the mean value of the normally distributed population of measurements of that quantity. To estimate that value, or to assert that it falls in a certain interval, is to make the corresponding claim about the mean μ of a normal distribution.

These examples suggest that the two probability distributions that are of the most importance to us in the context of inductive inference are the Bernoulli distribution and the normal distribution. The Bernoulli distribution is characterized by a single parameter p, representing a class ratio or a long run frequency or a propensity. The normal distribution is characterized by two parameters, its mean μ and standard deviation σ; again, these are characteristic of some real or hypothetical empirical population. An important part of our effort in uncertain or inductive inference is directed at evaluating such parameters as these by finding bounds within which they may confidently be believed to fall.

Definition 8.4 (Confidence Interval). *An interval $[\underline{\theta}(e), \overline{\theta}(e)]$ that is a function of the observed results e is a p-confidence interval for the parameter m, in an experimental setup E, if and only if the long run relative frequency with which that interval covers m is at least p.*

The method of confidence intervals is an approach closely allied to the testing of hypotheses. Suppose we consider an urn of black and white balls, and a sampling method that we have reason to believe produces each ball equally often in the long run. Let p be the proportion of black balls. Consider a test of $p = \underline{\theta}_0$ against the alternative $p > \underline{\theta}_0$, based on a sample of n draws. The hypothesis tested will be rejected at, say, the $\alpha/2$ level if there are too many black balls in it. Of course, any sample that leads to the rejection of $p = \theta_0$ will also lead to the rejection of $p = \theta_0'$ where $\theta_0' < \theta_0$.

The same story can be told about testing the hypothesis $p = \bar{\theta}_0$ against $p < \bar{\theta}_0$. We will commit an error of the first kind no more than $\alpha/2$ of the time if we reject the hypothesis tested when there are too few black balls in the sample.

How do we put flesh on the bones of "too many" and "too few"? Suppose that rf is the relative frequency of black balls in the sample. It is clear that we can find k and j such that *in the long run*, on our assumptions, whatever p may be, rf $- k$ will be less than p no more than $\alpha/2$ of the time and rf $+ j$ will be greater than p no more than $\alpha/2$ of the time. To see this, note that for a fixed value of p we can choose j and k so that $p - j > $ rf no more than $\alpha/2$ of the time, and $p + k < $ rf no more than $\alpha/2$ of the time, and these inequalities are equivalent to those just mentioned.

One-sided confidence intervals are constructed in a similar way. We obtain a lower bound on the parameter p by choosing k so that for any value of p the long run relative frequency that rf $- k$ will exceed p is less than α. This yields a one-sided lower $(1 - \alpha)$-confidence interval for p: we can have confidence $1 - \alpha$ that p is at least rf $- k$.

Example 8.2. *We would like to know a lower bound for the frequency with which items produced by a certain process pass inspection. Let us take a sample of 52 items. Suppose the observed relative frequency of passing inspection in this sample is 0.75. Let us take α to be 0.05. The rejection region for $p = \underline{\theta}$ against $p < \underline{\theta}$ is the set of samples with relative frequencies rf between 0 and a critical value f_c, where the sum*

$$\sum_{f=0}^{f_c} \binom{n}{nf} p^{nf}(1 - p)^{n-nf}$$

is constrained by the requirement that it be less than or equal to 0.05. We seek that value of $p = \underline{\theta}$ such that for any smaller value, the sum is less than 0.05: i.e., it should be very improbable that we should observe a relative frequency of 0.75 or less.

Because relative frequencies (like other sums) are approximated by the normal distribution, we can also use the normal distribution to approximate the lower bound on rejectable values of p. Note that 0.05 corresponds to the integral of the standardized normal distribution from $-\infty$ to -1.654. The hypothesis under test has a mean of $\underline{\theta}$ and a variance of $\underline{\theta}(1 - \underline{\theta})/52$. If we take $\underline{\theta}$ to be 0.60, the standard deviation is 0.063, and the rejection region is that of relative frequencies in the sample that are more than 0.704. If we take $\underline{\theta}$ to be 0.65, the standard deviation is 0.066, and the rejection region is that of sample frequencies more than 0.759. A value of $\underline{\theta}$ of about 0.638 yields a test of size 0.05 that just includes 0.75 in the rejection region. We can be 95% confident in the application of this test.

Just as we can obtain a lower bound for the parameter p by employing a test of size α, so we can obtain an upper bound in the same way. If we obtain a lower bound of size $\alpha/2$ and an upper bound of size $\alpha/2$, then we have obtained an *interval* that with probability $1 - \alpha$ captures the true value of the parameter p.

But be careful! From a classical point of view we cannot construe this *confidence* as a probability. In the example, we can say that the *probability* is 0.95 that the interval we calculate will cover the true value of the frequency with which items pass inspection; but to apply this to the case at hand would be to apply a probability to a particular event like the next toss or the last toss of a coin, and as we have seen, this makes no sense on a frequency interpretation of probability. The probabilities that we

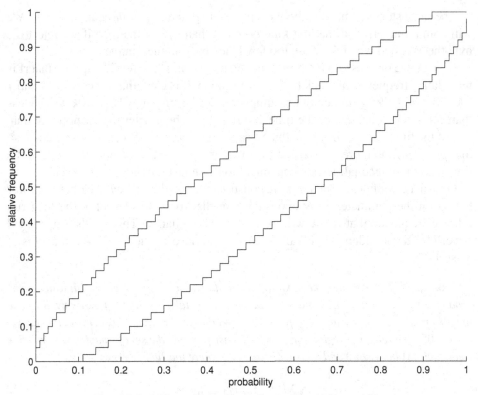

Figure 8.1: 95% confidence limits for the binomial, $n = 100$.

compute cannot be applied to the specific instances at hand: probabilities only apply to classes or sequences.

Graphs can be constructed to represent this form of inference. Such graphs appeared for the first time in 1934 in *Biometrika* [Clopper & Pearson, 1934]. Each graph deals with a certain sample size, and a certain level of confidence. Consider a 95% confidence diagram for $n = 100$. For every possible frequency p within the population under test we may compute the frequency $r_l(p)$ such that at most 2.5% of the samples will have a frequency below $r_l(p)$, and a frequency $r_u(p)$ such that at most 2.5% of the samples will have a frequency above $r(p)$. The graphs of these functions are not continuous: There are only a 101 possible values of $r_l(p)$. The graph thus constructed looks like Figure 8.1.

The graph is used by entering at the value of rf corresponding to the observed sample frequency, and inferring that the population frequency lies between the values indicated by the upper and lower curves.

For example, suppose we have a quality control problem like the one previously considered, but that we have a sample of 100 items, of which 85 pass inspection. What is the long run relative frequency of items that pass inspection? We enter the graph at 0.85, and find that we can be 95% confident that between 73% and 96% of the items will pass inspection in the long run.

Even simpler confidence procedures are available for estimating the mean of a normal distribution, in view of the fact that the difference between the mean μ of a normal distribution and the mean of a sample itself has a normal distribution with mean 0. Let Q be a random quantity normally distributed with a known variance of 1, but an unknown mean of μ. Let us draw a sample s consisting of one item—a sample of size 1. From the tables for the standard normal distribution, we can observe that the probability that an observed value of Q will be 2.575 standard deviations less than the mean of Q is (exactly) 0.005, the probability that it will exceed the mean of Q by 2.575 standard deviations is 0.005, and therefore the probability that the random interval $Q \pm 2.575$ will *fail* to cover the mean μ is 0.010, or 1%. In a particular case, having observed $Q(s)$, we say that the 99% confidence limits for μ are $Q(s) \pm 2.575$. We must, we are told, carefully avoid speaking of probability in this last case.

The same sort of argument will go through for a normal distribution of any known standard deviation.

Ronald A. Fisher proposed a form of inference that is much like the method of confidence limits just discussed. This is the method of *fiducial inference*. It applies only to continuous distributions, and then only when it is possible to identify a *pivotal quantity*, as there is in the present case. A pivotal quantity is a quantity with a known distribution that relates a property of a sample to the parameter whose value we seek.

This may be illustrated for the normal distribution. Let the quantity Q have a normal distribution $N(\mu, \sigma)$ with known standard deviation σ and unknown mean μ. The pivotal quantity $\mu - Q$ relates the parameter μ to the observational quantity Q. It has a normal distribution, because any linear function of a normally distributed quantity has a normal distribution. Furthermore, it is not hard to see that this quantity has a mean of 0.0, and a standard deviation of σ. Thus, the distribution of the pivotal quantity $\mu - Q$ is completely known: it is normal with mean 0 and standard deviation σ. Given a value of Q, $Q(s)$, we may invert the process and treat the pivotal quantity $\mu - Q(s)$ as giving a *fiducial* distribution for μ.[4] It does not give a classical frequency distribution, since μ and $Q(s)$ are both fixed constants and their difference is therefore not a random quantity. For any real z the fiducial probability that $\mu - Q(s) \leq z$ is formally a distribution function, and gives rise to fiducial probabilities.

An important application of fiducial probability, which can also be thought of as an application of confidence methods, is pervasive in the theory of measurement. It is generally supposed that errors of measurement are distributed normally about the true value of the quantity being measured. Thus, a well-calibrated balance yields results (in its design range) that are normally distributed about the true value of the weight being measured. Furthermore, we often know enough about the procedure of measurement to claim to know the standard deviation characteristic of the procedure. Of course, to the extent that the normal distribution can be taken to characterize the errors of measurement, an error of any size is *possible*; there is no limit to the magnitude of the error we might make. But an interval estimate of the quantity being measured at a given level of confidence can thus be constructed in accord with either the fiducial approach or the confidence approach (which in this context come to the same thing).

Example 8.3. *Suppose that we know (on the basis of previous uncertain inferences) that the distribution of errors of power measurement characterizing the dynamometer we use for testing tractors is $N(0 \text{ hp}, 2.5 \text{ hp})$. We test tractor 4106973 and obtain a reading of*

---

[4]This is Fisher's term; the fiducial distribution gives rise to fiducial probabilities, which Fisher distinguished from proper, long-run frequency distributions.

137.3 hp. *How powerful is this tractor? We don't know for sure. But we can say something with confidence or with fiducial probability: namely, that the 95% confidence interval for the power of the tractor is* [137.3 − 6.25, 137.2 + 6.25] = [131.05, 143.85], *because a normally distributed quantity lies within 2.5 standard deviations of its mean 95% of the time. Or, we could say that the* fiducial *probability that the true value lies in this interval is* 0.95.

Fisher's approach has the benefit that under the appropriate circumstances, we can calculate the fiducial probability that μ falls in any interval we please (or indeed any Borel set[5]) and not just the specific intervals picked out by confidence methods. It has the drawback that in any but the simplest cases the existence of pivotal quantities is controversial (even Fisher may have got it wrong on occasion[6]), and the validity of the method depends on the principle that " ...no subset of the entire set, having a different proportion, can be recognized" [Fisher, 1956, p. 109]. This principle is one to which we will return in the next chapter; it is not always clear how to apply it, especially in the face of vague or approximate data. We shall attempt to provide a precise and usable counterpart of the principle.

These characteristics have rendered fiducial inference suspect and controversial, though no one doubts Fisher's fundamental contributions to the discipline of classical statistics. Note that the interpretation of a fiducial interval is quite different from that of a confidence interval. The confidence interval grows out of a concern for controlling long run frequencies of error. The fiducial interval arises from direct inference concerning a pivotal quantity. According to Fisher, no such inference is possible for "discontinuous cases, such as the binomial distribution, ... [in which] a single observational value corresponds ... to a whole range in the values of the pivotal quantity" [Fisher, 1956, p. 118]. As we have seen, confidence methods do apply to the parameter p of a binomial distribution.

The interpretation and application of confidence estimates is also a matter of considerable controversy in practice. First, it might be thought that the fact that we obtain our sample first, before calculating the upper and lower confidence bounds for the parameter in question, somehow vitiates the inference procedure. In fact, this is not the case, even on the most conservative and classical views. The description of the confidence interval that we should focus on is the *general* description, in which the bounds are given *as a function of the observed frequency*. What we can correctly say is that a 95% confidence interval has a 0.95 probability—by which we mean a long run frequency or a propensity—of *covering* the target parameter, provided the conditions for making such an inference are satisfied. Alternatively, to focus on the avoidance of error, the 95% confidence interval has a chance of no more than 0.05 of being in error: of failing to cover the parameter.

On the classical view, however, one cannot say that the probability is 0.95 that a *particular* confidence interval—one obtained after looking at the data—includes the parameter being estimated. One can no more say this than one can say that the probability of heads on the very next toss of this coin is a half.[7] Neither the next toss nor

---

[5]A Borel set is an element of an algebra that includes all intervals and is closed under complementation and countable unions.

[6]The specific problem at issue, referred to as the Behrens–Fisher problem, concerns the development of a pivotal quantity for the parameters of a bivariate normal distribution [Bartlett, 1937].

[7]Alan Birnbaum once remarked somewhere that while one can learn not to *say* this, it is very hard to learn not to think it.

the interval estimate of the parameter p is the kind of thing that, on the classical view, can *have* a probability. We can only use probabilities to describe mass phenomena: long run frequencies, propensities, and the like do not characterize particular events in the right way. (Propensities do apply to single cases; that was what they were invented for. But they do not differentiate between observed and unobserved cases. They do not take account of our knowledge. The propensity of the last toss of the coin to yield heads is 0.5, just as the propensity of the next toss of the coin to yield heads is 0.5. But for assessing evidence, we need to be able to take account of what we know.)

Second, there is the question of the meaning of the statement inferred. To say that the frequency with which items produced by a certain process will pass inspection is at least 0.638 seems to suppose that the manufacturing process is frozen in time—that no improvements can be made. Populations have a tendency to change their character under our very eyes, and in many cases we have the power to change the populations by our actions.

Third, there is the question of the premises of the inference. These premises are often called "assumptions", but to call a premise an "assumption" doesn't eliminate the need for it to be justified in order for it to yield a warranted conclusion. Thus, all standard formulations of confidence interval inference suppose that the population being sampled is indefinitely large, that the sampling is being done "at random",[8] and that the conclusion of the inference applies to that same indefinitely large population. All standard formulations of confidence interval estimation assume that the *form* (binomial, normal) of the underlying distribution is known.

In most cases, not only are these things not known, but in fact we know that they are false. No actual population is literally infinite. There may be philosophers who would dispute that generalization (though there are and will be at most a finite number of them), but surely any natural population is bounded in space and time. There are no longer any dinosaurs, and the time will come when there are no more people. The boundaries of such populations may be fuzzy, but however these boundaries are construed, the cardinalities of the populations are finite.

This has the consequence that no random quantity can have a continuous distribution. Far from being the universal solvent for all statistical problems, the normal distribution *never* applies. To be sure, it may come very close to describing an actual population, and in fact that may be all we need ask of it. We may nevertheless ask for an explanation of its usefulness in particular cases.

Not only is it the case that ordinary sampling procedures are not known to be random, it is known for sure that they are not random. There is one exception. If we have a finite population, severely constrained in time and space, we may be able to assign numbers to each of its members, which can then be sampled with the help of table of random numbers. This is often presented as the "ideal" of which real sampling procedures are more or less pale imitations. But of course the numbers in the table of random numbers are not really random: they are pseudorandom—generated by a deterministic mathematical function.

With regard to the adequacy of this procedure, we should note that, even in the finite case, there is a finite probability that the sampling procedure so conducted

---

[8]This is one of the most difficult concepts in statistics. We shall return to it in the next chapter.

will yield an obviously inappropriate sample: for example, in sampling smokers to evaluate the carcinogenicity of smoking, we must face the probability that we will get a sample of smokers each of whom is the child of parents who both died early of some form of cancer. Any sensible researcher would reject this sample; but to reject it in this case is exactly to reject random sampling, and to undermine the frequency justification of error analysis.

Furthermore, when it comes to stochastic processes, such as the tosses of a coin or the drawing of cards from a deck, there is no way in which we can sample the process in such a way that each equinumerous sample has the same probability of being chosen: we must accept the cards as they fall.

In general, in doing science, we are interested in more than a bounded finite population. Even if we can randomize our selection from part of the population, we will not be selecting at random from a general population dispersed in time and space. Our sample cannot contain individuals remote from us in either space or time.

What we can know is that our sampling procedure is, while not random, not biased in certain respects. This reflects actual substantive knowledge about the sampling procedure. It is with biases that serious discussions about sampling are concerned.

All of these objections are objections only to the rhetoric that accompanies discussions of classical statistics. Given that we know how to handle uncertain inference, there is no reason that statistical inference, like other forms of inference, should not have empirical premises. If we can point to data that makes it reasonable to treat a certain population as (approximately) binomial, there is no reason for us not to take that claim as a premise. If we have evidence that shows that a certain sampling procedure is biased in a significant way, we should accept that conclusion, and admit that inferences based on that procedure are unwarranted unless that bias can be taken into account. There is more to be said on this topic, and we will return to it in the next few chapters.

On the positive side, the inferences that are discussed in classical statistics are exactly the kind of nonmonotonic, inductive, uncertain inferences that we want to be able to employ in learning from experience, from experiments, and from our peers (who usually—frequently—tell the truth). Classical statistical inference yields conclusions that expand our knowledge at the risk of incorporating falsehoods in our body of knowledge. This is all right, because increasing evidence will lead to new accepted statements that will replace the erroneous ones. Classical inference also seeks to find ways to tie probability to this process of inference, via the consideration of the probability of various kinds of errors. A recent book, *Error and the Growth of Experimental Knowledge* [Mayo, 1996], seeks to exploit these insights in developing a complete philosophy of science.

8.3   Bayesian Statistics

Bayesian statistics constitutes an alternative approach to the problems of statistics. This approach depends on the exploitation of Bayes' theorem. Recall that Laplace sought to found inverse inference on the principle of indifference. Lacking both a clear and consistent formulation of the principle of indifference, and grounds for employing the principle even if we could formulate it adequately, most statisticians before 1930 abandoned the hunt for a way to employ inverse inference in statistics.

Whereas classical statistics depended on an empirical concept of probability, the approach to statistics that came to be known as Bayesian (founded by Bruno de Finetti [de Finetti, 1980] and made popular by L. J. Savage [Savage, 1954]) took probability to be the expression of opinion: it adopted the subjective interpretation of probability first defended by Frank Ramsey [Ramsey, 1931]. The objects in the domain of the probability function, on this view (as on the logical view), are propositions: exactly the kinds of things for which we would like (sometimes) to have high probabilities. The probability, for an agent, of a proposition represents the degree of belief that that agent has in the proposition. The drawback of inverse inference was that it required prior probabilities. On the subjective interpretation of probability, these prior probabilities not only make sense, but are accessible: they merely represent the actual opinions of the agent.

Discovering the actual opinions of the agent—the agent's actual degrees of belief—can be difficult, and we have little reason to think that, if explored in detail, these degrees of belief would satisfy the probability calculus, as they are required to do. In practice, many statistical problems are characterized by special distributions (adopted as prior probabilities) that seem "reasonable" for particular problems and are at the same time mathematically tractable. The view may thus be seen to merge imperceptibly into the view that takes prior probabilities to be determined by the logical structure of the problem at hand.

Bayesians differ according to the factors they take into account when devising the prior distributions on which, on their view, all statistical inference is based. (I. J. Good, a well-known adherent of Bayesian statistics, has argued that there are 46,656 varieties of Bayesians [Good, 1971].) Some writers come close to the camp of the logical theorists, differing only in that they suppose that the appropriate language for a given problem may vary according to the problem at issue [Bacchus, 1992]. Others find it reasonable to derive a prior probability distribution from the inferential problem itself, by beginning with a "least informative prior"—a prior distribution that maximizes the informational entropy, perhaps in the sense of Shannon [Shannon & Weaver, 1949]. For example, in the case of a die we minimize the negative sum $\sum_{i=1}^{6} p_i \log p_i$, where p_i is the probability that i dots will result from a toss. This leads to the same result as the principle of indifference in this case; but while the principle of indifference leads to no solution in computing a prior probability distribution for a die with a known expectation of dots of, say, 3.0, we can maximize entropy subject to a constraint to get an answer in this case. The procedure is not uncontroversial [Seidenfeld, 1986], but at least it does yield results that its adherents can agree on. This approach is primarily due to Jaynes [Jaynes, 1958], but it is also followed by many writers on artificial intelligence, e.g., Cheeseman [Cheeseman, 1985].

One interesting feature of the subjectivist approach to statistics is that whereas the classical approach makes much use of (indeed, one may say, is essentially limited to) the study of independent and identically distributed (iid) quantities, the Bayesian approach must work around the lack of them. The outcome of this toss of a coin is relevant to the existence of bias in the coin, and thus relevant to the outcome on the next toss. Thus, the introduction of *exchangeable* quantities introduced in Chapter 4 was needed precisely because the outcomes of a set of coin tosses, on the subjective view, are not binomially distributed: knowing one outcome has an effect on the next. The outcomes are not independent of one another, and the distribution of the quantity representing the outcome on the next trial is different from the distribution of the quantity representing the outcome of the trial after next, given the outcome of the next trial. This fact is not a criticism of the Bayesian approach, but it does make the Bayesian approach somewhat more complicated mathematically, as we saw.

The attractiveness of the Bayesian approach is partly due to the fact that it gives us just what the classical approach to statistics denies us: probabilities for particular hypotheses. We can calculate the probability of the hypothesis H_1 that the long run frequency of heads in tosses of this coin is in the interval $[p, q]$.

More than this: What the classical approach gives us in the way of rejecting or accepting statistical hypotheses is, from the subjective point of view, unnecessary. What do we want statistical inference for? To bridge the gap between observed frequencies, on the one hand, and decisions based on the maximization of expectation on the other. Once we have probabilities—the probabilities of specific events, which the classical view denies us—the acceptance and rejection of hypotheses is gratuitous. Acceptance and rejection serve no function, or at best an intermediary function, in choosing among actions.

Furthermore, it can be argued that acceptance and rejection of hypotheses serve no epistemological function either. We have looked at such arguments in Chapter 4. The idea is that rationality consists in assigning to each proposition its appropriate degree of belief, which, in the view of both many subjectivists and many logicists, should always be less than 1.0. It should be observed that this view usually depends on a sharp distinction between "hypotheses" and "observations"; we update our degree of belief in hypotheses by conditionalizing, via Bayes' theorem, on our observations. But our *observations* are simply "accepted". Similarly, in most practical situations there is a large body of background knowledge that is simply accepted ("presupposed") without question, or taken for granted. In real life, then, the idea that every proposition should be assigned its appropriate degree of belief less than 1.0 is less than universally honored.

One of the most serious drawbacks to the subjectivist approach to statistical inference is the dependence of conclusions on the prior distributions with which one starts, about which there may be little agreement. It is argued that this is of little moment [Skyrms, 1966], in view of the fact that evidence will push two people toward agreement as the amount of evidence increases. But this argument depends on two premises: in the simplest case that the two people regard a sequence of events as exchangeable, and that neither has an "extreme" view: i.e., regards the outcome in question as certain or impossible, or (what is sometimes forgotten) regards the occurrences in the sequence as *independent*. This argument also depends on the acceptance of conditionalization as a mechanism for updating degrees of belief, and we have noted that the arguments for conditionalization are weak [Kyburg et al., 1990]. Furthermore, there is the problem that there is no reason that the agent cannot scrap his entire prior probability, or the part of it that leads, via the evidence, into conflict with his present beliefs. The normative element in the Bayesian approach is very delicate and must be handled gently. We want people caught in a "contradiction" in their degrees of belief to back out of the conflict gracefully, but we don't want them to do so by adopting beliefs that conflict with our own.

From the present point of view, these tensions are symptoms of a broader conflict. As we pointed out in Chapter 4, probabilistic inference, in the sense of inference about probabilities, is not uncertain inference, but deductive inference with a certain kind of subject matter, viz., uncertainty. Because Bayesian inference—inference based on Bayes' theorem—is a special case of probabilistic inference, this observation holds *a fortiori* for Bayesian inference. Important computational methods, such as the methods involving Bayesian networks, fall under this heading as well. Though such techniques are important and valuable, and though they have a lot to do with uncertainty, they do not represent any kind of uncertain *inference*, and it is that that is our proper topic.

Classical statistics does involve leaping to conclusions (uncertain inference) that may be subject to refutation and withdrawal (nonmonotonic inference). It includes at least some forms of what has been traditionally thought of as inductive inference— inference from particular observations and premises to a conclusion that is not entailed by those premises but is more general than the premises. Bayesian statistics does not involve any of these things directly.

Now there is one way in which the Bayesian approach to statistics can impinge on our topic of uncertain inference. If high probabilities are necessary for uncertain or inductive inference, Bayesian calculations could be an extremely important ingredient in arriving at these probabilities [Korb, 1992]. Relatively few Bayesians make an effort to accommodate induction. Most, like Carnap [Carnap, 1968], take the rational thing to do to be to assign to every statement its appropriate degree of belief, and never to *accept* any statement not entailed by the evidence. To the extent that rational agents appear to make inductive inferences, accepting hypotheses (for example, the atomic theory of matter) that are not entailed by evidence, this "acceptance" is to be construed as a loose way of speaking, representing what is more accurately described as the assignment of a very high probability to the hypothesis in question. In general, then, but with some exceptions, those of Bayesian persuasion do not discuss the acceptance of uncertain conclusions.

It thus seems that if uncertain inference is possible at all, we are more likely to get illumination from classical statistics than from Bayesian statistics. Nevertheless, Bayesian approaches, particularly in the form of robust inference [Huber, 1977], may be important in handling uncertainty, in part because we are often prepared to ignore the source of the required prior distributions. Calling them "subjective" allows us to treat them simply as given and thus (in some sense) immune to demands for justification. On the other hand, it is important to remember that the mathematical and computational machinery developed by Bayesian writers is perfectly acceptable from a classical point of view, provided only that the prior probabilities can be given an empirical (frequency or propensity) interpretation.

8.4 Summary

Statistical inference reveals in its foundational controversies many of the problems of inductive or uncertain inference. *Direct inference* is the inference to the composition of a sample of a population of a premise concerning the whole population. Direct inference may have two forms. One is strictly deductive, as the inference from "the coin is fair" to "the probability is high that about half of the next thousand tosses will exhibit heads". The other form of direct inference would assert, from the same premise, "about half of the next thousand tosses will exhibit heads," without any qualification, but with the understanding that the inference may need to be withdrawn in the light of further evidence.

Inverse inference is the inference from a part of a population to the population as a whole. It may be similarly divided: from the fact that about half of a long sequence of tosses have landed heads, we may infer that the coin is *probably* fair; or from the same evidence we may infer, subject to further information, that the coin *is* fair.

Direct inference, in the first form, is deductively valid: Given that the coin (and tossing apparatus) is fair, it is deductively certain that the probability that about half of a long sequence of tosses yields heads is very high.

Inverse inference carries no such deductive guarantee. To obtain that guarantee requires additional premises. Believing that such additional premises could not be justified, the founders of classical statistical inference—Fisher, Neyman, and Pearson—concentrated on the (probable) prevention of errors in the tests of statistical hypotheses.

Thus, significance testing: we test the null hypothesis of no effect. If what we observe is incredible (very improbable) *given* that hypothesis, we reject the hypothesis of no effect, which is to accept the hypothesis that there is some effect, with small risk of error.

A *test* of a statistical hypothesis is a rule that specifies circumstances under which a hypothesis is to be rejected. For example: Examine a random sample of a shipment of widgets; if more than 10% of the sample are defective, reject the hypothesis that less than 5% of the shipment are defective.

The theory of testing statistical hypotheses introduced the consideration of two *kinds* of error: Type I error, consisting in the rejection of the hypothesis under test, given that it was true; and Type II error, consisting in the failure to reject the hypothesis under test given that it was false—i.e., given that some contrary hypothesis was true.

In the case of testing one simple statistical hypothesis (the long run relative frequency of white balls is 0.30) against another simple statistical hypothesis (the long run relative frequency of white balls is 0.60), these two conditional probabilities can readily be calculated. In other circumstances their upper and lower bounds can provide guidance.

Although this classic approach provides for the (tentative) rejection—and, ipso facto, acceptance—of inductive conclusions, it does not provide any way of assigning probabilities to these conclusions. Although we can be 90% confident that a given confidence interval includes its target parameter, we cannot say that the probability that it does so is at least 0.90.

The Bayesian approach to statistics does lead to such probabilities. It does so by providing the additional premises that the classical theorists claimed to be unjustified. These premises are the *prior probabilities* of the relevant hypotheses. Because for most Bayesians these prior probabilities represent personal opinions or the distribution over the space of hypotheses that is "minimally informative", no justification (on their view) is needed.

What are we to think of all this? First, that the house of statistics is divided according to the interpretation of probability that is adopted. We need to settle on an interpretation of probability in order to settle on a global approach to statistical inference. This is what we shall do in the next two chapters. Given the approach to probability that we settle on, we shall see that each of these approaches to statistics

8.5 Bibliographical Notes

It is interesting that the field of statistics is close enough to its origins that there are usable textbooks written by the founding fathers. Thus, Neyman's *First Course in Probability and Statistics* [Neyman, 1950] falls in this category, as does Fisher's *Design of Experiments* [Fisher, 1971]. Cramér's text [Cramér, 1955] and my own [Kyburg, 1969] are more recent contributions, the first reflecting a frequency view, the second a logical view rooted in relative frequencies. For a strongly subjectivistic text, see [Lindley, 1965]. For a thorough treatment of the classical approach to statistical inference, [Lehman, 1959] is excellent and up to date, despite its years. An excellent book written from a philosophical point of view that pays close attention to the issues involved in statistical inference is [Baird, 1992]. For more general discussions of the issues raised by subjectivism, see the Bibliographical Notes of Chapter 4.

8.6 Exercises

(1) Describe the most powerful test of the hypothesis that a biased coin yields 75% heads, against the hypothesis that the coin yields 25% heads, at the level of 0.5, making use of a sample of 15 flips. What assumptions are involved?

(2) Why not use Laplace's rule of succession to generate a mortality table for humans—given that a baby survives its first day, the probability that it will survive its second is $\frac{2}{3}$, that it will survive its third is $\frac{3}{4}$, . . .?

(3) Suppose that blue and pink flowers are simply inherited in sweet peas, and that blue (B) dominates over pink (b). This means that a blue flowering pea is either (Bb) or (BB), but a pink flowering pea is always (bb). Let us take the blue flowering offspring of a hybrid cross ((Bb) with (Bb)), and cross it with itself. We obtain five blue flowering plants and no pink flowering plants. Our original plant is a hybrid if and only if the long run of flowers it produces is 75% blue and 25% pink. What is the probability of this statistical hypothesis? What assumptions are involved? Do we have good grounds for them? What is the nature of their grounds? Is this inference classical or Bayesian?

(4) Suppose you want to test the hypothesis that classical music makes plants grow better. Formulate the hypothesis in a testable form. Describe how you would set up a test of this hypothesis. What are the two possible outcomes? How would you interpret them?

(5) (a) When a measurement is made in the laboratory, it is done by reading an instrument. Instrument readings do not give the true values of the quantities measured: every measurement is subject to error. The standard view of measurement error (which we neither endorse, nor reject, nor attempt to explain) is that random errors are distributed normally, with a mean of 0 and a variance that is characteristic of the method of measurement. What does this standard view *mean*?

   (b) You make a measurement, and read 8.65 units. You conclude with 0.99 confidence that the quantity measured lies between 8.50 and 8.80 units. What is going on? What is the inference, and what justifies it? And what is the standard deviation of the error of measurement characteristic of the

(6) Assume the same standard view of measurement error. Suppose you work for a hematology laboratory that makes thousands of BPH determinations a day. Of the million or so determinations made in a year, how many would you expect to be in error? As a responsible physician, how do you treat the results from the laboratory? What are the issues raised by this example?

(7) "Because the individuals in the experiment were assigned treatments arbitrarily rather than at random, we can draw no conclusion from the experimental results." Explain the distinction at issue here, and explain where it fits in the two major schools of statistical inference, the classical and the subjective Bayesian. Note that exactly the same assignment of treatments to individuals might have been made under a randomized procedure. Why should anyone think it makes a difference?

(8) Why do many statisticians reject the idea of prior probability in statistical inference? How can other statisticians insist that we always have to make assumptions, so assuming a prior distribution is plausible? Is there any hope of reconciliation?

(9) What is the relation between statistical inference and other forms of nonmonotonic inference? What role does acceptance play?

Bibliography

[Bacchus, 1992] Fahiem Bacchus. *Representing and Reasoning with Probabilistic Knowledge*. The MIT Press, 1992.

[Baird, 1992] Davis Baird. *Inductive Logic: Probability and Statistics*. Prentice-Hall, Englewood Cliffs, NJ, 1992.

[Bartlett, 1937] M. S. Bartlett. Properties of sufficiency and empirical tests. *Proceedings of the Royal Society*, 160:268–282, 1937.

[Carnap, 1968] Rudolf Carnap. Inductive logic and inductive intuition. In Imre Lakatos, editor, *The Problem of Inductive Logic*, pp. 268–314. North Holland, Amsterdam, 1968.

[Cheeseman, 1985] Peter Cheeseman. In defense of probability. In *IJCAI 1985*, pp. 1002–1009. IJCAI, Morgan Kaufman, 1985.

[Clopper & Pearson, 1934] C. J. Clopper and E.S. Pearson. The use of confidence or fiducial limits illustrated in the case of the binomial. *Biometrika*, 26:404–413, 1934.

[Cramér, 1955] Harald Cramér. *The Elements of Probability Theory*. Wiley, New York, 1955.

[de Finetti, 1980] Bruno de Finetti. Foresight: Its logical laws, its subjective sources. In Henry E. Kyburg, Jr., and Howard Smokler, editors, *Studies in Subjective Probability*, pp. 53–118. Krieger, Huntington, NY, 1980.

[Fisher, 1922] Ronald A. Fisher. On the mathematical foundations of theoretical statistics. *Philosophical Transactions of the Royal Society Series A*, 222:309–368, 1922.

[Fisher, 1930] Ronald A. Fisher. Inverse probability. *Proceedings of the Cambridge Philosophical Society*, 26:528–535, 1930.

[Fisher, 1956] Ronald A. Fisher. *Statistical Methods and Scientific Inference*. Hafner, New York, 1956.

[Fisher, 1971] Ronald A. Fisher. *The Design of Experiments*. Hafner, New York, 1971.

[Good, 1971] I. J. Good. 46656 varieties of bayesians. *American Statistician*, 25:62–63, 1971.

[Huber, 1977] Peter J. Huber. *Robust Statistical Procedures*. SIAM, Philadelphia, 1977.

[Jaynes, 1958] E. T. Jaynes. Probability theory in science and engineering. *Colloquium Lectures in Pure and Applied Science*, 4:152–187, 1958.

[Korb, 1992] Kevin Korb. The collapse of collective defeat: Lessons from the lottery paradox. In *PSA92*. Philosophy of Science Association, 1992.

[Kyburg, 1969] Henry E. Kyburg, Jr. *Probability Theory*. Prentice Hall, Englewood Cliffs, NJ, 1969.

[Kyburg et al., 1990] Henry E. Kyburg, Jr., Fahiem Bacchus, and Mariam Thalos. Against conditionalization. *Synthese*, 85:475–506, 1990.

[Laplace, 1951] Pierre Simon Marquis de Laplace. *A Philosophical Essay on Probabilities*. Dover, New York, 1951.

[Lehman, 1959] E. L. Lehman. *Testing Statistical Hypotheses*. Wiley, New York, 1959.

[Lindley, 1965] Dennis V. Lindley. *Introduction to Probability and Statistics*. Cambridge University Press, Cambridge, U.K., 1965.

[Mayo, 1996] Deborah Mayo. *Error and the Growth of Experimental Knowledge*. University of Chicago Press, Chicago, 1996.

[Neyman, 1950] Jerzy Neyman. *First Course in Probability and Statistics*. Henry Holt, New York, 1950.

[Ramsey, 1931] F. P. Ramsey. *The Foundations of Mathematics and Other Essays*. Humanities Press, New York, 1931.

[Savage, 1954] L. J. Savage. *Foundations of Statistics*. Wiley, New York, 1954.

[Seidenfeld, 1986] Teddy Seidenfeld. Entropy and uncertainty. *Philosophy of Science*, 53:467–491, 1986.

[Shannon & Weaver, 1949] Claude E. Shannon and Warren Weaver. *The Mathematical Theory of Communication*. University of Illinois Press, Urbana, 1949.

[Skyrms, 1966] Brian Skyrms. *Choice and Chance: An Introduction to Inductive Logic*. Dickenson, Belmont, CA, 1966.

9

Evidential Probability

9.1 Introduction

The idea behind evidential probability is a simple one. It consists of two parts: that probabilities should reflect empirical frequencies in the world, and that the probabilities that interest us—the probabilities of specific events—should be determined by everything we know about those events.

The first suggestions along these lines were made by Reichenbach [Reichenbach, 1949]. With regard to probability, Reichenbach was a strict limiting-frequentist: he took probability statements to be statements about the world, and to be statements about the frequency of one kind of event in a sequence of other events. But recognizing that what concerns us in real life is often decisions that bear on specific events—the next roll of the die, the occurrence of a storm tomorrow, the frequency of rain next month—he devised another concept that applied to particular events, that of *weight*. "We write $P(a) = p$ thus admitting individual propositions inside the probability functor. The number p measures the weight of the individual proposition a. It is understood that the weight of the proposition was determined by means of a suitable reference class, ... " [Reichenbach, 1949, p. 409]. Reichenbach appreciated the problem of the reference class: "... we may have reliable statistics concerning a reference class A and likewise reliable statistics for a reference class C, whereas we have insufficient statistics for the reference class $A \cdot C$. The calculus of probability cannot help in such a case because the probabilities $P(A, B)$ and $P(C, B)$ do not determine the probability $P(A \cdot C, B)$" [Reichenbach, 1949, p. 375]. The best the logician can do is to recommend gathering more data.

Another strand leading to the idea of evidential probability is the recognition that we do not know the precise values of most limiting frequencies or class ratios. We have adequate evidence to the effect that coin tosses yield heads *about* half the time, that live human births result in *a little over* 50% males, etc. Taking seriously the fact that our knowledge of probabilities is not precise is relatively recent [Kyburg, 1961; Smith, 1961; Good, 1962]; a thorough exploration of interval-valued probabilities is [Walley, 1991]. If we admit *approximate* statistical knowledge, then, unless we arbitrarily limit what we count as an approximation, we always have statistical knowledge about the occurrence of one kind of event (object) in a sequence of other events (set of objects), if only the vacuous knowledge that it occurs between 100% of the time and 0% of the time. Incorporating this strand into our notion of evidential probability only aggravates the problem of the reference class, of course: it means that every class to which an object or event belongs must be considered as potentially contributing to its probability.

This brings up the third strand: in order to accept even approximate statistical knowledge into our bodies of knowledge, we need an inductive or nonmonotonic logic: a system of rules spelling out the conditions under which uncertain statements can be accepted as a basis for probabilities, and the conditions under which those statements must cease to be accepted, and be replaced by other statements.

The fourth strand is taken from the logical and subjective conceptions of probability: If probabilities are to be assigned to statements, then equivalent statements should be assigned the same probability. This is certainly the case for *logically* equivalent statements; we shall also claim that it holds for statements *known* or *maximally believed* to be equivalent.

Here, then, is the framework that sets the problems we want a theory of evidential probability to solve: Every statement in our formal language is believed to be equivalent to a large number of membership statements. Many of these membership statements can, potentially, be assigned probabilities on the basis of direct inference making use of known (generally empirical and approximate) statistical statements. We must clarify and articulate the structure of direct inference. In any given case, there may be a number of competing and conflicting direct inferences that can be made. These may involve different classes to which an object is known to belong, or they may involve different objects belonging to different classes, by way of the principle that equivalent statements should have the same probability. We must lay down principles according to which we can adjudicate these conflicts. If probabilities are to be based on empirical knowledge, we must establish that we can *have* such knowledge. It is particularly pressing to show that a plausible form of nonmonotonic or inductive inference based on high probability can account for this knowledge without circularity.

These strands come together in a large and highly integrated package. There are a number of problems that arise at each step in this agenda. In the present chapter we shall examine the basic structure of direct inference. We will deal with the problem of choosing the right reference class, given a body of knowledge. In Chapter 10, we shall be concerned with the semantics of probability, and explore the connections between probabilities and frequencies in the world. In Chapter 11, we will offer a treatment of nonmonotonic or inductive logic based on probability, and show that empirical evidence can render approximate statistical generalizations acceptable without circularity and without ungrounded assumptions. Finally, in Chapter 12, we will consider a variety of applications of uncertain inference to scientific inference in general, and argue that they are adequately explained by evidential probability construed as providing an inductive logic.

9.2   Background Issues and Assumptions

What do we need probability for? We want probability as a guide in life. For it to function as a *guide*, we require that it be objective rather than subjective. Two agents, sharing the same evidence—that is, agreeing on the objective *facts* about the world—should assign the same probabilities. (We do not assume that they assign the same utilities, and thus we do not assume that they make the same decisions.)

We focus on *argument*: As in the case of first order logic, we take the formulation of standards of argumentation as the first function of inductive logic. Because arguments take place in a language, it is natural to focus on syntax. We can only *give* an argument in a specific language. We can't regulate argument by first looking at all the worlds in which the premises are true and then checking the truth of the conclusion. Propositional logic is a bit misleading in this respect, because there we *can* look at the possible worlds—the possible truth assignments to the atomic sentences in the premises—and then check that each one of these assignments leads to the truth of the conclusion.

There is another reason for focusing on syntax rather than semantics: We want to tie probability to what we know about frequencies, and there are too many things we know about frequencies. Every object, every event, belongs to many sets of things, and among these sets of things a given property may have (objectively) any given frequency. For example, the next toss of this coin is a member of the set of tosses that yield heads or are identical to the next toss; and also of the set of tosses that yield tails or are identical to the next toss. Almost all the members of the first set land heads; almost all the members of the second set land tails; the next toss is a member of each.

Similarly, given a class of events the frequency with which a given member of that class exhibits a particular property can have any value, according to the way that property is characterized. Thus, almost all emeralds are green; but this emerald is green if and only if it belongs to the set of emeralds that are now green and examined, or unexamined and blue.[1]

One way to deal with the plethora of frequencies is to limit the number of classes that are to be considered relevant. This is what is done by John Pollock when he limits his considerations to "projectible" properties [Pollock, 1990, p. 81 ff]. Unfortunately, despite the best efforts of many writers, the notion of projectibility remains obscure. Characterizing the set of projectible properties is a hard job, and one that no one seems to have managed to do successfully. Even if it is done successfully, however, what counts in determining probability is what is *known* about these properties, and, in virtue of what we take to be the function of probability, what can be said about them in the language we speak. Thus, we need to pick out a set of *terms* of our language; knowledge expressed in these terms will serve to ground our statements of probability.

We do not assert that there is no way of picking out such a set of terms on a semantic basis. Furthermore, if it can be done, it will presumably lead to sets of terms that satisfy the constraints that we shall lay down in the next section. These constraints will be given in terms of the syntax of a given first order language. Thus, the statistical knowledge we have that is relevant to our probabilities will also be relativized to that language. On our syntactical view of the probability relation, probabilities may depend on what language we adopt. Thus, it is important to provide an analysis of the basis on which one language is to be preferred to another as a medium for embodying our knowledge. This is, as already recognized by Goodman [Goodman, 1955], a part

---

[1]This puzzle, or rather an elaboration of it, is due to Nelson Goodman [Goodman, 1955].

of the problem of induction. It is already clear that this choice will depend on what we can defensibly say within the two languages. We must come up with a principled way of choosing among languages.[2] Chapter 12 will be concerned with this inductive issue, among others.

Syntax will give us much of what we want in the way of a framework for criticizing and evaluating arguments, and may even supply us with an algorithm for calculating degrees of partial entailment. But syntax is not the end of the matter, in probability any more than it is in deductive argument. In first order logic it is important to show that every valid argument can be captured by our syntactical rules of inference, and that every consequence drawn from true premises by our rules of inference will also be true. Similarly, we take it to be important to show the connection between probability, as we construe it syntactically, and frequencies of truth among the models of our language. This is the focus of the following chapter.

9.3 The Syntax of Statistical Knowledge

Let us begin by giving a syntactical characterization of the object language in which our knowledge of the world is to be expressed, and in which the terms corresponding to reference classes (or to potentially relevant properties) and the terms corresponding to useful target properties are to be found.

Like Bacchus [Bacchus, 1992] and Halpern [Halpern, 1990], not to mention Kyburg [Kyburg, 1961], we begin with a standard first order language. We allow for names, functions, predicates of any arity (i.e., functions taking any number of arguments and predicates and relations of any degree). We adopt the technique used by both Halpern and Bacchus of using a two-sorted language, with separate variables for objects (or events) and for "field" terms, intended to be interpreted as real numbers. The domain of the object variables (as opposed to the mathematical variables) is to be finite. The numerical terms have their standard interpretation. We include the constants 0 and 1, and the standard operations and relations for real numbers. We include a countable set $\Re$ of canonical terms representing real numbers between 0 and 1, inclusive (for example decimals). These terms are *canonical* in the sense that they will be given the same interpretation in every model; they are to be rigid designators.

The only syntactic novelty in the language is the inclusion of a primitive variable binding operator "%" that connects formulas and real numbers: $\ulcorner \%\bar{\eta}(\tau, \rho, p, q) \urcorner$. The first two places of the four place operator are to be filled by formulas, and the last two by number variables or constants of canonical form; $\bar{\eta}$ represents a sequence of variables.

For example, "$\%x(\text{heads}(x), \text{toss}(x), 0.5, 0.6)$" would be taken to assert that among tosses, between 0.5 and 0.6 yield heads; "$\%x, y, z(L(x, y), L(x, z) \wedge L(z, y), 0.95, 0.99)$" would be taken to assert that almost always when z is judged longer than y and x is judged longer than z, x is judged longer than y. Note that the reference and target terms in this instance are two-place predicates, though the statistical quantifier binds three variables.

The expression $\ulcorner \%\bar{\eta}(\tau, \rho, p, q) \urcorner$ may be given various informal interpretations. Most straightforwardly, it may be taken to assert that the proportion of objects in

---

[2] Some tentative suggestions are to be found in [Kyburg, 1990].

the object domain satisfying the open reference formula ρ that also satisfy the open target formula τ is between p and q.[3] The formula $\ulcorner\%\overline{\eta}(\tau, \rho, p, q)\urcorner$ may also be taken to represent a propensity: the propensity or tendency of objects satisfying ρ also to satisfy τ. We may also take the formula to represent the strength of an analogy: the interval $[p, q]$ representing the strength of the analogy between the subject of the closed formula τ and the subject of the closed formula ρ. We will give formal truth conditions for these formulas in the next chapter.

Formulas of this form represent just the sort of statistical or quasistatistical knowledge that we can realistically suppose ourselves to possess; it is the sort of knowledge that can be obtained from sampling, or from experience or from a general theory. That this kind of knowledge can be inductively obtained on the basis of the same kind of inference we are discussing will be illustrated later, in Chapter 11.

Not all formulas can usefully occur in the second place of the "%" operator—the place holding the reference formula. For example, consider again the fact that almost all the events in {coin tosses that yield heads} $\cup$ {the next toss} are tosses that yield heads. We would never want to say, on this account, that the next toss was almost certain to yield heads. And of course almost none of the events in {coin tosses that yield tails} $\cup$ {the next toss} yield heads. Neither of these subsets of the set of coin tosses should be allowed to prevent our taking the set of coin tosses to be the right reference class for the statement that the next toss will yield heads.

Similarly, the target formula τ must be restricted. The next toss will land "heads" just in case the next toss lands "preads", where "preads" applies to those tosses that land heads in the temporal interval including the next toss, and land at all at any other time. Almost all coin tosses land preads, but we don't want to take that as a basis for attributing a high probability to heads on the next toss.[4]

Furthermore, the constraints on ρ and τ are connected, since we shall often want to conditionalize on an instantiation of τ: We may want to consider pairs of tosses and the probability that the second toss lands heads, *given* that the first toss lands heads.

The most straightforward way to deal with the problem of distinguishing potential reference formulas and potential target formulas in the language is to take our object language to be partly characterized by its set of *target formulas* and its set of *reference formulas*. We then stipulate that a formula $\ulcorner\%\overline{\eta}(\tau, \rho, p, q)\urcorner$ is well formed only when τ is among the official target formulas and ρ among the official reference formulas. In the following section, we will discuss these formulas further.

It will be useful to have a set of rigidly designating real number terms, that is, terms that denote the same number in every (standard) model. For example, "$\frac{3}{7}$" is understood to denote the same number in every model; "the proportion of planets in our solar system that have more than one moon" is a perfectly good term, and

[3]Bacchus [Bacchus, 1992] introduces a similarly motivated notation. "$\%x(\psi, \phi, p, q)$" corresponds to "$p \leq [\psi, \phi]_x \leq q$". This has also been adopted by [Bacchus et al., 1992; Halpern, 1990], and in other places. Note that "$[\psi, \phi]_x$" is a term, while "$\%x(\psi, \phi, p, q)$" is a formula; we can constrain the *expressions* p and q to have certain forms, which is useful. The term "$[\psi, \phi]_x$" is a real number expression, and is accessible by quantifiers and substitution, which seems awkward: "The table is $[\psi, \phi]_x$ meters long."

[4]The philosophically sophisticated reader will find here reverberations of "grue" and "emeroses." Something is "grue" if it is green before the year 2000, and blue afterwards; "emeroses" are things that are emeralds existing before 2000, or roses existing after. The former predicate is due to [Goodman, 1955]; the latter to [Davidson, 1966].

denotes a perfectly well-defined rational number, but *which* number it denotes varies from model to model. It will also be useful to limit the precision of rigidly designated numbers to some specific granularity. We choose the granularity corresponding to N decimal places.

Definition 9.1. Rigid(t) *iff* $\exists s, i(N\text{-sequence}(s) \wedge \text{Range}(s) \subseteq \text{digits} \wedge t = \ulcorner \sum_{i=1}^{N} s(i) \times 10^{-i} \urcorner)$.

9.4 Reference Classes and Target Classes

When we attribute a probability of landing heads of about a half to the next toss of this coin, we are taking some class of coin tosses as the *reference* class, and the subclass of coin tosses landing heads as the *target* class. Generically, we might say: The probability that the object o is in the class T, given that it is in the class R, is about p. In this formula, we take R to be the reference class, and T to be the target class. But, as we just argued, what really concerns us is the set of formulas in our language that can be construed as determining reference classes (formulas that can take the position of R) and the set of formulas in our language that determine target classes (formulas that can usefully appear in the position of T).

9.4.1 Reference Formulas

Our formal characterization of reference formulas and target formulas will be designed to support such considerations as these, and to avoid the kind of morass we might be led to by artificially constructed reference formulas and target formulas. Whereas we cannot accommodate all classes, or even all the classes determined by formulas, we want a rich enough supply of reference classes to meet our needs in science and in everyday life. The game is to provide enough, but not too many. Because this construction is frankly language relative, we may both avoid the knotty metaphysical problems of characterizing projectibility and (more important) be able to modify our characterization of reference formulas should the need arise.

We begin with atomic formulas. Atomic formulas in general consist of predicates followed by an appropriate sequence of terms. For our purposes we restrict our concerns to atomic formulas in which the sequence of terms consists of (object) variables and terms containing no free variables.

Definition 9.2. ϕ *is a restricted atomic formula of* $\mathcal{L}$, $\text{At}^r_{\mathcal{L}}(\phi)$, *if and only if* ϕ *is the concatenation of an n-place predicate of* $\mathcal{L}$ *with a sequence of n terms of* $\mathcal{L}$, *and the only variables free in* ϕ *are among those n terms.*

Thus, the atomic formulas "Taller-than(x, y)," "Taller-than(John, y)," and "Taller-than(the blonde sister of John, y)" are all restricted atomic formulas, but "Taller-than(the mother of(x), y)" is not, since "x" is free in the formula, but is not one of the n terms following the predicate. What might we want to say about the set of pairs $\langle x, y \rangle$ such that the mother of x is taller than y? Ingenuity could perhaps provide some answers, but practically it seems an artificial set of pairs.

It is clear that we need reference classes other than those determined by restricted atomic formulas. We continue the characterization of reference formulas by introducing two kinds of considerations that serve to enlarge the class of reference formulas obtainable from the restricted atomic formulas.

First, we need to take into account collections of predicates, such as color predicates, that generate a partition of the objects they apply to. A partition of a set S is a collection of subsets $S_1, \ldots, S_n$ that exhaust S ($S \subset \bigcup_{i=1}^{n} S_i$) and that are mutually exclusive ($S_i \cap S_j = \emptyset$ when $i \neq j$). A disjunction of partition predicates may be taken as a constraint in a reference formula. We relativize these notions to an open formula ϕ, and we include an unsatisfiable formula.

Definition 9.3. Partition-constraint $(\tau$ in $\psi)$ *if and only if*

$$\exists \phi_1, \ldots, \phi_n (\vdash^{\ulcorner} \psi \supset (\phi_1 \vee \phi_2 \vee \cdots \vee \phi_n)^{\urcorner} \wedge \vdash^{\ulcorner} i \neq j \wedge \psi \supset (\phi_i \supset \neg \phi_j)^{\urcorner}$$
$$\wedge \, \exists i_1, \ldots, i_j (i_1 \leq n \wedge \cdots \wedge i_j \leq n \wedge \tau =^{\ulcorner} \phi_{i_1} \vee \cdots \vee \phi_{i_j}^{\urcorner} \vee \tau =^{\ulcorner} \phi_{i_1} \wedge \neg \phi_{i_1}^{\urcorner})).$$

In words, τ in ψ is a partition constraint if τ is a disjunction of formulas picked out of a set of formulas that form a partition of the objects satisfying ψ or a contradictory formula. A set of formulas forms a partition of ψ if it is a theorem that everything that satisfies ψ satisfies exactly one of these formulas. Theoremhood of the disjunction and logical incompatibility are represented by the turnstiles in the definition.

Example 9.1. *An urn example serves to illustrate the idea of a partition constraint. Suppose we have an urn, filled with balls of various colors, each of which bears a numeral. Take ψ to denote the set of balls in the urn. As noted, the set of color predicates gives rise to a collection of restricted atomic predicates: "is-blue", "is-red", etc. We ensure that each of the balls in the urn satisfies one or another of these predicates by including among them the predicate "is-colorless", which we interpret to mean that the object in question cannot be characterized by one of the specific colors. Thus, we ensure that it is a theorem that every ball satisfies at least one of these color predicates, and that it is a theorem (a consequence of the rules or "meaning postulates" of the language) that no more than one of these predicates can apply to a ball.*

We can now speak of the frequency with which red balls bear the numeral 3 or the frequency with which the numeral 3 appears among balls that are either red or white or blue. We can thus use these classes as reference classes for generating probability statements in answer to such questions as "What is the probability that a red ball will exhibit the numeral 3?"

Second, we need formulas based on quantitative functions or quantities as constraints on reference formulas. We may be interested, for example, in the frequency with which people less than 6 ft 1in. in height contract the flu or bump their heads. We begin by defining atomic terms as we defined atomic formulas:

Definition 9.4. f *is a restricted atomic term of $\mathcal{L}$, $\mathrm{Fn}_{\mathcal{L}}^r(f)$, if and only if f is the concatenation of an n-place functor of $\mathcal{L}$ with n terms of $\mathcal{L}$, and the only variables free in f are among those n terms.*

In general, we may take a primitive quantity f to generate a partition of the objects satisfying a formula ψ, just as the set of colors generates a partition of the set of extended objects. The following definition captures one variety of quantitative term that may reasonably be taken to constrain reference formulas:

Definition 9.5. function-constraint (ρ in ψ) iff

$$\exists f, b(\text{Fn}_{\mathcal{L}}^r(f) \wedge \vdash^\ulcorner \psi \supset \text{Range}(f) \subseteq \mathfrak{R}^\urcorner \wedge \vdash^\ulcorner b \in \text{Borel set}^\urcorner \wedge \rho = {^\ulcorner f \in b^\urcorner}).$$

Some remarks on this brief definition are in order.

(1) The range of a function f is the set of values of $f(a)$ for objects a; the turnstile says that it is to be a rule of the language that the range of f is included in the set of reals.

(2) A set is a Borel set if it can be obtained from open intervals of reals by a finite or countable number of unions and complements.

(3) We could eliminate the need for partition constraints by reconstruing partition formulas as functions: Thus, we could replace 100 color predicates by a single function, "color-of", that would have a hundred numerical values. We refrain from doing this on grounds of simplicity and familiarity.

Example 9.2. *Consider the urn described in Example 9.1. Let N be a function from objects to numbers represented by numerals inscribed on the objects: thus if ball a bears the numeral "3", $N(a) = 3$. We may now employ such statistical statements as: "The proportion of red balls among those whose N-value is 3 is between 0.4 and 0.5", "The proportion of red-or-green balls among the odd-numbered balls is more than 0.8", etc. These can be used to found the corresponding probability statements in the appropriate circumstances.*

Example 9.3. *Important everyday functions include "weight of" and "length of". In a biological experiment, we may be interested in the fertility (frequency of live births) among the members of a species that weigh at least 500 grams. We may be interested in the distribution (that is, the probability distribution) of weights, given (i.e., as a function of) the length of the tibia in a certain species. In each case, we need to be able to take a subset of a population for which the quantity has a certain value, or satisfies a certain inequality, as a suitable reference class.*

The general set of reference formulas is the smallest set including the restricted atomic formulas closed under conjunction, and constrained by function constraints and partition constraints, *but not closed under disjunction.*[5]

Definition 9.6.

$$\mathcal{R} = \bigcap \{ S : \forall \rho (\text{At}_{\mathcal{L}}^r(\rho) \supset \rho \in S)$$
$$\wedge \forall \rho_1, \rho_2(\rho_1, \rho_2 \in S \supset {^\ulcorner (\rho_1 \wedge \rho_2)^\urcorner} \in S)$$
$$\wedge \forall \rho(\rho \in S \supset \forall \rho'(\text{function-constraint}(\rho' \text{ in } \psi)) \vee \text{partition-constraint}$$
$$(\rho' \text{ in } \rho) \supset {^\ulcorner \rho \wedge \rho'^\urcorner} \in S)) \}.$$

For our purposes the set $\mathcal{R}$ of reference formulas consists of the restricted atomic formulas of our language, conjunctions of these formulas, and functional and partition restrictions of these formulas. Note that this also gives us formulas corresponding to

---

[5] The difficulties engendered by disjunctive reference classes are well known: [Kyburg, 1961; Davidson, 1966; Pollock, 1990].

sequences of objects satisfying any of these formulas: $\ulcorner \rho(x_1) \wedge \rho(x_2) \wedge \cdots \wedge \rho(x_n) \urcorner \in \mathcal{R}$. For example, if "coin-toss" is a one-place predicate in our object language, "coin-toss(x)" is a reference term, as is "coin-toss(x_1) $\wedge$ coin-toss(x_2) $\wedge \cdots \wedge$ coin-toss(x_n)".

Similarly, it gives us multidimensional quantitative functions: We can use the statistics about members of the college who are female, and both are more than five feet tall and weigh less than 120 lb, to generate a probability.

Our object here has been to provide as rich a collection of reference formulas as we need for ordinary purposes without admitting the troublesome cases that give rise to arbitrary statistics. Note that *within* a reference class we may even have disjunctions: thus we can consider the frequency of 3s among the balls in the urn that are either red or green.

9.4.2   Target Formulas

The target formulas are those corresponding to the projectible properties, properly so-called.[6] We will specify four different kinds of target properties.

To obtain a set $\mathcal{T}(\rho)$ of target formulas appropriate to a reference formula ρ, we begin with partition constraints. Given a reference property, the set of appropriate target properties should form a field, since in probability theory one expects target properties to form a field. We will call the sets of formulas corresponding to these fields *target fields*. A target field emerges naturally in the case where a family of atomic formulas $\{\tau_1, \ldots, \tau_k\}$ corresponds to a partition of the objects satisfying the reference formula ρ; for example, a set of color predicates would correspond to a partition of any set of extended objects. The corresponding field takes as atoms the formulas $\{\tau_1, \ldots, \tau_k\}$, and as generic elements the sets corresponding to the disjunctions of some of the formulas. The disjunction of all the formulas corresponds to the universal set, and the intersection of any distinct pair of atoms yields the empty set. What we must avoid is the construction of arbitrary algebras—that is the fatal trick behind "grue."

In this case, to be an acceptable target formula is to be a partition constraint in some reference formula:

Definition 9.7. $\mathcal{T}_1(\rho) = \{\tau : \text{partition-constraint}(\tau \text{ in } \rho)\}$.

Example 9.4. *Consider an urn containing balls that may be any of fifteen different colors. The partition of the set of balls into fifteen categories corresponding to their colors leads to a field: the atoms are the formulas like "x is blue"; the union of the set of colors in two elements of the field is the set of colors that belong to either of the two elements (e.g., {"x is red," "x is blue"} $\cup$ {"x is red", "x is green"} is the set of atomic formulas {"x is red," "x is blue," "x is green"}); the intersection of these two elements is the singleton {x is red}.*

A similar definition will capture the formulas of target fields based on the values of quantitative functions. Because the set of Borel sets is a field, the set of formulas of the form $\tau = \ulcorner f = b \urcorner$, where b is a Borel set, is a field.

Definition 9.8. $\mathcal{T}_2(\rho) = \{\tau : \text{function-constraint}(\tau \text{ in } \rho)\}$.

The reference formulas themselves constitute one sort of target term. These give rise to what we shall call target fields in a perfectly straightforward way. For example, if "is-green" is in $\mathcal{R}$, then the field whose members are {"is-green," "is-not-green," "is-green-and-is-not-green," "is-green-or-is-not-green"} is a perfectly reasonable target field, and all the formulas corresponding to elements in that algebra are acceptable target formulas.

Definition 9.9. $T_3(\rho) = \{\tau : \exists \rho'(\rho' \in \mathcal{R} \wedge (\tau = \rho' \vee \tau = \ulcorner \neg \rho' \urcorner \vee \tau = \ulcorner \rho' \wedge \neg \rho' \urcorner \vee \tau = \rho))\}$.

Note that "is-Green" and "is-examined-before-2000" are elements of perfectly acceptable fields, but that there is no provision for combining target fields to obtain a new target field. This is what allows us to avoid projecting "grue."

Finally, in the sequel we need one special form of target formula: that which characterizes the relative cardinality of the set picked out by a target term of one of the first three kinds, and the corresponding reference term.

Definition 9.10.

$$T_4(\rho) = \left\{ \tau : \rho \in \mathcal{R} \right. $$

$$\wedge \exists \tau', b \left(\tau' \in T_1 \cup T_2 \cup T_3 \wedge b \in \text{Borel set} \right.$$

$$\left. \left. \wedge \tau = \ulcorner \frac{|\{x : \tau'(x) \wedge \rho(x)\}|}{|\{x : \rho(x)\}|} \in b \urcorner \right) \right\}.$$

These considerations lead to the following definition of $T(\rho)$, the set of target formulas appropriate to the reference formula ρ.

Definition 9.11. $T = T_1(\rho) \cup T_{2\in}(\rho) \cup T_3(\rho) \cup T_4(\rho)$.

There are thus four kinds of target terms: those coming from predicate partitions of the reference formula ρ; those derived from primitive quantitative functions in the language—the function partitions of ρ; those corresponding to reference formulas themselves; and those representing the ratio of the cardinality of the set of things satisfying a target formula and a reference formula to that of those satisfying the reference formula alone.

For conditionalization to work, we must be sure that the conjunction of an ordinary target formula and a reference formula will yield a reference formula.

Theorem 9.1. $\rho \in \mathcal{R} \wedge \tau \in (T_1(\rho) \cup T_2(\rho) \cup T_3(\rho)) \supset \tau \wedge \rho \in \mathcal{R}$.

The formation rules of the language $\mathcal{L}$ are to include a clause concerning statistical statements in addition to the standard ones for first order logic: If $\rho \in \mathcal{R}$ and $\tau \in T(\rho)$ and $p, q \in \text{Vbls}$ or $\text{Rigid}(p) \wedge \text{Rigid}(q)$ and $\overline{\eta}$ is a sequence of object variables, then $\ulcorner \%\overline{\eta}(\tau, \rho, p, q) \urcorner$ is a well-formed formula.

9.5 Prima Facie Support

In order to approach the issue of computing probabilities in the simplest way, let us first focus on what John Pollock [Pollock, 1990] has called *indefinite probabilities*, and then consider the more complicated issues surrounding probabilities that refer to

9.5.1   Indefinite Probabilities

There is one ordinary English locution involving probability (or one of its cognates) that often wears its reference class on its sleeve. This is the locution employing the indefinite article "a." When we say that the probability of *a* live human birth being the birth of a male is about 0.51, it is pretty clear that we are talking about the set of live human births, and that the intended reference formula is "*x* is a live human birth." Of course, this set doesn't have sharp boundaries: we're talking about "modern" times, and not paleolithic times. By the same token we are talking about the practical future—the next few hundred or thousands of years—and not the geological future of millions of years.[7] Being precise about the scope of this set is no more essential for practical purposes than being precise about the meaning of "about."

We are justified in asserting that the probability of a live human birth being the birth of a male is about 0.51 just in case we are justified in believing that about 51% of the class of live human births result in the birth of a male. What Pollock [Pollock, 1990] calls "indefinite" probability statements are, as he observes, equivalent to statements about classes or sets or kinds.

Consider the set of balls in urn A. Suppose that we know that between 0.2 and 0.3 of them are white, that between 0.4 and 0.6 of them are black, and that the rest are red. (Thus, between 0.1 and 0.4 are red.) We also know that the colors are exclusive. Identifying the probability with the interval, rather than with some unknown point within the interval, we may say that the probability that *a* ball in A is red is [0.1, 0.4]. Because we may also infer that the proportion of non-red balls in the urn is in [0.6, 0.9], we may say that the probability that a ball in A is not red is [0.6, 0.9]. Similarly, we may infer that the proportion of red-or-white balls is in [0.3, 0.7]. Whereas this is true, we may also note that the proportion of red-or-white balls is the same as the proportion of nonblack balls, and that the latter is under sharper constraints: the proportion of nonblack balls lies in [0.4, 0.6] ⊂ [0.3, 0.7]. Thus, we would want to say that the probability that *a* ball in A is red-or-white is the same as the probability that it is nonblack, which is [0.4, 0.6].

Indefinite probabilities obtained by direct inference reflect the properties of proportions.

The same is true of conditional probabilities: The probability that *a* ball is red, given that *it* is not black, is determined by the interval information we have. The lower bound on the conditional probability is 0.25 (corresponding to the minimum frequency of red balls, combined with the maximum frequency of white balls), and the upper bound is 0.67, corresponding to the opposite case.

If we know that between 0.4 and 0.6 of the balls are nonblack, then of course we also know that between 0.3 and 0.7 of the balls are nonblack. The former interval is the one we would want to use to generate probabilities to guide our actions, since it is more precise.

9.5.2   Definite Probabilities

To pass from the indefinite article "a" to definite descriptions like "*the* next toss," "*the* set of bets made in a certain casino on a certain day," "*the* amount of error made

---

[7]In fact, we'd better limit the future relatively severely: when parents can determine the sex of their offspring,

in a particular measurement" is complicated. The indefinite article leads you to a single reference class with no problem; but that very fact means that the probability so obtained is useless as a guide in life. To guide our response to a particular case, we must take account of all that we know. We never have to do with the outcome of a coin toss that is *merely* a coin toss, and not the toss of a coin on some particular occasion, an applicant for insurance who is *merely* a 40-year-old male, and not some unique individual, a measurement error that is *merely* a measurement error, and not the error made in some specific measurement. But as soon as the *mere* event or object becomes clothed with particularity, there are unlimited possibilities for choosing a reference class. The coin toss we are concerned with is the toss of a particular coin, at a certain place and time, performed by a certain person, in a certain way. What collection of properties of this particular toss should we select as determining the class whose proportion of heads we should take as guiding our degree of belief in heads?

We cannot say "all" properties of this toss, or even all namable properties, since that yields a class containing one member—that's the point of definite descriptions! If we knew the frequency in that unit class, we would have no need of probability. If we don't know the frequency—that is, if we know only that it is one of the two extreme frequencies in $[0, 1]$—then we get no guidance at all. We want to say "the *relevant* properties," but that is exactly the idea we need to make precise.

We face a unique situation in each case in which we confront the problem that though we wish we knew whether the sentence S was true or not, we simply don't know which. We start with a background of things we do know: Γ. Various items of knowledge in Γ bear on the truth of S in various ways. Some items obscure the import of others. What we are seeking to do is to sort out the way in which a collection of knowledge bearing on S can be integrated.

The degree of belief a body of knowledge Γ supports in a sentence S may be construed in a number of ways, according to the way in which "%" is construed. It is our conviction, however, that the epistemic bearing of a body of knowledge on a sentence may be cashed out in terms of statistical knowledge on the one hand, and knowledge about individuals on the other.

We will write $\Delta(S, \Gamma)$ for the set of statistical statements—statements of the form $\ulcorner\%\overline{\eta}(\tau, \rho, p, q)\urcorner$—in Γ that bear on S:

Definition 9.12. $\Delta(S, \Gamma) = \{\sigma : \exists p, q, \alpha, \tau, \rho, \overline{\eta}(\ulcorner\rho(\alpha)\urcorner \in \Gamma$
$\wedge \ulcorner S \equiv \tau(\alpha)\urcorner \in \Gamma \wedge \sigma = \ulcorner\%\overline{\eta}(\tau, \rho, p, q)\urcorner)\}$.

Now we will define what it is for Δ to give support of $[p, q]$ to S, by means of the triple of terms α, τ, ρ, with background knowledge Γ. We must take account of the specific terms $\alpha, \tau,$ and ρ in which that knowledge is expressed. The third clause of the definition allows us to ignore information about τ and ρ that is merely weaker than the strongest information we have about those terms.

Definition 9.13 (Support). $\text{Support}(\Delta, S, \Gamma, \alpha, \tau, \rho) = [p, q]$ *iff*
$\ulcorner\%\overline{\eta}(\tau, \rho, p, q)\urcorner \in \Delta \ \wedge \ \Delta \subseteq \Delta(S, \Gamma) \wedge \forall p', q'(\ulcorner\%\overline{\eta}(\tau, \rho, p', q')\urcorner \in \Delta$
$\supset [p, q] \subseteq [p', q']) \wedge \ulcorner\rho(\alpha)\urcorner \in \Gamma \wedge \ulcorner S \equiv \tau(\alpha)\urcorner \in \Gamma.$

A set Δ of statistical statements supports the sentence S relative to background knowledge Γ via the terms $\alpha, \tau,$ and ρ to the degree $[p, q]$ just in case $\ulcorner\%\overline{\eta}(\tau, \rho, p, q)\urcorner$

represents the most precise knowledge we have about τ and ρ. The terms p and q are among our canonical real number terms, since the only other possibility (according to our syntactic rules) is that they are variables, and then $\ulcorner \%\overline{\eta}(\tau, \rho, p, q) \urcorner$ would not be in Γ, which contains only closed formulas.

In what follows, we will look at three grounds for ignoring information in Δ. We have already mentioned that some information may obscure the import of other information. We need principles by which we may judiciously disregard some of our evidence.

The following example shows how a database with more statistical data can be less informative about a sentence S than one with less statistical data.

Example 9.5. *Consider a body of knowledge concerning death rates among 50-year-old females in the city of Rochester, and in the United States as a whole, and concerning the characteristics of an individual j (Mrs. L. W. Q. Jones). It is plausible to suppose that the complete body of knowledge includes (as a result of statistical inference from the actuarial observations) something like*

$$\{\text{``}\%x(L(x), F(x) \wedge \text{US}(x), 0.945, 0.965)\text{''},$$
$$\text{``}\%x(L(x), F(x) \wedge \text{US}(x) \wedge \text{Roch}(x), 0.930, 0.975)\text{''}\}$$

where "$L(x)$" means that x lives for two more years, and "$F(x)$" means that x is a 50-year-old female. The discrepancy between [0.945, 0.965] and [0.930, 0.965] reflects—or may reflect—the fact that the sample on which the first interval is based is larger than the sample on which the second interval is based.

Among the sentences of Γ are "$F(j) \wedge \text{US}(j)$" and "$\text{Roch}(j) \wedge F(j) \wedge \text{US}(j)$", in addition to the mentioned statistical statements.[8]

We have both Support$(\Delta, \text{``}L(j)\text{''}, \Gamma, \text{``}j\text{''}, \text{``}L(x)\text{''}, \text{``}F(x) \wedge \text{US}(x)\text{''}) = [0.945, 0.965]$ *and* Support$(\Delta, \text{``}L(j)\text{''}, \Gamma), \text{``}j\text{''}, \text{``}L(x)\text{''}, \text{``}F(x) \wedge \text{US}(x) \wedge \text{Roch}(x)\text{''}) = [0.930, 0.975]$.

Because there is no conflict between these two constraints—one is just more precise than the other—we can see that the more complete body of evidence does not contribute to constraining our beliefs about "$L(j)$." If anything, it merely weakens what we can say about "$L(j)$". A more precise anticipation is provided by taking Δ to be {"$\%x(L(x), F(x) \wedge \text{US}(x), 0.945, 0.965)$"}.

9.6 Sharpening

We say that one body of evidence is more informative than, or *sharpens*, another with respect to a sentence S, given total knowledge Γ, if it provides an interval more relevant to the truth of S. How do we construe "more relevant"? A narrower interval is surely better (more useful) than a wider one; but at the same time a more complete body of evidence is better (more informative) than a truncated one. These considerations can conflict, as we have seen.

There are two ways in which intervals can disagree. One may be included in the other, for which there is a convenient standard notation. Alternatively, neither may

---

[8]For the purposes of our examples, we shall take Γ to be logically closed. In general, however, there may be good reason for taking Γ to consist of a finite number of closed theories.

be included in the other. We introduce a notation in the object language for this
case:

Definition 9.14. $[p, q]\#[p'q']$ *iff* $\neg[p, q] \subseteq [p', q'] \wedge \neg[p', q'] \subseteq [p, q]$.

There are three ways in which a larger body of information may contain statements
that merely obscure the force of a smaller body of information, and thus three ways
in which a smaller body of evidence can sharpen a larger one. These are *precision*,
specificity, and *richness*. These are described below.

9.6.1 Precision

The more inclusive knowledge represented by Δ may include an item of knowledge
$\ulcorner \%\overline{\eta}(\tau, \rho, p, q)\urcorner$ that supports S to the degree $[p, q]$, where $[p, q]$ includes all the prima
facie degrees of support given by $\Delta - \ulcorner \%\overline{\eta}(\tau, \rho, p, q)\urcorner$ to S. This is illustrated by the
story of Mrs. Jones, where intuitively we would wish to use the more precise mortality
derived from the national statistics as a basis for decision and action, since it does not
conflict with the statistics obtained from Rochester. It makes sense to disregard the
Rochester data in this case, for the same reason that we would want to disregard the
very vague data that might be derived from a very small sample of people who live
in the same apartment house as Mrs. Jones, or the ultimately vague data embodied
in "$\%x(L(x), x = j, 0.0, 1.0)$". We shall call this sharpening by precision. It does not
depend on the structure of the support provided by Δ'_1 that is not provided by Δ_2, but
other forms of sharpening do, so we include the reference to α', ρ', and τ'.

Definition 9.15. Δ_1 *sharpens-by-precision* Δ_2 *in the context* S, Γ *if and only if*
$\Delta_2 \subset \Delta(S, \Gamma) \wedge \exists p', q', \alpha', \tau', \rho', \overline{\eta}(\Delta_1 = \Delta_2 - \{\ulcorner \%\overline{\eta}(\tau', \rho', p', q')\urcorner\}$
$\wedge \forall \alpha, \tau, \rho\, (\text{Support}(\Delta_1, S, \Gamma, \alpha, \tau, \rho) \subseteq [p', q']))$.

Sharpening by precision is illustrated by Example 9.5.

It will be useful to have a characterization of the result of sharpening a set of statistical statements (in the context S, Γ) that cannot be further sharpened by considerations
of precision.

Definition 9.16. $Pr(\Delta, S, \Gamma) = \{\ulcorner \%\overline{\eta}(\tau, \rho, p, q)\urcorner : \ulcorner \%\overline{\eta}(\tau, \rho, p, q)\urcorner \in \Delta$
$\wedge \neg\exists\Delta'(\Delta'$ *sharpens-by-precision* Δ *in the context* S, $\Gamma)\}$

We read "$Pr(\Delta, S, \Gamma)$" as the precisification of Δ. It is clearly unique. One may
think of Δ as divided into tree structures, in which each parent has as children the
statements whose mentioned interval is included in the interval mentioned in the
parent. (Because a child may have more than one parent, this is really a directed
acyclic graph.) The set $Pr(\Delta, S, \Gamma)$ is the set of roots of these trees.

9.6.2 Specificity

More general data may be also rendered irrelevant by more specific data derived from
the background knowledge Γ. Suppose we add to our knowledge of Mrs. Jones that
she is of Polish extraction ("PE(j)"), and suppose we know that the two year survival
rate of 50-year-old U.S. women of Polish extraction is in the interval [0.920, 0.950].
This data conflicts with both the statistics concerning Rochester and the statistics
concerning the U.S. Intuitively (as we shall see in detail shortly), this alone would

undermine the use of either the Rochester or the U.S. interval. There are frequencies allowed by this interval that are not allowed by [0.930, 0.975] : [0.920, 0.950] # [0.930, 0.975].

Example 9.6. *If we add the sentences "PE(j)" and "%x($L(x)$, $F(x) \wedge$ Roch(x) $\wedge$ US(x) $\wedge$ PE(x), 0.920, 0.950)" to the sentences* Γ *of Example 9.5, we get*

$$\Delta("L(j)", \Gamma) = \{ "%x(L(x), F(x) \wedge US(x)0.945, 0.965)",$$
$$"%x(L(x), F(x) \wedge US(x) \wedge Roch(x), 0.930, 0.975)",$$
$$"%x(L(x), F(x) \wedge Roch(x) \wedge US(x) \wedge PE(x), 0.920, 0.950)" \}.$$

Delete "%x($L(x)$, $F(x) \wedge$ US(x)0.945, 0.965)". Then

$$\Delta_1 = \{ "%x(L(x), F(x) \wedge US(x) \wedge Roch(x), 0.930, 0.975)",$$
$$"%x(L(x), F(x) \wedge Roch(x) \wedge US(x) \wedge PE(x), 0.920, 0.950)" \}.$$

Delete "%x($L(x)$, $F(x) \wedge$ US(x) $\wedge$ Roch(x), 0.930, 0.975)". Then

$$\Delta_2 = \{ "%x(L(x), F(x) \wedge US(x) \wedge PE(x), 0.920, 0.950)" \}.$$

Intuitively, Δ_2 *contains the statistical knowledge we want.*

Despite the fact that the interval in the third statement is broader than the other intervals, we should take the specificity of that statistical statement into account, because it *conflicts* with the other statistical statements, and assign a probability of [0.920, 0.950] to "$L(j)$."

Specificity and precision may interact, however. Add to the previous case knowledge that Mrs. Jones is a medical doctor ("MD(j)"), and that the two year survival rate for 50-year-old women of Polish extraction who are medical doctors is [0.925, 0.985]. Because this both conflicts with the statistics for women of Polish extraction (PE) and fails to conflict with the statistics for U.S. women in general, this *undoes* the influence of being of Polish extraction, and leads us back to the U.S. interval.[9]

Schematically, we have

$$%x(L(x), F(x) \wedge US(x) \wedge PE(x) \wedge MD(x), 0.925, 0.985). \qquad (9.1)$$
$$%x(L(x), F(x) \wedge US(x) \wedge F(x) \wedge PE(x), 0.920, 0.950). \qquad (9.2)$$
$$%x(L(x), F(x) \wedge US(x), 0.945, 0.965). \qquad (9.3)$$
$$%x(L(x), F(x) \wedge US(x) \wedge Roch(x), 0.930, 0.975). \qquad (9.4)$$

Given a choice between (9.2) and (9.3), (9.2) should be preferred as a basis. Given a choice between (9.1) and (9.2), it seems natural that (9.1) should be preferred. But given a choice between (9.1) and (9.3), (9.3) should be preferred.

Should (9.3) still be preferred to (9.1) in the presence of (9.2)? We think so, since (9.2) loses to (9.1) by specificity, and could be disregarded; (9.3) does not conflict with (9.1) but merely offers higher precision.

On the other hand, if there were a further refinement that conflicted with (9.1), say "%x($L(x)$, US(x) $\wedge$ $F(x)$ $\wedge$ PE(x) $\wedge$ MD(x) $\wedge$ Roch(x), 0.0.920, 0.960)", but did *not* conflict with (9.2), we would not want to take (9.3) as a basis. The definition of partial proof in Section 9.7 will take this into account.

---

[9]This represents a change from [Kyburg, 1974; Kyburg, 1983]. It was proposed by Bulent Murtezaoglu in

We have embodied this intuition in the following definition of "sharpens-by-specificity."

Definition 9.17. Δ_1 *sharpens-by-specificity* Δ_2 *in the context* S, Γ *if and only if*

$$\exists \bar{\eta}, \alpha', \tau', \rho', p', q'(\Delta_2 \subseteq \Delta(S, \Gamma) \wedge \Delta_1 = \Delta_2 - \ulcorner \% \bar{\eta}(\tau', \rho', p', q') \urcorner$$
$$\wedge \exists \rho''(\ulcorner \forall \bar{\eta}(\rho'' \supset \rho') \urcorner \in \Gamma \wedge \ulcorner \rho''(\alpha') \urcorner \in \Gamma$$
$$\wedge \ \mathrm{Support}(S, \Delta_1, \Gamma, \alpha', \tau', \rho'') \# [p', q']$$
$$\wedge \forall \rho (\ulcorner \forall \bar{\eta}(\rho \supset \rho'') \urcorner \in \Gamma \wedge \ulcorner \rho(\alpha') \urcorner \in \Gamma)$$
$$\supset \neg \mathrm{Support}(S, \Delta_1, \Gamma, \alpha', \tau', \rho)$$
$$\# \mathrm{Support}(S, \Delta_1, \Gamma, \alpha', \tau', \rho''))).$$

In words, this says that we can sharpen a set of statements Δ_2 by specificity in the context S, Γ by deleting the statistical statement by which Δ_2 supports S to degree $[p', q']$, when Γ includes the knowledge that the ρ'''s are a subset of the ρ''s and α' satisfies ρ'', *provided* there is no further formula ρ leading to conflict with the support given by Δ_1 to S via ρ''.

Let us see how this definition supports our intuitions with respect to our ongoing example.

Example 9.7. *Let us collect the ingredients of earlier examples. Let us define*

$s_1 = $ "$\% x(L(x), F(x) \wedge \mathrm{US}(x), 0.945, 0.965)$"
$s_2 = $ "$\% x(L(x), F(x) \wedge \mathrm{US}(x) \wedge \mathrm{Roch}(x), 0.930, 0.975)$"
$s_3 = $ "$\% x(L(x), F(x) \wedge \mathrm{Roch}(x) \wedge \mathrm{US}(x) \wedge \mathrm{PE}(x), 0.920, 0.950)$"
$s_4 = $ "$\% x(L(x), F(x) \wedge \mathrm{Roch}(x) \wedge \mathrm{US}(x) \wedge \mathrm{PE}(x) \wedge \mathrm{MD}(x), 0.925, 0.985)$".

The general body of background knowledge Γ *consists of these four statements, together with* "$F(j)$", "$\mathrm{US}(j)$", "$\mathrm{Roch}(j)$", "$\mathrm{PE}(j)$", "$\mathrm{MD}(j)$", *and* "$\forall x(\mathrm{Roch}(x) \supset \mathrm{US}(x))$", *together with logical truths such as* "$\forall x(\phi \wedge \psi \supset \phi)$."

$\Delta(\text{"}L(j)\text{"}, \Gamma) = \{s_1, s_2, s_3, s_4\}$
$\Delta_1 = \Delta(\text{"}L(j)\text{"}, \Gamma) - $ "$\% x(L(x), F(x) \wedge \mathrm{Roch}(x) \wedge \mathrm{US}(x) \wedge \mathrm{PE}(x), 0.920, 0.950)$"
$\quad = \{s_1, s_2, s_4\}$

because of specificity;

$\Delta_2 = \Delta_1 - s_2$,

because of precision;

$\Delta_3 = \Delta_2 - s_4 = \{$"$\% x(L(x), F(x) \wedge \mathrm{US}(x), 0.945, 0.965)$"$\}$

because of precision.

Of course, if we take $\Delta(\text{"}L(j)\text{"}, \Gamma)$ to be just $\{s_1, s_2, s_3\}$, then $\{s_3, s_2\}$ sharpens-by-specificity $\{s_1, s_2, s_3\}$, and $\{s_3\}$ sharpens-by-specificity $\{s_3, s_2\}$.

Theorem 9.2. $\forall \Delta'(\neg \Delta'$ *sharpens-by-specificity* Δ *in the context* $S, \Gamma)$
$\supset \forall \rho', \rho, \alpha, \tau(\ulcorner \forall \bar{\eta}(\rho' \supset \rho) \urcorner \in \Gamma \supset \neg \mathrm{Support}(S, \Delta, \Gamma, \alpha, \tau, \rho))$
$\# \mathrm{Support}(S, \Delta, \Gamma, \alpha, \tau, \rho'))$.

If Δ is a set of statistical statements relevant to the context S, Γ, and it cannot be (in the sense of #) from that due to ρ. Of course, if some superset of ρ conflicted with ρ it would already have been sharpened away. So in any chain of reference formulas, there is one formula that embodies the most relevant statistical knowledge. We pick these most specific formulas by the following definition:

Definition 9.18. $Sp(\Delta, S, \Gamma) = \{\ulcorner\%\overline{\eta}(\tau, \rho, p, q)\urcorner : \ulcorner\%\overline{\eta}(\tau\rho, p, q)\urcorner \in \Delta$
$\wedge \neg\exists\Delta'(\Delta' \text{ sharpens-by-specificity } \Delta \text{ in the context } S, \Gamma)\}.$

9.6.3   Richness

Third, our evidence may include knowledge of a marginal distribution that is also more fully articulated in our total body of evidence $\Delta(S, \Gamma)$. Under these circumstances we do well to ignore the marginal distribution in favor of the full distribution. An example will make this clear.

Example 9.8. *Suppose we have three urns: the first contains three black and two white balls; the second and third contain seven white and three black balls. (A more realistic example will be discussed in connection with Bayes' theorem in Chapter 11.) We choose an urn u, and then a ball b from the urn. The statement S whose support interests us is "Bl(b)": the chosen ball is black. Below are the relevant statements, where $E(\langle x, y\rangle)$ is the set of compound (two stage) experiments just described, consisting of the selection of urn x and then ball y from urn x; $Bl^*(\langle x, y\rangle)$ is the set of experiments in which the ball ultimately chosen is black; B is the set of balls; and Bl is the set of black balls. We may readily calculate that the chance of getting a black ball from an instance $\langle u, b\rangle$ is $\frac{1}{3} \times \frac{3}{5} + \frac{2}{3} \times \frac{3}{10} = 0.4$. We have*

$$\Delta(\text{``}Bl(b)\text{''}, \Gamma): \begin{cases} \%x, y(Bl^*(\langle x, y\rangle), E(\langle x, y\rangle), 0.4, 0.4) \\ \%x\left(Bl(x), B(x), \frac{9}{25}, \frac{9}{25}\right) \end{cases}$$

$$\Gamma - \Delta(\text{``}Bl(b)\text{''}, \Gamma): \begin{cases} E(\langle u, b\rangle) \\ Bl^*(\langle u, b\rangle) \equiv Bl(b) \\ B(b) \\ \forall x, y(E(\langle x, y\rangle) \supset B(y)). \end{cases}$$

We want the probability of a black ball to be given by $\%x, y(Bl^*(\langle x, y\rangle), E(\langle x, y\rangle), 0.4, 0.4)$ rather than $\%x(Bl(x), B(x), \frac{9}{25}, \frac{9}{25})$.

A useful way to visualize this situation is to look at a table:

	u_1	u_2	u_3	
B	3	3	3	9
W	2	7	7	16
	5	10	10	25

If we just look at the numbers in the right hand margin, we obtain $\frac{9}{25}$; but this neglects some of the information we have about the experiment.

When we have full information about a compound experiment, we should use it, and ignore marginal frequencies. Note that "specificity" as we have formulated it does not cover this case. There is no subset of the set of balls involved that is a better guide to our beliefs. The issue is really that of giving grounds for ignoring some of the statistical information in Δ. Of course, we don't *always* want to ignore it. If we have no information or only very vague information about the frequency with which the various urns are chosen, it might be the case that the relevant statistics are given by $\%x(\text{Bl}(x), B(x), \frac{9}{25}, \frac{9}{25})$.

We will call the relation that allows us to ignore irrelevant marginal frequencies *sharpens-by-richness*; it is the opposite of simplifying by marginalization. It is this condition that allows Bayesian arguments to take precedence over arguments based on marginal frequencies, *when they are appropriate*, that is, when the resulting intervals *conflict* with the intervals that emerge from the marginal distributions.[10] As in the case of *specificity*, we need to be sure that it is not the case that there is statistical knowledge of yet higher dimension that can undermine the higher dimensional set in the same way. The definition of partial proof in Section 9.7 will ensure this.

Definition 9.19. Δ_1 sharpens-by-richness Δ_2 *in the context* S, Γ *if and only if*

$$\Delta_2 \subseteq \Delta(S, \Gamma) \wedge \exists \overline{\eta}, \alpha', \tau', \rho', p', q'(\Delta_1 = \Delta_2 - \{\ulcorner \%\overline{\eta}(\tau', \rho', p', q')\urcorner\}$$
$$\wedge \exists \alpha, \tau, \rho \, (\text{Support}(S, \Delta_1, \Gamma, \alpha, \tau, \rho) \# [p', q'] \wedge \exists p''(\ulcorner \forall \overline{\eta}, \overline{\eta}'(\rho(\overline{\eta}, \overline{\eta}') \supset \rho'(\overline{\eta}) \wedge \rho''(\overline{\eta}'))\urcorner \in \Gamma)$$
$$\wedge \forall \rho_1(\ulcorner \forall \overline{\eta}, \overline{\eta}'(\rho_1(\overline{\eta}, \overline{\eta}') \supset \rho_1(\overline{\eta}) \wedge \rho(\overline{\eta}'))\urcorner \in \Gamma)$$
$$\supset \neg \exists \alpha'', \tau'' \, (\text{Support}(S, \Delta_1, \Gamma, \alpha'', \tau'', \rho_1) \# \text{Support}(S, \Delta_1, \Gamma, \alpha, \tau, \rho)))).$$

To see how this works, take ρ to be "$E(\langle x, y \rangle)$", τ to be "$\text{Bl}^*(\langle x, y \rangle)$", p and q to be 0.4, and α to be "$\langle u, b \rangle$"; the statement is "$\text{Bl}(b)$." All the requisite formulas are in Δ; it is not the case that $[\frac{9}{25}, \frac{9}{25}] \subseteq [0.4, 0.4]$; and of course every support provided for S by Δ_1 is included in $[0.4, 0.4]$, since that is the only support there is. The general idea is that when the statistics of a full joint distribution are available but conflict with marginal statistics, we are entitled to ignore the marginal distribution.

As before we get a tree structure; the relation of child to parent is that of a distribution to a marginal distribution. If moving to a marginal distribution yields nothing different in our state of knowledge, we should not do it. Thus, from the point of view of richness, we may disregard everything below the reference formula that represents the last sharpening by richness on any branch. As before, what is above that on the branch can be sharpened away. Thus, we have:

Theorem 9.3. $\forall \Delta'(\neg \Delta'$ *sharpens-by-richness* Δ *in the context* S, Γ).
$$\supset \forall \rho, \rho', \rho'', \alpha, \alpha', \tau, \tau'(\ulcorner \forall \overline{\eta}, \overline{\eta}'(\rho'(\overline{\eta}, \overline{\eta}') \supset \rho''(\overline{\eta})) \wedge \rho(\overline{\eta}')\urcorner \in \Gamma$$
$$\supset \neg \, \text{Support}(S, \Delta, \Gamma, \alpha, \tau, \rho) \# \text{Support}(S, \Delta, \Gamma, \alpha', \tau', \rho')).$$

It is handy to define the enrichment of a set of statistical statements:

Definition 9.20. $Ri(\Delta, S, \Gamma) = \{\ulcorner \%\overline{\eta}(\tau, \rho, p, q)\urcorner : \ulcorner \%\overline{\eta}(\tau, \rho, p, q)\urcorner \in \Delta$
$$\wedge \neg \exists \Delta' (\Delta' \textit{ sharpens-by-richness } \Delta \textit{ in the context } S, \Gamma)\}.$$

9.6.4 Sharpens

We can now combine these three definitions into a definition of what it means for one body of evidence to sharpen another. It means that the first body of evidence sharpens every *different* support provided by the other in one of the three ways just defined.

---

[10]Of course, this is what the Bayesians have been telling us for 50 years. Our point, however, is that it is relevant only when we have the statistical knowledge that makes it relevant.

Definition 9.21. Δ_1 sharpens Δ_2 *in the context S,* Γ *iff*

$$\exists \overline{\eta}, \alpha, \tau, \rho, p, q \ (\text{Support}(\Delta_2, S, \Gamma, \alpha, \tau, \rho) = [p, q]$$
$$\wedge \neg \text{Support}(\Delta_1, S, \Gamma, \alpha, \tau, \rho) = [p, q])$$
$$\supset \exists \Delta (\Delta_1 \subset \Delta \subset \Delta_2$$
$$\wedge (\Delta \ sharpens\text{-}by\text{-}precision \ \Delta_2 \ in \ the \ context \ S, \ \Gamma$$
$$\vee \ \Delta \ sharpens\text{-}by\text{-}specificity \ \Delta_2 \ in \ the \ context \ S, \ \Gamma$$
$$\vee \ \Delta \ sharpens\text{-}by\text{-}richness \ \Delta_2 \ in \ the \ context \ S, \ \Gamma)).$$

One set of sentences, Δ_1, sharpens another set of sentences, Δ_2, in the context S, Γ, just in case every support provided by Δ_2 and not provided by Δ_1 satisfies one of three conditions: (1) it is less precise than any support provided by Δ_1; (2) it is undermined by conflict with a set of sentences in Δ_1 which makes use of a more specific reference class and is not undermined by a yet more specific reference class; or (3) it is undermined by conflict with a set of sentences included in Γ that makes use of a Bayesian construction whose marginal distribution yields the (misleading) support provided by Δ_2. In other words, by expanding our evidence to include an element of $\Delta_2 - \Delta_1$ we do not learn anything new and useful about S. Put the other way about, we lose nothing and gain accuracy (but not necessarily a narrower interval) by pruning the extra statistical statements from Δ_2.

The following theorem suggests a useful algorithm for obtaining a body of statistical statements that cannot be sharpened:

Theorem 9.4. $Pr(Sp(Ri(\Delta(S, \Gamma))))$ *sharpens* $\Delta(S, \Gamma)$ *in the context* S, Γ).

Whereas this theorem has intuitive appeal, and does lead to an unsharpenable set of statistical statements, it is arbitrary: we could arrive at different set of statements by changing (or mixing) the order in which we sharpen statements from the initial set $\Delta(S, \Gamma)$.

The following trivial theorems provide convenient principles for applying these definitions in particular cases.

Theorem 9.5. *If* $\Delta \subseteq \Delta(S, \Gamma)$ *and* $\ulcorner \%\overline{\eta}(\tau, \rho, p, q) \urcorner \in \Delta$ *and* $\forall \rho'(\ulcorner \overline{\eta}(\rho' \supset \rho) \urcorner \in \Gamma \supset \forall \alpha, \tau, \neg \text{Support}(\Delta, S, \Gamma, \alpha, \tau, \rho') \# [p, q])$ *then* $\neg \exists \Delta''(\Delta'' \ sharpens\text{-}by\text{-}specificity \ \Delta)$.

In other words, if no known subset of ρ provides support that differs from that provided by ρ, then nothing can sharpen the corresponding statistical statement by specificity.

Theorem 9.6. *If* $\Delta'' \subseteq \Delta \subseteq \Delta'$ *and* Δ'' *sharpens-by-precision* Δ *in the context S,* Γ, *and* $\neg \exists \Delta'''(\Delta''' \ sharpens\text{-}by\text{-}specificity \ or \ sharpens\text{-}by\text{-}richness \ \Delta'$ *in the context S,* Γ), *then* $(\exists \Delta'''')(\Delta'''' \ sharpens\text{-}by\text{-}precision \ \Delta'$ *in the context S,* Γ).

In the absence of arguments from specificity or richness, precision alone may be used to prune the set of relevant statistical statements.

Theorem 9.7. *If* $\neg (\exists \Delta)(\Delta \ sharpens\text{-}by\text{-}specificity \ or \ sharpens\text{-}by\text{-}richness \ \Delta'$ *in the context S,Γ), and* $\ulcorner \%\overline{\eta}(\tau, \rho, p, q) \urcorner \in \Delta'$, *and* $\ulcorner \%\overline{\eta}(\tau, \rho', r, s) \urcorner \in \Delta'$, *then if* $[r, s] \subset [p, q]$ *then* $\Delta' - \{\ulcorner \%\overline{\eta}(\tau, \rho', p, q) \urcorner\}$ *sharpens-by-precision* Δ' *in the context S,* Γ.

In general, in the absence of specificity or richness, we may simply prune by

The following lemma, which we will call the *pruning lemma*, will prove important. It shows that in the process of sharpening, we shall always succeed in reducing the number of statements in the set of statistical supporting statements that is sharpened:

Lemma 9.1 (The Pruning Lemma). *If Δ_1 sharpens Δ_2 in the context S, Γ, then there is a statistical statement σ that is in Δ_2 and not in Δ_1.*

Proof: By Definition 9.21, if the hypothesis holds, there is a set of statistical statements Δ that includes Δ_1 and that sharpens Δ_2 in the context S, Γ, either by precision, by specificity, or by richness. In any of the three cases, there is a statistical statement that is in Δ_2 but not in Δ. Because Δ_1 is included in Δ, that statement cannot appear in Δ_1 either.   ∎

9.7   Partial Proof

Finally, we come to the formal definition of evidential probability, which corresponds to a notion of *partial proof*. The general idea is that we collect in Δ the statistical statements seriously relevant to S (in general this will not be all the statistical statements in Γ) and take the *degree of proof* of S provided by Γ, our total knowledge, to be the lower bound and the upper bound of the degrees of support provided by Δ. This is what we shall mean by "evidential probability": the evidential probability of a statement S, relative to a body of knowledge Γ, is the interval $[p, q]$ determined by the unsharpenable *evidence* Δ contained in Γ bearing on S. The interval $[p, q]$ is the *total support* given S by Δ with total background Γ.

Pursuing the analogy with first order deductive logic, we shall write "$\Gamma \vdash_{[p,q]} S$" to mean that Γ provides a partial proof of S bounded by p and q.

Definition 9.22. $\Gamma \vdash_{[p,q]} S$ *iff*

$$\exists \Delta (p = \min\{t : \exists u, \alpha, \tau, \rho(\text{Support}(\Delta, S, \Gamma, \alpha, \tau, \rho) = [t, u])\}$$
$$\wedge\, q = \max\{u : \exists t, \alpha, \tau, \rho(\text{Support}(\Delta, S, \Gamma, \alpha, \tau, \rho) = [t, u])\}$$
$$\wedge\, \forall \Delta'(\neg \Delta' \text{ sharpens } \Delta \text{ in the context } S, \Gamma)$$
$$\wedge\, (\exists \alpha, \tau, \rho)(\text{Support}(\Delta', S, \Gamma, \alpha, \tau, \rho) \# [p, q] S)$$
$$\supset \Delta \text{ sharpens } \Delta' \text{ in the context } S, \Gamma))).$$

In words: Γ provides a partial proof of S to the degree $[p, q]$ just in case there is a set of statistical sentences Δ included in Γ such that the minimum and maximum of the degrees of support provided by Δ are p and q, respectively, no alternative Δ' sharpens Δ, and any alternative Δ' that contains statistics conflicting (in the sense of #) with Δ is sharpened by Δ.

Note particularly the last clause of this definition. Except for the last clause, we would be unable to show that partial proof was unique. If, as is tempting, we replace the existential quantification in this definition by $Pr(Sp(Ri(\Delta(S, \Gamma))))$, we easily get uniqueness, but at the cost of arbitrariness. The down side of the approach we have adopted is that there may be circumstances in which for some Γ and S there may be no partial proof.

It is important that there should be only one interval of partial support, that is, that support, as we have defined it, is unique:

Theorem 9.8 (Uniqueness). *If* $\Gamma \vdash$ . . . *S and* $\Gamma \vdash$ . . . *S then* . . .

Proof: Let the sets of statements called for by Definition 9.22 be Δ_1 and Δ_2. By Definition 9.22, instantiating Δ' in turn to Δ_1 and Δ_2, we have

$\neg\Delta_1$ sharpens Δ_2 in the context S, Γ, and
$\neg\Delta_2$ sharpens Δ_1 in the context S, Γ.

But we also have, from the same instantiations, using the last clause of Definition 9.22,

$\neg\exists\alpha, \tau\rho(\text{Support}(\Delta_2, S, \Gamma, \alpha, \tau, \rho)\#[p, q]$, and
$\neg\exists\alpha, \tau\rho(\text{Support}(\Delta_1, S, \Gamma, \alpha, \tau, \rho)\#[r, s]$.

No support in Δ_1 can include a point outside the interval $[r, s]$, unless it includes the whole interval; but then Δ_2 would sharpen Δ_1 (by precision), so $[p, q] \subseteq [r, s]$. Similarly, $[r, s] \subseteq [p, q]$, and $[r, s] = [p, q]$.   ∎

Finally, the evidential probability of a sentence S relative to a body of evidence and knowledge Γ, $\text{EP}(S, \Gamma)$, is just the interval that represents the degree to which Γ provides a partial proof of S:

Definition 9.23. $\text{EP}(S, \Gamma) = [p, q]$ *iff* $\Gamma \vdash_{[p,q]} S$.

Complementation holds in the obvious way for evidential probability, but that is the only relation among evidential probabilities that can be obtained without more explicit consideration of the contents of Γ.

Theorem 9.9. $\text{EP}(S, \Gamma) = [p, q]$ *if and only if* $\text{EP}(\neg S, \Gamma) = [1 - q, 1 - p]$.

In many applications of probability, we are concerned with *conditional* probabilities: the probability, for example, of such and such a result, *given* that we perform such and such an action. The extension of evidential probability to the case in which we take account of hypothetical evidence is straightforward, so long as the hypothetical evidence is possible according to our body of knowledge. When the hypothetical evidence contradicts what we have reason to believe, the situation is more complicated. Rules must be provided—which may depend heavily on context—that will tell us how the body of knowledge must be modified to allow for the hypothetical evidence.

9.8   Extended Example

Suppose that a company is insuring bicycles against theft and collision. The company has records concerning 6537 bicycles of various colors, types, and provenances. The records may be summarized as in Table 9.1.

A number of people have submitted forms, filled out with varying degrees of carelessness, to obtain insurance. We will use the names $b_1, \ldots, b_{12}$ for the bicycles we are being asked to insure. Naturally, we do not want to expose ourselves to greater losses than are implicit in providing insurance; but if we are to be competitive, we need to quote the lowest rates we can. Thus, we want to estimate the probability of loss as accurately as possible.

We want to use the numbers in the table as a basis for these probabilities, but of course the relative frequencies derivable from the table are not themselves probabilities concerning events in the future. (For example, the relative frequency with which imported yellow mountain bikes get stolen is, according to the table, 100%; but that does not give us a reasonable rate for an insurance premium.) Thus, we use

Table 9.1: Bicycle data.

Color	Type	Provenance	Number	Stolen	Crashed
Red	Racing	Imported	350	4	20
		Domestic	500	6	32
	Touring	Imported	250	0	5
		Domestic	600	10	10
	Mountain	Imported	0	0	0
		Domestic	800	50	20
Yellow	Racing	Imported	100	0	2
		Domestic	150	1	2
	Touring	Imported	400	10	10
		Domestic	300	5	10
	Mountain	Imported	2	2	0
		Domestic	700	45	15
Blue	Racing	Imported	100	5	3
		Domestic	120	10	3
	Touring	Imported	500	20	10
		Domestic	1000	25	15
	Mountain	Imported	215	10	15
		Domestic	450	15	20

the data in Table 9.1 to support general statistical hypotheses about the long run frequencies of various losses. We will adopt a standard of certainty of 0.95, and so use a 0.95 confidence interval to represent our knowledge about the various categories; we assume only that the variance is less than $\frac{1}{2}$, as it is for every relative frequency.[11]

The statistical generalizations in Γ of this level are presented in Table 9.2, in which t_l and t_u represent the bounds for frequencies of stolen bikes, and c_l and c_u represent the bounds for frequencies of crashed bikes. The most specific inferences we can make are given in Table 9.3.

In many contexts, including the one under discussion here, the reference formulas form a tree, in which each node branches into more specific formulas, and all our statistical knowledge concerns the reference formulas in the tree. In calculating the degree to which such a tree of knowledge supports a statement, it is convenient to display the relevant branch of the underlying statistical tree with the potential reference classes ordered from most specific to most general. For example, consider b_1, a red, imported racing bike. What we know relevant to its likelihood of being stolen can be represented as follows:

Red racing imported	[0.0, 0.0638]
Red racing	[0.0, 0.0454]
Red imported	[0.0, 0.0467]
Imported racing	[0.0, 0.0582]
Red	[0.0084, 0.0476]
Racing	[0.0, 0.0467]
Imported	[0.0042, 0.0490]
All	[0.0212, 0.0455]

---

[11]For simplicity we use the normal approximation to the binomial.

Table 9.2: Inferred frequencies.

Class	t_l	t_u	c_l	c_u
All	0.0212	0.0455	0.173	0.0415
Red	0.0084	0.0476	0.0152	0.0544
Yellow	0.0140	0.0622	0.0000	0.0477
Blue	0.0156	0.0557	0.0076	0.0477
Domestic	0.0217	0.0506	0.0131	0.0419
Imported	0.0042	0.0490	0.0115	0.0563
Racing	0.0000	0.0467	0.0200	0.0739
Touring	0.0052	0.0407	0.0019	0.0374
Mountain	0.0352	0.0774	0.0113	0.0534
Red touring	0.0000	0.0454	0.0000	0.0513
Red racing	0.0000	0.0454	0.0276	0.0948
Red mountain	0.0279	0.0971	0.0000	0.0596
Red imported	0.0000	0.0467	0.0017	0.0817
Red domestic	0.0123	0.0572	0.0101	0.0551
Yellow touring	0.0000	0.0585	0.0000	0.0656
Yellow racing	0.0000	0.0660	0.0000	0.0780
Yellow mountain	0.0300	0.1039	0.0000	0.0584
Yellow imported	0.0000	0.0676	0.0000	0.0676
Yellow domestic	0.0154	0.0732	0.0000	0.0524
Blue touring	0.0047	0.0553	0.0000	0.0420
Blue racing	0.0021	0.1343	0.0000	0.0933
Blue mountain	0.0000	0.0756	0.0146	0.0906
Blue imported	0.0086	0.0773	0.0000	0.0687
Blue domestic	0.0071	0.0566	0.0000	0.0489
Touring imported	0.0000	0.0550	0.0000	0.0506
Touring domestic	0.0000	0.0453	0.0000	0.0409
Racing imported	0.0000	0.0582	0.0037	0.0872
Racing domestic	0.0000	0.0574	0.0127	0.0834
Mountain imported	0.0000	0.1218	0.0026	0.1357
Mountain domestic	0.0342	0.0786	0.0060	0.0504

The frequency of theft of imported red racing bikes lies between 0.0 and 0.0638. Does this differ from the frequency for any higher node in the statistical tree? No. Consider red racing bikes. This frequency conflicts with that for red bikes, so it sharpens by specificity any body of knowledge that contains red. We may delete the information about red bikes. Red imported conflicts with imported, and so rules that class out as part of the evidence. Finally, *All* conflicts with its subset red racing, and so is ruled out. We are left with the following candidates, no two of which differ:

Red racing imported	[0.0, 0.0638]
Red racing	[0.0, 0.0454]
Red imported	[0.0, 0.0467]
Imported racing	[0.0, 0.0582]

There is no Δ that sharpens any of the corresponding set of statistical statements by specificity. Now we may apply Theorem 9.7, noting that these items of statistical knowledge can be ordered by precision: the sole class remaining is red racing, and Δ is

$$\{``\%x(\text{Stolen}(x), \text{ReRaBike}(x), [0.0, 0.0454]"\}$$

Table 9.3: Most specific inferences.

Class	t_l	t_u	c_l	c_u
Red racing imported	0.0000	0.0638	0.0048	0.1095
Red racing domestic	0.0000	0.0558	0.0202	0.1078
Red touring imported	0.0000	0.0620	0.0000	0.0820
Red touring domestic	0.0000	0.05672	0.0000	0.0567
Red mountain imported	0.0000	1.0000	0.0000	1.0000
Red mountain domestic	0.0279	0.0971	0.0000	0.0596
Yellow racing imported	0.0000	0.0980	0.0000	0.1180
Yellow racing domestic	0.0000	0.0867	0.0000	0.0933
Yellow touring imported	0.0000	0.0740	0.0000	0.0740
Yellow touring domestic	0.0000	0.0732	0.0000	0.0899
Yellow mountain imported	0.3070	1.0000	0.0000	0.6930
Yellow mountain domestic	0.0272	0.1013	0.0000	0.0585
Blue racing imported	0.0000	0.1480	0.0000	0.1280
Blue racing domestic	0.0000	0.1728	0.0000	0.1145
Blue touring imported	0.0000	0.0838	0.0000	0.0638
Blue touring domestic	0.0000	0.0560	0.0000	0.0460
Blue mountain imported	0.0000	0.1133	0.0029	0.1366
Blue mountain domestic	0.0000	0.0795	0.0000	0.0906

yielding partial proof, relative to our whole knowledge base, of degree [0.0, 0.0454]. This is the evidential probability that bike b_1 will be stolen.

Consider bike b_2, a red domestic racing bike:

Red racing domestic	[0.0, 0.0558]
Red racing	[0.0, 0.0454]
Red domestic	[0.0123, 0.0572]
Domestic racing	[0.0, 0.0574]
Red	[0.0084, 0.0476]
Racing	[0.0, 0.0467]
Domestic	[0.0217, 0.0506]
All	[0.0212, 0.0455]

The frequency of theft in the most specific class lies between 0 and 0.0558. This conflicts with the interval [0.0123, 0.0572] associated with red domestic bikes. Thus, any Δ containing both reference structures will be sharpened by a Δ containing only the more specific structure; we may leave red domestic bikes out of account. Next, consider red racing bikes, with the associated interval [0.0, 0.0454], and domestic racing bikes, with the interval [0.0, 0.0574]. The former is included, the latter not; but in either event there is no conflict. Red racing conflicts with red; red would bow to specificity. Finally, red racing conflicts with *All*. Resolving all conflicts in terms of specificity that we can, we are left with:

Red racing domestic	[0.0, 0.0558]
Red racing	[0.0, 0.0454]
Domestic racing	[0.0, 0.0574]
Domestic	[0.0217, 0.0506]

Notice that this does not mean that we have no more conflicts: red racing and domestic conflict, but since neither is a subset of the other, we must retain both. Now we apply considerations of precision: red racing provides the smallest of the nested intervals, but we must also take account of the conflicting general class of domestic bikes. The relevant relation is that of inclusion; this allows us to rule irrelevant both red racing domestic and domestic racing. The evidential probability of "Stolen (b_2)" is the interval cover [0.0, 0.0506].

Bike b_3 is a red imported touring bike; what is the probability that it will be stolen? We have:

Red touring imported	[0.0, 0.0620]
Red touring	[0.0, 0.0454]
Red imported	[0.0, 0.0467]
Touring imported	[0.0, 0.0550]
Red	[0.0084, 0.0476]
Touring	[0.0052, 0.0407]
Imported	[0.0042, 0.0490]
All	[0.0212, 0.0455]

Nothing in our data conflicts with the frequency limits imposed by the most specific class. Red touring, however, conflicts with red and with *All*, and red imported conflicts with imported. Taking account of considerations of specificity, we have:

Red touring imported	[0.0, 0.0620]
Red touring	[0.0, 0.0454]
Red imported	[0.0, 0.0467]
Touring imported	[0.0, 0.0550]
Touring	[0.0052, 0.0407]

Applying Theorem 9.7, we obtain the single reference class touring, and the probability [0.0052, 0.0407].

Here is a final example: Bike b_4 is a red, imported mountain bike. The list of potential reference classes is:

Red mountain imported	[0.0, 1.0]
Red mountain	[0.0279, 0.0971]
Red imported	[0.0, 0.0467]
Mountain imported	[0.0, 0.1218]
Red	[0.0084, 0.0476]
Mountain	[0.0352, 0.0774]
Imported	[0.0042, 0.0490]
All	[0.0212, 0.0455]

Note that we have 0 imported red mountain bikes reported in our database; but in the same vein, we have 0 members of $\{b_4\}$ in our database, so by itself that cuts no ice. We proceed as before: red mountain conflicts with red and with *All*; red imported conflicts with imported; we have:

Red mountain imported	[0.0, 1.0]
Red mountain	[0.0279, 0.0971]
Red imported	[0.0, 0.0467]
Mountain imported	[0.0, 0.1218]
Mountain	[0.0352, 0.0774]

Applying Theorem 9.7, we obtain:

Red imported	[0.0, 0.0467]
Red mountain	[0.0279, 0.0971]
Mountain	[0.0352, 0.0774]

The cover of these three intervals, none of which can be eliminated, leads to a probability of [0.0, 0.0971].

9.9 A Useful Algorithm

The procedure we followed in each of these cases was the same. We considered all of the possible reference classes to which the entity we are concerned with is known to belong. We listed the intervals corresponding to our statistical knowledge about each class. We then eliminated each potential reference set that could be eliminated on grounds of specificity: that is, we sharpened the large set, step by step, by specificity. This yielded a set Δ of statistical statements that cannot be sharpened by specificity: if there are two statistical statements in Δ that conflict in the sense of #, neither concerns a subclass of the other.

Now we can proceed to sharpen Δ by precision. We can do so, step by step, until we have statistical knowledge that cannot be further reduced. The probability is then given by the cover of the remaining intervals.

Note that this algorithm is perfectly applicable to instances about which we have only partial knowledge; whatever we know in the tree of knowledge can be first sharpened by specificity, and then sharpened by precision.

We have not discussed sharpening by richness in these examples. We will consider some examples later on that make use of this relation. It may be incorporated into the algorithm easily enough: sharpen the initial set of statistical statements, as far as possible, by richness, and then proceed as we have suggested.

9.10 Relations to Other Interpretations

According to the frequentists, all probabilities *are* frequencies or measures. Probabilities concern sets of cases or kinds of cases, not particular cases. Evidential probability takes probabilities to be *based on* known frequencies. There are a lot of empirical facts and generalizations that we may properly claim to know; whether one wishes to call Γ in a particular case a set of items of knowledge or a set of justified knowledge claims seems irrelevant to its use and a mere terminological matter. In this sense we know (defeasibly, of course) a lot of general statements, including statistical generalizations. That knowledge in this sense depends on probability is no problem: If we are concerned with the life expectancy of Dr. Jones, we take for granted the inferences we may have made from the mortality tables. If we are concerned with the statistics appropriate to a class of individuals, that is a different sort of question, and different background knowledge may be taken for granted.

One difference between our view and the classical limiting frequency or measure views is that on the latter views the reference sets are generally taken to be infinite. But we know of no infinite sets, except in mathematics. We can be quite sure that the set of humans (past, present, and future) is finite, as is the set of coin tosses. Whereas it is mathematically convenient to have infinite sets, it is possible to regard unbounded collectives to be merely the idealization of practically large collectives. Because we do not suppose that *real* (as opposed to hypothetical) probabilities are ever known with perfect precision, there is little to be lost in eschewing infinities of

In common with the logical views of probability, evidential probability shares the conventionality of the choice of a language, including the choice of primitive functions and predicates. And like the logical views of Carnap [Carnap, 1950] and Keynes [Keynes, 1952], it construes probability as legislative for rational belief. What does this mean?

It is not altogether clear what belief is, or how it is to be measured. Ramsey [Ramsey, 1931] proposed the strict behavioristic interpretation of belief discussed previously. But it isn't at all clear that degrees of belief should be measured by real numbers. If degrees of belief are fuzzy, then interval constraints seem to make more sense, although they would still represent an idealization. Probability could impose *bounds* on belief (as suggested, for example, by Shafer's theory of belief functions), which may partially constrain betting behavior. We consider that to be enough of a normative connection between evidential probability and belief. Given what an agent has in the way of evidence and background knowledge, the evidential probability of any statement S is determined for him, and his "degree of belief"—his propensities to bet or to take chances—*ought* to be constrained by that interval.

The subjective view allows any degree of belief in any statement. The constraints of the subjective view involve only the *relations* among degrees of belief in related statements. Many writers have found this view of probability too liberal, and have sought to strengthen it with principles akin to the principle of indifference. These views (for example, the view of Jaynes [Jaynes, 1958] are objective in a local sense, but lack the global objectivity sought by Carnap and Keynes. In contrast, there is nothing subjective in evidential probability except that the relevant evidence may vary from individual to individual—from subject to subject. Being confident that something is so does not make it probable.

There is no dependence either on precise "degrees of belief" or on a logical measure defined over the sentences of a language. This last point deserves special emphasis. On the view developed in this chapter no prior distributions are defined or needed. Probability is dependent on the language we speak, through its dependence on the classes of reference and target terms. But, first, this choice of a language does not commit one to any probabilities at all: with no empirical background knowledge, the a priori probability of any empirical statement is [0, 1]. And, second, we will argue later that our interests and our background knowledge will lead us, not to an "ideal" language, but to the grounded choice between two suggested languages. We will argue that this entire procedure, from the determination of probabilities to the choice between two languages, is objective in the sense that any two rational agents sharing the same evidence will be bound by the same interval constraints.

In the next chapter we will develop a corresponding semantic approach, which will clarify the relation being captured by evidential probability and also justify the use we will make of the notion in the chapter after next.

9.11 Summary

Evidential probability is based on the assumption that it is legitimate to base our assessments of probability on defeasible statistical knowledge. We formalize this notion with the help of a special notation: $\ulcorner \%\overline{\eta}(\tau, \rho, p, q)\urcorner$, where this statement spells out the claim that the proportion of objects (perhaps tuples) satisfying the *reference* formula ρ that also satisfy the *target* formula τ lies between the value denoted by the canonical term p and that denoted by the canonical term q.

To make sense of this, we must stipulate what are to count as the reference terms of our language, and what are to count as the target terms, as well as what terms will be taken to be the canonical real number terms.

Once this is done, we proceed to characterize statistical relevance. There are three ways in which formulas in a set of potentially relevant statistical formulas can be eliminated as actually irrelevant: precision, specificity, and richness. Precision corresponds to the inclusion relation between the intervals of statistical statements. Specificity corresponds to an inclusion relation between the domains of statistical statements. Richness concerns dimensionality; it allows us sometimes to disregard marginal statistics when we have adequate data concerning a full multidimensional distribution. These three ways of sharpening the focus of a body of potential statistical data as it bears on a given statement are all that are needed.

Once such a body of data has been sharpened as much as it can be, probability is taken to be the interval cover of the intervals mentioned in that body of data. Thus, the value of an evidential probability is, in general, an interval.

The definitions presented in this chapter give rise to an algorithm that can be used to obtain evidential probabilities in simple cases, including the typical case in which the properties relevant to the objects we are concerned with form a tree, and all of our statistics concern these properties.

9.12   Bibliographical Notes

This conception of probability was given, in its main outlines, in two papers presented to the Association for Symbolic Logic [Kyburg, 1959a; Kyburg, 1959b], and in more detail in [Kyburg, 1961]. It was called "epistemological probability" in [Kyburg, 1971]. A more accessible version, focused primarily on statistical inference, was presented in [Kyburg, 1974]. The system, in essentially its present form, appeared as [Kyburg, 1997].

Several efforts have been made to render this notion of probability computable. The two that have been most extensive are those of Ron Loui [Loui, 1986] and Mucit [Kyburg & Murtezaoglu, 1991; Murtezaoglu, 1998]. Much relevant material, though written from a different viewpoint, will be found in [Walley, 1991; Walley, 1995; Walley, 1996], as well as [Kyburg & Pittarelli, 1996].

9.13   Exercises

(1)   What are the chances of b_5, a yellow, imported touring bike, being stolen?

(2)   What are the chances of b_6, a blue, domestic mountain bike, being stolen?

(3)   What are the chances of b_7, a blue, imported racing bike, being stolen?

(4)   Use table 9.2 to compute the chances of b_8, a red, imported racing bike, being crashed.

(5)   What are the chances of b_9, a red racing bike, being stolen? (The applicant for insurance neglected to fill in the blank for "domestic" or "imported".)

(6)   What are the chances of b_{10}, a blue touring bike, being stolen?

(7)   What are the chances of b_{11}, a yellow mountain bike, being crashed?

(8)   What are the chances of b_{12}, a green, imported racing bike, being stolen?

(9)   What are the chances of b_{13}, a red bike, being crashed?

(10)   What are the chances of b_{14}, a green bike, being crashed?

Bibliography

[Bacchus, 1992] Fahiem Bacchus. *Representing and Reasoning with Probabilistic Knowledge*. The MIT Press, 1992.

[Bacchus et al., 1992] F. Bacchus, A. J. Grove, J. Y. Halpern, and D. Koller. From statistics to degrees of belief. In *AAAI-92*, pp. 602–608, 1992.

[Carnap, 1950] Rudolf Carnap. *The Logical Foundations of Probability*, University of Chicago Press, Chicago, 1950.

[Davidson, 1966] Donald Davidson. Emeroses by other names. *Journal of Philosophy*, 63:778–780, 1966.

[Good, 1962] I. J. Good. Subjective probability as a measure of a nonmeasurable set. In Patrick Suppes, Ernest Nagel, and Alfred Tarski, editors, *Logic, Methodology and Philosophy of Science*, pp. 319–329. University of California Press, Berkeley, 1962.

[Goodman, 1955] Nelson Goodman. *Fact, Fiction, and Forecast*. Harvard University Press, Cambridge, MA, 1955.

[Halpern, 1990] Joseph Y. Halpern. An analysis of first-order logics of probability. *Artificial Intelligence*, 46:311–350, 1990.

[Jaynes, 1958] E. T. Jaynes. Probability theory in science and engineering. *Colloquium Lectures in Pure and Applied Science*, 4:152–187, 1958.

[Keynes, 1952] John Maynard Keynes. *A Treatise on Probability*. Macmillan, London, 1952.

[Kyburg, 1959a] Henry E. Kyburg, Jr. Probability and randomness I, abstract. *Journal of Symbolic Logic*, 24:316–317, 1959.

[Kyburg, 1959b] Henry E. Kyburg, Jr. Probability and randomness II, abstract. *Journal of Symbolic Logic*, 24:317–318, 1959.

[Kyburg, 1961] Henry E. Kyburg, Jr. *Probability and the Logic of Rational Belief*. Wesleyan University Press, Middletown, 1961.

[Kyburg, 1971] Henry E. Kyburg, Jr. Epistemological probability. *Synthese*, 23:309–326, 1971.

[Kyburg, 1974] Henry E. Kyburg, Jr. *The Logical Foundations of Statistical Inference*. Reidel, Dordrecht, 1974.

[Kyburg, 1983] Henry E. Kyburg, Jr. The reference class. *Philosophy of Science*, 50:374–397, 1983.

[Kyburg, 1990] Henry E. Kyburg, Jr. Theories as mere conventions. In Wade Savage, editor, *Scientific Theories*, Volume XIV, pp. 158–174. University of Minnesota Press, Minneapolis, 1990.

[Kyburg, 1997] Henry E. Kyburg, Jr. Combinatorial semantics. *Computational Intelligence*, 13:215–257, 1997.

[Kyburg & Murtezaoglu, 1991] Henry E. Kyburg, Jr., and Bulent Murtezaoglu. A modification to evidential probability. In *Uncertainty in AI—91*, pp. 228–231, 1991.

[Kyburg & Pittarelli, 1996] Henry E. Kyburg, Jr. and Michael Pittarelli. Set based bayesianism. *IEEE Transactions on Systems, Man, and Cybernetics*, 26:324–339, 1996.

[Loui, 1986] Ronald P. Loui. Computing reference classes. In *Proceedings of the 1986 Workshop on Uncertainty in Artificial Intelligence*, pp. 183–188, 1986.

[Murtezaoglu, 1998] Bulent Murtezaoglu. Uncertain Inference and Learning with Reference Classes, PhD thesis, The University of Rochester, 1998.

[Pollock, 1990] John L. Pollock. *Nomic Probability and the Foundations of Induction*. Oxford University Press, New York, 1990.

[Ramsey, 1931] F. P. Ramsey. *The Foundations of Mathematics and Other Essays*. Humanities Press, New York, 1931.

[Reichenbach, 1949] Hans Reichenbach. *The Theory of Probability*. University of California Press, Berkeley and Los Angeles, 1949.

BIBLIOGRAPHY   229

[Smith, 1961] C. A. B. Smith. Consistency in statistical inference and decision. *Journal of the Royal Statistical Society B*, 23:1–37, 1961.
[Walley, 1991] Peter Walley. *Statistical Reasoning with Imprecise Probabilities*. Chapman and Hall, London, 1991.
[Walley, 1995] Peter Walley. Inferences from multinomial data: Learning about a bag of marbles. *Journal of the Royal Statistical Society B*, 57, 1995.
[Walley, 1996] Peter Walley. Measures of uncertainty in expert systems. *Artificial Intelligence*, 83 pp. 1–58, 1996.

10

Semantics

10.1 Introduction

Uncertain reasoning and uncertain argument, as we have been concerned with them here, are reasoning and argument in which the object is to establish the credibility or acceptability of a conclusion on the basis of an argument from premises that do *not* entail that conclusion. Other terms for the process are inductive reasoning, scientific reasoning, nonmonotonic reasoning, and probabilistic reasoning. What we seek to characterize is that general form of argument that will lead to conclusions that are worth accepting, but that may, on the basis of new evidence, need to be withdrawn.

What is explicitly excluded from uncertain reasoning, in the sense under discussion, is reasoning from one probability statement to another. Genesereth and Nilsson [Nilsson, 1986; Genesereth & Nilsson, 1987], for example, offer as an example of their "probabilistic logic" the way in which constraints on the probability of Q can be established on the basis of probabilities for P and for $P \rightarrow Q$. This is a matter of deduction: as we noted in Chapter Five, it is provable that any function prob satisfying the usual axioms for probability will be such that if $\text{prob}(P) = r$ and $\text{prob}(P \rightarrow Q) = s$ then $\text{prob}(Q)$ must lie between $s + r - 1$ (or 0) and s. This deductive relation, though often of interest, is not what we are concerned with here. It has been explored by Suppes and Adams [Suppes, 1966; Adams, 1966] as well as Genesereth and Nilson.

We are concerned with the argument from a set of sentences Γ to a particular sentence ϕ, where that argument is not deductive: that is, the case in which Γ does not *entail* ϕ. Keep in mind that the statement ϕ is *not* a statement of probability. It is a categorical statement, such as "Mary's child will be a male," "The casino will make a profit of at least \$10,000 Friday night," "The sample of bullfinches we have observed exhibits typical behavior in regard to territoriality." Arguments of this sort have been considered at length in philosophy, but as we have seen, no consensus regarding their treatment has emerged. They have also been considered in various guises in artificial intelligence default logic, nonmonotonic logic, etc. We have touched on these views earlier. Again, no consensus has emerged.

Although the focus of our interest is argument that leads to acceptance via high (evidential) probability, we examine here the general semantics underlying low evidential probabilities as well as high ones.

When axioms and rules of inference are offered for deductive relations, whether in simple first order logic or in a modal extension of first order logic, it is conventional to present *soundness* and *completeness* proofs for the system. A soundness proof establishes that if the premises of an argument are true in a model, the conclusion is also true. A completeness proof shows that if the conclusion is true in *every* model in which the premises are true, then the conclusion is derivable from the premises in the system considered.

Obviously soundness and completeness depend as much on the characterization of the models as on the premises and rules of inference of the system. What is proved is a "fit" between the system and the models of the system. In order to do this for modal logics, we turn to models that consist of sets of possible worlds, together with an accessibility relation. The tendency among nonmonotonic logicians has been modeled, as we have seen, on the practices of modal logicians.

In Chapter 9, we offered a syntactical characterization of a process that we called *partial proof*. The ideas of soundness and completeness do not capture what we want in the case of inductive logic as we have characterized it in the last chapter. Clearly soundness is asking too much: We are perfectly willing to allow that our conclusions will *sometimes* be wrong—we only ask that they not be wrong too often. And completeness may similarly be too strong a requirement: to ask that every statement that is almost always true in the models of our premises, where our "premises" are construed as a collection of background knowledge together with evidence, be partially entailed by our premises is asking a lot.

We will offer below a semantics for logics that purport to be inductive, or probabilistic, or uncertain, or nonmonotonic, or defeasible. It should be remembered that inference of the sort in question is not truth preserving. Thus, we do not want to apply conventional semantics directly: we are not concerned with the characterization of some form of truth preserving inference. (Note that inference within the framework of the probability calculus—from $P(A) = p$ and $P(B) = q$ to infer $P(A \vee B) \leq p + q$— *is* truth preserving.) We are not concerned with the logical relations between one statement concerning uncertainty and another, however that may be construed. We are concerned with inference from a set of sentences Γ to a single sentence ϕ, where Γ does not entail ϕ, but in some sense *justifies* it, and where ϕ need not concern uncertainty, though it may involve frequencies or dispositions. We will impose minimal constraints on Γ: we will allow it to contain statements representing general knowledge, as well as observational data, since we often make use of background knowledge as well as data in arriving at inductive conclusions.

One argument undermining this effort is the argument that such inference is "subjective." We will take for granted, but also attempt to illustrate, that inconclusive arguments can be just as "objective" as conclusive deductive arguments, and thus that inductive, probabilistic, or uncertain argument as much admits of characterization in semantic terms as does deductive argument. It is true that each individual may base his or her assessment on a unique body of evidence; this suggests that each individual may arrive at slightly different conclusions; it does not suggest that the relation between evidence and the sentence supported by it is not as objective as the relation of entailment.

David Israel [Israel, 1983] once argued that "real" inference—the kind of inference we need to understand—is not like "logic" at all, where he included both classical first order logic and the modern nonmonotonic variants under the rubric "logic." Typically, the standards of adequacy for inference, and so for logic, include soundness and completeness. Neither criterion is plausible for the reasonings and arguments that we need in the real world.

The standard we would like to suggest for real inference is that a conclusion is warranted to some degree when it is true in most (or almost all) of the models in which the premises are true. In contexts where to speak of premises is misleading, the conclusion should be true in most of the models in which the sentences of our background knowledge and evidence are true. We will have to give more details about what a model is, and we will want to restrict ourselves to "relevant" models. We will also have to be specific about the meaning of "most," and we will want to generalize this criterion slightly to allow for bodies of knowledge that are not fully consistent and therefore admit of no models.

10.2   Models and Truth

A model m of our language is exactly the usual sort of thing: a pair $\langle \mathcal{D}, \mathcal{I} \rangle$ where $\mathcal{D}$ is a domain of individuals and $\mathcal{I}$ is an interpretation function. We restrict the class of models that interest us to those in which the field terms take their values in the domain of the reals, and the relations and functions applicable to field terms have their standard interpretations. Thus, we divide the domain into a domain of mathematical objects, $\mathcal{D}_m$, and a finite domain of empirical objects, $\mathcal{D}_e$. Similarly, we construe the interpretation function $\mathcal{I}$ as the union of two partial functions: one whose domain is the set of mathematical terms, and one whose domain is the set of empirical terms. We assume that these two classes of expressions are exhaustive and exclusive.[1]

Henceforth, we will focus on $\mathcal{D}_e$ and $\mathcal{I}_e$ and take $\mathcal{D}_m$ and $\mathcal{I}_m$ for granted. We will drop the subscripts, and write $\mathcal{D}$ and $\mathcal{I}$ for $\mathcal{D}_e$ and $\mathcal{I}_e$. Because we want to talk about the ratios of the cardinalities of models, and since our concerns are primarily practical, we will fix on a particular domain $\mathcal{D}$ for all of our models. The set of all models with this domain we denote by "$\mathcal{M}^{\mathcal{D}}$." We may think of $\mathcal{D}$ as "the" domain of empirical objects, though exactly what goes into such a domain—physical objects, time slices of physical objects, sets of parts of physical objects, events, spatiotemporal segments, or a combination of various of these items—we may leave open.

The interpretation function $\mathcal{I}$ maps individual (empirical) terms into members of $\mathcal{D}$, predicates into subsets of $\mathcal{D}$, n-term relations into $\mathcal{D}^n$, function terms into functions from $\mathcal{D}^n$ to $\mathcal{D}$, etc. The only oddity, for such a rich language, is that we shall take $\mathcal{D}$ to be finite (though we need not give its cardinality an upper bound). Thus, we are committed to the existence of a first crow and a last goldfish. This seems no more odd than to be committed to an infinite number of crows and goldfish.

Truth is defined, in the standard way, relative to a model m and a variable assignment v. An open one-place formula is true in the model m, under the variable assignment v, just in case the object assigned to the variable is in the set constituting the interpretation of the predicate. Note that the variable assignment must assign objects both to mathematical variables and to empirical variables.

---

[1] Note that a function (e.g., "height in inches of") may be applied to an empirical object (e.g., Sally) to yield a mathematical object (e.g., 66.3), without requiring a mixture of the domains.

The only novelty here is the procedure for assigning truth to the special statistical statements of the form $\ulcorner \%\overline{\eta}(\tau, \rho, p, q)\urcorner$.

The *satisfaction set*[2] of an open formula ϕ, whose only free variables are n empirical variables, relative to a model $m = \langle \mathcal{D}, \mathcal{I} \rangle$—in symbols, $SS_m(\phi)$—is the subset of $\mathcal{D}^n$ that satisfies ϕ. In particular, if ϕ is a formula containing n free empirical variables $x_1, x_2, \ldots, x_n$, then the satisfaction set of ϕ is the set of n-tuples $\langle a_1, a_2, \ldots, a_n \rangle$ of $\mathcal{D}$ such that in the model m, under any variable assignment v that assigns a_1 to $x_1, \ldots,$ a_n to x_n, the formula ϕ is true. Formally,

Definition 10.1. $SS_m(\phi) = \{\langle a_1, \ldots, a_n \rangle \in \mathcal{D}^n : v_{x_1,\ldots,x_n}^{a_1,\ldots,a_n}$ satisfies ϕ in $m\}$, where ϕ contains n free object variables and no mathematical variables, and m is a model whose domain is $\mathcal{D}$.

If it is required that some of the variables are to remain free, we must replace the variable assignment function by a partial assignment function that specifies the variables whose assignment is to be left open. Thus, we write $v[x_1, \ldots, x_k]_{x_{k+1},\ldots,x_n}^{a_{k+1},\ldots,a_n}$ for the variable assignment function that is like v except that (1) it leaves the assignment to the variables $x_1, \ldots, x_k$ open, and (2) assigns $a_{k+1}, \ldots, a_n$ to the variables $x_{k+1}, \ldots, x_n$. Thus, the satisfaction set of the formula ϕ, in a certain model m, and under a variable assignment v, when we leave the assignment to $x_1, \ldots, x_k$ untouched, $SS_{m,v}(\phi)[x_1, \ldots, x_k]$ depends on the variable assignment to $x_1, \ldots, x_k$ as well as the interpretation.[3]

Definition 10.2. $SS_{m,v}(\phi)[x_1, \ldots, x_k] = \{\langle a_{k+1}, \ldots a_n \rangle \in \mathcal{D}^{n-k} : v_{x_{k+1},\ldots,x_n}^{a_{k+1},\ldots,a_n}$ satisfies ϕ in m under the assignment $v\}$, where ϕ contains at least n free object variables and m is a model whose domain is $\mathcal{D}$.

If ϕ is a primitive n-place predicate followed by n variables, the satisfaction set of ϕ is just its interpretation in the model m. If ϕ is a formula in $\mathcal{R}$ or $\mathcal{T}(\rho)$ having k distinct free empirical variables, the satisfaction set of ϕ is the set of objects $\langle a_1, a_2, \ldots, a_k \rangle \in \mathcal{D}^k$ such that the assignment of those objects to variables of ϕ renders ϕ true in the model m. Thus, $SS_m(\text{"Taller}(y, \text{Father}(a))\text{")}$ in $m = \langle \mathcal{D}, \mathcal{I} \rangle$ would be the set of individuals in $\mathcal{D}$ who are taller than the father of a, in the particular model m; $SS_m(\text{"Taller}(y, \text{Father}(x))\text{")}$ $[x]$ in $m = \langle \mathcal{D}, \mathcal{I} \rangle$ would be the set of individuals in $\mathcal{D}$ who are taller than the father of x, for a given assignment to x, in the particular model m.

In general the number of free variables may be less than the number of terms. If ϕ and ψ contain exactly the same free variables, the satisfaction set of $\ulcorner \phi \wedge \psi \urcorner$ in a model m is the intersection of the satisfaction sets of ψ and of ϕ in m. If ϕ and ψ contain different numbers of free variables, the intersection of their satisfaction sets is empty, though the satisfaction set of their conjunction may be nonempty. For example, if "L" stands for "longer than," then $SS_m(\text{"}L(x, y) \wedge L(y, z)\text{"}) \cap SS_m(\text{"}L(x, z)\text{"}) = \emptyset$, but $SS_m(\text{"}L(x, y) \wedge L(y, z) \wedge L(x, z)\text{"})$ is a perfectly good set of triples. If ϕ contains

---

[2]Mendelson [Mendelson, 1979, pp. 53–54] refers to this as "the relation (or property) of the interpretation associated with" the open formula ϕ. Cursory examination has failed to turn up prior use of the natural-sounding expression "satisfaction set", but it must have occurred somewhere.

[3]We shall not need this generality here, but as Rolf Eberle has pointed out to me in conversation, it is needed in order to say "In 30% of the first grade classes, from 60% to 70% of the students are out with measles": $\%y(\%(M(x), S(x, y), 0.6, 0.7), FG(y), 0.3, 0.3)$.

n free variables, the satisfaction set of $\ulcorner \neg \phi \urcorner$ is the complement of the satisfaction set of ϕ in $\mathcal{D}^n$. We will take the satisfaction set of a closed formula ϕ to be $\mathcal{D}$ if $\mathcal{I}$ makes ϕ true, and the empty set $\emptyset$ if ϕ is false under $\mathcal{I}$.

We say that $\ulcorner \%\overline{\eta}(\tau, \rho, p, q) \urcorner$ is true in model m, under a variable assignment v, if and only if

(1) p and q are mathematical variables or members of the set of canonical numerical terms (required by well-formedness),

(2) ρ and τ are open formulas (for the quasiquoted sentence to be well formed, they must be in $\mathcal{R}$ and $\mathcal{T}(\rho)$ respectively), τ contains no free variables that are not free in ρ (other variables may be free in ρ), and all the variables of $\overline{\eta}$ occur in ρ,

(3) $|\mathrm{SS}_{m,v}(\rho)| > 0$, and

(4) The ratio of the cardinality of the satisfaction set in m, v of $\ulcorner \rho \wedge \tau \urcorner$ to the cardinality of the satisfaction set in m, v of ρ is between the real number assigned to or denoted by p and the real number assigned to or denoted by q.

Definition 10.3 (Statistical Truth). $\ulcorner \%\overline{\eta}(\tau, \rho, p, q) \urcorner$ *is* true *in model m, under a variable assignment v, if and only if* $|\mathrm{SS}_{m,v}(\rho)| > 0$ *and* $p \leq \mathrm{SS}_{m,v}[\overline{\eta}_1] \ulcorner \tau \wedge \rho \urcorner / \mathrm{SS}_{m,v}[\overline{\eta}_1]\rho \leq q$, *where all the variables of $\overline{\eta}$ occur in ρ, and all variables in τ occur in ρ, and $\overline{\eta}_1$ is the sequence of variables free in ρ but not bound by $\overline{\eta}$.*

As an illustration, let "$L(x, y)$" be interpreted to mean that x seems longer than y. It is plausible to suppose of most triples of objects that if the first seems longer than the second, and the second seems longer than the third, then the first will seem longer than the third; we may take as true "$\%(L(x, z), L(x, y) \wedge L(y, z), 0.90, 1.0)$." Of the triples x, y, z of objects in $\mathcal{D}$ satisfying "$L(x, y) \wedge L(y, z)$" almost all satisfy "$L(x, z)$".

The symbol "$\%\overline{\eta}$" thus represents a variable binding operator (it binds the empirical variables $\overline{\eta}$ in its scope) and represents relative frequency in a finite model.[4] If p or q is a field variable, then the truth of a measure statement in which it occurs depends on the variable assignment; otherwise (since then it is a real number term in canonical form, i.e., a rigid designator of a number), it does not.

We now have truth defined for all the well-formed formulas of our language relative to models and variable assignments for the intended finite object domain $\mathcal{D}$. We take as basic the notion of the truth of a sentence (a formula with no free variables) in a model m: $\models_m \phi$. We extend this in the obvious way to sets of sentences Δ:

Definition 10.4. $\models_m \Delta$ *iff* $\forall \phi (\phi \in \Delta \supset \models_m \phi)$.

We need to talk about the set of models with domain $\mathcal{D}$ that makes a set Δ of sentences true:

Definition 10.5. $\mathcal{M}_{\mathcal{D}}(\Delta) = \{\langle \mathcal{D}, \mathcal{I} \rangle : \forall \phi (\phi \in \Delta \supset \models_{\langle \mathcal{D}, \mathcal{I} \rangle} \phi)\}$.

When the subscript $\mathcal{D}$ is suppressed we intend to refer to all the (finite) models with the intended domain satisfying Δ. We will sometimes refer to these models

---

[4]This does not mean that we are committed to a statistical interpretation of "$\%\overline{\eta}$": $\ulcorner \%\overline{\eta}(\tau, \rho, p, q) \urcorner$ might represent that the force of the *analogy* between the target formula τ and the reference formula ρ is to be measured by a number between p and q. This is a possibility we will not explore further here.

as "possible worlds", meaning no more than that these models represent the way things (i.e., the objects in our canonical universe $\mathcal{D}$) might be, so far as that can be characterized within the constraints of our language.

10.3   Model Ratios

Whereas soundness for valid deductive inference consists in the fact that the conclusion is true in *every* model in which the premises are true, partial entailment need only meet a weaker condition. The conditions for partial entailment are not quite a simple weakening, which would require that the conclusion be true in almost all of the models in which the premises are true. Rather, as we saw in the previous chapter, we must allow for the existence of obfuscating or irrelevant data in the knowledge base, and require that the conclusion be true in *almost all* of the models in which the *informative* or *relevant* premises are true.

In what follows, we will assume that all the free variables of ρ are bound by $\bar{\eta}$, so that we need not take account of variable assignments in determining satisfaction sets. To allow for free empirical variables in $\ulcorner \% \bar{\eta}(\tau, \rho, p, q) \urcorner$ introduces little that is new, but would involve some notational inconvenience.

The first auxiliary notion is that of the *model ratio* $r_m(\tau, \rho)$ of one formula τ to another ρ in a model m.

Definition 10.6. *Let* $r_m(\tau, \rho) = |SS_m(\tau \wedge \rho)|/|SS_m(\rho)|$ *if m is a model with the canonical domain* $\mathcal{D}$, $|SS_m(\rho)| > 0$, ρ *is in* $\mathcal{R}$, *and τ is in* $T(\rho)$; *otherwise let* $r_m(\tau, \rho)$ *be zero.*

In our example of satisfaction sets concerning length, we claimed that in the model w that corresponds to the real world, $r_w("L(x, z)","L(x, y) \wedge L(y, z)") \geq 0.9$, i.e., we claimed that the model ratio of "$L(x, z)$" to "$L(x, y) \wedge L(y, z)$" was at least 0.9. Put otherwise, at least 90% of the time, if one thing seems longer than another, and that thing seems longer than a third, the first will seem longer than the third.

We use this notion as the basis for constructing equivalence classes of models with a given domain. These equivalence classes play an important role in what follows. Let us define the *set* of model ratios of τ to ρ permitted by models of the set of sentences Δ having domain $\mathcal{D}$:

Definition 10.7. $r_{\mathcal{M}_{\mathcal{D}}(\Delta)}(\tau, \rho) = \{r_m(\tau, \rho) : m \in \mathcal{M}_{\mathcal{D}}(\Delta)\}$.

The equivalence classes we are interested in constitute a partition of this set given by specifying the value of a model ratio:

Definition 10.8. $[r_{\mathcal{M}_{\mathcal{D}}(\Delta)}(\tau, \rho) = i/j] = \{m \in \mathcal{M}_{\mathcal{D}}(\Delta) : r_m(\tau, \rho) = i/j\}$.

These sets constitute a partition of any set of models of Δ with domain $\mathcal{D}$. The set $[r_{\mathcal{M}_{\mathcal{D}}(\Delta)}(\tau, \rho) = 0]$ includes all those models in which the satisfaction set $SS_m(\rho)$ is empty, as well as those in which the satisfaction set $SS_m(\rho \wedge \tau)$ is empty.

In our example, we supposed that we had reason to believe that from 90% to 100% of the triples $\langle x, y, z \rangle$ such that x appears longer than y, and y appears longer than z, are also such that x appears longer than z. If we include the corresponding % statement, "$\%x, y, z(L(x, z), L(x, y) \wedge L(y, z), 0.90, 1.0)$", in Δ, then the models of Δ with given domain $\mathcal{D}$ are partitioned by the equivalence classes just defined. Thus, if

$|\mathcal{D}^3| < 10$, there can be only one element in the partition, $[r_{M_{\mathcal{D}(\Delta)}}(\text{``}L(x, z)\text{''}, \text{``}L(x, y) \wedge L(y, z)\text{''}) = 1]$—all the models have a model ratio of "$L(x, z)$" to "$L(x, y) \wedge L(y, z)$" equal to 1. If $|\mathcal{D}^3| = 27$, there are three, corresponding to the rationals $\frac{25}{27}, \frac{26}{27}$, and $\frac{27}{27}$. If $|\mathcal{D}| = 10$, so that $|D^3| = 1000$, there are a hundred and one, given by $i/j \in \{0.900, \ldots, 1.000\}$.

Note that here we are calculating the numbers of models subject only to the constraint imposed by the statistical statement. We are not even worrying about such "analytic" constraints as that nothing can appear longer than itself. We simply count the triples in $\mathcal{D}$, so that $|\mathcal{D}^3| = |\mathcal{D}|^3$. Further constraints beyond the statistical constraint cited could reduce the number of models in each partition set, though on pain of inconsistency at least one partition set must be nonempty.

10.4   Relevant Models

It is all very well to say that we will count models to evaluate the degree of validity of an uncertain inference, but we need to specify *which* models. Answer: The relevant ones. In this section we will attempt to characterize the relevant models for partial entailment between a set of sentences Γ and a sentence S. We will write the model-theoretic relation we are concerned with as $\Gamma \models_{[p,q]} S$, which says that S is *partially entailed* to the degree $[p, q]$ by the premises (evidence, body of knowledge) Γ. The partial entailment relation will be represented in terms of the model-theoretic relation just described: S must be true in a fraction of the *relevant* models that is at least p, and at most q. The *relevant* models are those picked out by certain subsets of Γ. (For inductive logic we may settle for "at least p", but for generality we are well be advised to keep track of the upper limit q as well.)

Both the words "support" and "entailment" may be a bit misleading, since both suggest that the evidence Γ is positively related to S. Of course, the case of positive relations, where p is pretty close to 1.0, is the case that interests us most. But we will use these same words when p and q are close to 0.0—that is, when the evidence concerning S is largely negative—as well as when the interval representing the degree of entailment is the whole interval $[0, 1]$, which tells us that Γ gives us no useful information about S at all.

Given an empirical domain $\mathcal{D}$, a sentence S, and evidence Γ, we will explicate the semantic relation of partial entailment in terms of *support sets* of models. A support set of models for a sentence S and a set of sentences Γ is a set of models of a set of three sentences corresponding (unsurprisingly) to the sets of sentences we focused on in the syntactic characterization of partial proof in the last chapter: $\{\ulcorner \rho(\alpha) \urcorner, \ulcorner S \equiv \tau(\alpha) \urcorner, \ulcorner \%\overline{\eta}(\tau, \rho, p, q) \urcorner\}$, where $\ulcorner \%\overline{\eta}(\tau, \rho, p, q) \urcorner$ is the *strongest* statement in Γ about τ and ρ.

Each of these support sets is a set of models picked out by the sentence S, the set of sentences Γ constituting evidence and background knowledge, and a triple of terms α, τ, ρ. There are at most a countable number of these triples of terms in the language, since we are constrained by the expressions of our language. Suppose that these triples of terms in the language are enumerated in some standard way: the ith triple is denoted by $\ulcorner \langle \alpha_i, \tau_i, \rho_i \rangle \urcorner$. Quantification over indices now does service for quantification over triples of formulas.

Let the support set $\Lambda_i(S, \Gamma, \mathcal{D})$ be the set of models corresponding to the ith triple $\{\ulcorner \rho_i(\alpha_i)\urcorner, \ulcorner S \equiv \tau_i(\alpha_i)\urcorner, \ulcorner \%\bar{\eta}(\tau_i, \rho_i, p, q)\urcorner\}$ for a given sentence S, background Γ, and empirical model domain $\mathcal{D}$:

Definition 10.9 (Support Set).

$$\Lambda_i(S, \Gamma, \mathcal{D}) = \{m : \exists p, q(\models_m \ulcorner S \equiv \tau_i(\alpha_i) \land \rho_i(\alpha_i) \land \%\bar{\eta}(\tau_i, \rho_i, p, q)\urcorner$$
$$\land \forall p', q'(\models_m \ulcorner \%\bar{\eta}(\tau_i, \rho_i, p, q)\urcorner \supset p \geq p' \land q \leq q'))\}$$

These are the basic model structures on which we will base uncertain validity. They depend on the language we adopt, but they do not depend on any notion of "objective chance" (although they are perfectly compatible with such notions), nor on any metaphysical indeterminism. They depend on the existence of statistical regularities—something that automatically follows from the finiteness of our model domain $\mathcal{D}$, since these regularities just represent frequency ratios.

They are also *objective*. We have pointed out that our syntactical constructions are objective in the sense that logic is objective: A sentence S is derivable from a set of premises Γ in first order logic whether or not anyone has performed the derivation, or even thought of it. Similarly, according to the preceding chapter, a set of sentences Γ partially proves the sentence S whether or not anyone has shown it, whether or not Γ is the evidence available to any agent, and whether or not it is even *possible* that Γ is someone's body of evidence.

But the support sets we have just defined are objective in another sense as well: they represent objective facts about the world. That the next ball to be drawn is (now) a ball in the urn is, if true, a fact. That the proportion of black balls among balls in the urn lies between 0.40 and 0.45 is, if true, a fact. That the next ball to be drawn is black if and only if the next ball to be drawn is black is certainly true; we may call it a fact by courtesy. The models in support sets purport to be models of the actual world.[5] There are many of them in each support set: in some of them the next ball to be drawn is black, in others it is white. The set of support models as a whole, however, provides objective constraints on the way the world can be.

It is interesting that so far this coincides with the intuitions of the frequentists. But the problem of the reference class is central—and insuperable, because, as we shall see, it has no *direct* solution.

For exactly the reasons adumbrated in the last chapter, we don't want to be concerned with *all* support sets $\Lambda_i(S, \Gamma, \mathcal{D})$, but only those that are *relevant*. Not all such models are relevant: the models corresponding to the fact that between none and all of the balls drawn today by Susan are black are not relevant, for example.

Let us call the set of relevant support sets $\Theta_{\mathcal{D}}(S, \Gamma)$; its members are the relevant support sets. For each i, Λ_i is a set of models; $\Theta_{\mathcal{D}}(S, \Gamma)$ is a set of these sets. We will now define this set. Not surprisingly, our considerations will parallel those of the previous chapter. To simplify our notation, we will take $\mathcal{D}$, S, and Γ to be fixed,

---

[5] Of course, we can reason in this way about *hypothetical* worlds; when we do so we are treating that hypothetical world as actual.

and simply write Λ_i for $\Lambda_i(\mathcal{D}, S, \Gamma)$. We define three conditions under which a set of support models Λ_i can be deleted from a set of sets of support models Θ with no loss. These are *weakness*, *generality*, and *marginality*.

To illustrate these conditions as we go along, let us consider the set of support sets $\Theta = \{\Lambda_1, \Lambda_2, \Lambda_3, \Lambda_4, \Lambda_5\}$ where the Λs are characterised by strongest statements:

$$\Lambda_1 : S \equiv \tau_1(\alpha_1) \wedge \rho_1(\alpha_1) \wedge \%\overline{\eta}(\tau_1, \rho_1, 0.3, 0.9).$$

$$\Lambda_2 : S \equiv \tau_2(\alpha_2) \wedge \rho_2(\alpha_2) \wedge \%\overline{\eta}(\tau_2, \rho_2, 0.5, 0.7).$$

$$\Lambda_3 : S \equiv \tau_3(\alpha_3) \wedge \rho_3(\alpha_3) \wedge \%\overline{\eta}(\tau_3, \rho_3, 0.6, 0.8).$$

$$\Lambda_4 : S \equiv \tau_4(\alpha_4) \wedge \rho_4(\alpha_4) \wedge \%\overline{\eta}(\tau_4, \rho_4, 0.55, 0.75).$$

$$\Lambda_5 : S \equiv \tau_5(\alpha_5) \wedge \rho_5(\alpha_5) \wedge \%\overline{\eta}(\tau_5, \rho_5, 0.60, 0.78).$$

In addition, we assume $\Gamma \models \ulcorner \forall \overline{\eta}(\rho_4 \supset \rho_3) \urcorner$ and $\Gamma \models \ulcorner \forall \overline{\eta}, \overline{\eta}'(\rho_5(\overline{\eta}, \overline{\eta}') \supset \rho_2(\overline{\eta}) \wedge \rho_6(\overline{\eta}')) \urcorner$.

Quite clearly, any set of support models Λ_i in which the model ratios include all the model ratios represented by other sets of support models in $\Theta_{\mathcal{D}}(S, \Gamma)$ adds nothing to the relevant constraints on the world. We define a relation between a support set Λ_i and a set of support sets Θ that holds when the support set Λ_i imposes no restrictions on Θ.

Definition 10.10 (Weakness). $W(\Lambda_i, \Theta)$ *iff* Θ *is a set of support sets and*
$\forall j (\Lambda_j \in \Theta \supset r_{M_{\mathcal{D}}(\Lambda_j)}(\tau_j, \rho_j) \subseteq r_{M_{\mathcal{D}}(\Lambda_i)}(\tau_i, \rho_i))$.

In our illustration, $W(\Lambda_1, \Theta)$.

If the constraints provided by a more specialized reference class conflict with the constraints provided by a less specialized reference class, then, provided that that reference class is not undermined by yet another, the less specialized can be ruled out as providing constraints. We use the symbol "#" introduced in the last chapter to express noninclusion between two sets of ratios.

Definition 10.11 (Generality). $G(\Lambda_i, \Theta)$ *if and only if* Θ *is a set of support sets and*

$$\exists j (\Lambda_j \in \Theta \wedge \Gamma \models \ulcorner \forall \overline{\eta}(\rho_j \supset \rho_i) \urcorner \wedge r_{M_{\mathcal{D}}(\Lambda_i)}(\tau_i, \rho_i) \# r_{M_{\mathcal{D}}(\Lambda_j)}(\tau_j, \rho_j)$$
$$\wedge \forall k (\Lambda_k \in \Theta \wedge \Gamma \models \ulcorner \forall \overline{\eta}(\rho_k \supset \rho_j) \urcorner \supset \neg r_{M_{\mathcal{D}}(\Lambda_j)}(\tau_j, \rho_j) \# r_{M_{\mathcal{D}}(\Lambda_k)}(\tau_k, \rho_k))).$$

In our illustration $G(\Lambda_3, \Theta)$.

Finally, if the ith support set is simply the marginal projection of a higher-dimensional set, as discussed in the last chapter, then in case of conflict it is appropriate to regard that marginal support set as irrelevant. When it is instructive to conditionalize, we should do so.

Definition 10.12 (Marginality). $M(\Lambda_i, \Theta)$ *if and only if* Θ *is a set of support sets and*

$$\exists j, \rho (\Lambda_j \in \Theta \wedge \Gamma \models \ulcorner \forall \overline{\eta}, \overline{\eta}'(\rho_j(\overline{\eta}, \overline{\eta}') \supset \rho_i(\overline{\eta}) \wedge \rho(\overline{\eta}')) \urcorner$$
$$\wedge r_{M_{\mathcal{D}}(\Lambda_i)}(\tau_i, \rho_i) \# r_{M_{\mathcal{D}}(\Lambda_j)}(\tau_j, \rho_j) \wedge \forall j', \rho [\Lambda'_j \in \Theta$$
$$\wedge \Gamma \models \ulcorner \forall \overline{\eta}, \overline{\eta}'(\rho_{j'}(\overline{\eta}, \overline{\eta}') \supset \rho_j(\overline{\eta}) \wedge \rho'(\overline{\eta}')) \urcorner \supset \neg r_{M_{\mathcal{D}}(\Lambda_i)}(\tau_{j'}, \rho_{j'}) \# r_{M_{\mathcal{D}}(\Lambda_j)}(\tau_j, \rho_j)]).$$

In our illustration $M(\Lambda_2, \Theta)$.

Now we can define the set of relevant support sets for S given Γ. Because we cannot be sure that there is only one set of support sets that cannot be improved in one of the three ways defined (by eliminating weakness, generality, or marginality), we define $\Theta_{\mathcal{D}}(S, \Gamma)$ to be the union of all such sets of sets. The union has the virtue the relevant frequency, and also the virtue (as we shall see) that the upper and lower bounds of the relevant model ratios are captured by our syntactical construction. The intersection of the set of relevant support sets is a set of models that includes the models of Γ and is included in each relevant support set Λ_i. It might be thought to provide a better guide to life, but note that if it could provide useful constraints, those constraints would be embodied in a relevant support set based on the intersection of reference classes used in other support sets, as is illustrated below in Example 10.1.

Definition 10.13 (Relevant Support Sets).

$$\Theta_D(S,\Gamma) = \{\Lambda_j(S,\Gamma,\mathcal{D}) : (\exists\Theta)(\forall i)(\Lambda_i \in \Theta \equiv (\neg W(\Lambda_i,\Theta) \wedge \neg G(\Lambda_i,\Theta) \wedge \neg M(\Lambda_i,\Theta)))\}.$$

Example 10.1. *Suppose that we know that 50% of red berries are good to eat, and that 70% of soft berries are good to eat, and all we know of the berry before us is that it is red and soft. We get no guidance from the set of all models of our knowledge: it is compatible with our knowledge base that all red and soft berries are good to eat, and compatible with our knowledge base that no red and soft berries are good to eat. The bounds on the proportion of models in which it is true that the berry is good to eat are 0 and 1, so far as our story goes. This provides us with no enlightenment. We may, however, expand the models we are considering beyond those that simultaneously satisfy all the constraints imposed by Γ. (We already know that in any single model, the berry in question is good to eat or is not good to eat. That's no help either.)*

In this example, one triple consists of the berry before us, the set of berries that are good to eat, and the set of red berries. The proposition that this berry (construed as a proper name) is good to eat is true in 50% of these models. Another triple consists of the berry before us, the set of berries that are good to eat, and the set of soft berries: that determines a support set of models in which 70% satisfy the proposition that this berry is good to eat. The models of each of these triples will be said to correspond to support sets. In addition, there is the support set based on this berry, the set of berries that are good to eat, and the set of soft and red berries. The bounds on the model ratios in this reference class are 0.00 and 1.00; these bounds, like the bounds on the reference class consisting of the singleton of the berry before us, include all the alternative bounds, and therefore may be disregarded. (Note that if we did have conflicting data about red-and-soft berries—e.g., "%x(Edible(x), Soft(x) \wedge Red(x), 0.2, 0.2)"—it would render the support sets associated with "red" and with "soft" irrelevant in virtue of their generality.) We have

$$\Theta_D(\text{``This berry is good to eat''}, \ \Gamma)$$
$$= \{\{m : \models_m \text{``Red (this berry)} \wedge \%x(Edible(x), Red(x), 0.5, 0.5)\text{''}\},$$
$$\{m : \models_m \text{``Soft (this berry)} \wedge \%x(Edible(x), Soft(x), 0.7, 0.7)\text{''}\}\}.$$

More realistically, we might know that between 40% and 60% of the red berries are good to eat, and that between 60% and 80% of the soft berries are good to eat. That would lead us, again, to two sets of models, one of which, Λ_1, contains models with model ratios of edibility falling between 0.40 and 0.60, and the other of which, Λ_2, contains models with model ratios of edibility falling between 0.60 and 0.80.

Pictorially, what we have for a given domain $\mathcal{D}$ looks like Figure 10.1. In the picture $\mathcal{M}_D\Gamma$ represents the set of models of the full body of evidence and background knowledge we have available, and $\bigcap \Theta_D(S,\Gamma) = \Lambda_1 \cap \Lambda_2$ represents the set of models compatible with the evidence that remain if we consider only the evidence that constrains "This berry

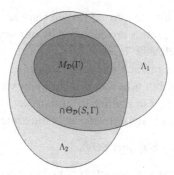

Figure 10.1: The models for berries.

red berries are good to eat will be pruned away.) The sets of models Λ_1 and Λ_2 represent the support sets of models corresponding to individual items of statistical evidence.

Each of the sets Λ_1 and Λ_2 in the example represents a support set of models, partitioned by model ratios. Each such set includes the set of models corresponding to Γ. Each set represents one useful way in which the impact of the evidence provided by Γ can be understood. Furthermore, there are no other useful constraints we can call on. The set of sets of models, $\Theta_{\mathcal{D}}(S, \Gamma)$, is what we have to go on. The problem is to represent the joint influence of these various items of evidence.

Whereas we know that the actual model—the world—lies in the intersection of the support sets derived from Γ, this information, like much of that reflected in Γ, is of little help, because we have no constraints on the frequency among those models of models in which S is true.

How can we capture the import of this collection of items of evidence not ruled irrelevant by considerations of weakness, specificity, and marginalization? We propose to consider $\Theta_{\mathcal{D}}(S, \Gamma)$ itself:

$$\Theta_{\mathcal{D}}(S, \Gamma) = \{\Lambda_{i_1}, \Lambda_{i_2}, \ldots \Lambda_{i_k}\}.$$

Although we know that the actual world is a world in the intersection of these sets of models (and indeed in the even smaller set of worlds Γ), we do not have statistics concerning this set of worlds on which to base our beliefs and actions. (We have no statistics concerning the edibility of berries that are both soft and red, much less "statistics" concerning the edibility of *this* berry, which is what really interests us.)

One thing we know about these constraints on the truth frequency of S in each of these support sets is that it is bounded by the largest and the smallest model ratio among the Λ_{i_j}. These objective constraints all agree in this. We therefore adopt this as our definition of *partial validity*.

10.5 Partial Validity

We define partial validity of S, given Γ, in our canonical domain $\mathcal{D}$ as follows, making use again of the canonical ordering of triples relevant to S, given Γ:

Definition 10.14. $\Gamma \models_{[p.q]} S$ *iff there is* $\Theta_{\mathcal{D}}(S, \Gamma)$ *such that*

$$p = \min\{r_{\mathcal{M}_{\mathcal{D}}}(\tau_i, \rho_i) : \Lambda_i \in \Theta_{\mathcal{D}}(S, \Gamma)\}$$
$$q = \max\{r_{\mathcal{M}_{\mathcal{D}}}(\tau_i, \rho_i) : \Lambda_i \in \Theta_{\mathcal{D}}(S, \Gamma)\}.$$

Validity in deductive logic requires that the conclusion of a valid argument be true in every model in which the premises are true. In the case of partial entailment characterized by the interval $[p, q]$, what corresponds to validity is that the conclusion should hold in a fraction between p and q of the set of models "corresponding" to the premises. As we have argued, this set is not the set of models of Γ, the total body of knowledge, nor even the larger set of models of $\bigcap \Theta_\mathcal{D}(S, \Gamma)$, but the *set* of sets of support provided by Γ for S itself. Each set of support models represents a possible constraint. It would clearly be a mistake, an objective mistake, to adopt a measure falling outside any of the constraints corresponding to the relevant support sets Λ_i as an appropriate measure by which to characterize S given Γ.

Lemma 10.1. *The proportion of models in $\Lambda_i(S, \Gamma, \mathcal{D})$ in which S holds lies between* $\min\{r_{\mathcal{M}_\mathcal{D}(\Lambda_i)}(\tau_i, \rho_i)\}$ *and* $\max\{r_{\mathcal{M}_\mathcal{D}(\Lambda_i)}(\tau_i, \rho_i)\}$.

Proof: Assume that the cardinality of the satisfaction set of ρ_i is n_i and that the cardinality of the satisfaction set of $\tau_i \wedge \rho_i$ is m_i. Suppose p_i and q_i are the minimal and maximal values asserted to exist by Definition 10.9. Then since $\ulcorner\%\overline{\eta}(\rho_i, \tau_i, p_i, q_i)\urcorner$ is satisfied in Λ_i we must have $p_i \leq m_i/n_i \leq q_i$. Because the only constraint that must be satisfied by the interpretation of α_i in Λ_i is that it be in the satisfaction set of ρ_i, the number of possible interpretations of α_i is exactly n_i. Of these n_i interpretations, m_i must also satisfy τ_i. Because the biconditional $\ulcorner S \equiv \tau_i(\alpha_i)\urcorner$ is true in every model, S must be true in the same models that make $\ulcorner\tau_i(\alpha_i)\urcorner$ true:

$$\min\{r_{\mathcal{M}_\mathcal{D}}(\tau_i, \rho_i)\} \leq \frac{|\Lambda_i(S, \Gamma, \mathcal{D}) \cap \mathcal{M}_\mathcal{D}(S)|}{|\Lambda_i(S, \Gamma, \mathcal{D})|} \leq \max\{r_{\mathcal{M}_\mathcal{D}}(\tau_i, \rho_i)\} \qquad ■$$

We must be careful here to distinguish between use and mention. The definition is in the metalanguage, in which p and q are variables ranging over real numbers. Inside the quoted sentence, of course, p and q must be thought of as real number *expressions*. Due to our syntactical constraints, however, they are real number expressions in canonical form (they obviously can't be variables), and thus they correspond directly to real numbers, and relations among them are decidable.

Our first and most important theorem extends the result of Lemma 1 to the set of sets of support models $\Lambda(S, \Delta)$ generated by Γ. It is this fundamental theorem that captures the kernel of our approach, and corresponds to the proof of soundness (deduction preserves truth) for deductive systems. It says that if Γ partially proves S to the degree $[p, q]$, then Γ partially entails S to the same degree, i.e., for any large $\mathcal{D}$, the proportion of relevant sets of models with domain $\mathcal{D}$ in $\Theta_\mathcal{D}(S, \Gamma)$ in which S is satisfied lies between p and q. Note that the number of models at issue is the sum of the numbers of models in *each* of the support sets, not the number of models in the union or intersection of the support sets, and that the number of models satisfying S is similarly the sum of the numbers of satisfying models in the support sets, rather than the number of satisfying models in their union or intersection.

Theorem 10.1. *If $\Gamma \vdash_{[p,q]} S$ then $\Gamma \models_{[p,q]} S$.*

Proof: We must show that

$$\forall i \left(p \leq \frac{|\Lambda_i(S, \Gamma, \mathcal{D}) \cap \mathcal{M}_\mathcal{D}(S)|}{|\Lambda_i(S, \Gamma, \mathcal{D})|} \leq q \right).$$

Suppose that $\Lambda_i(S, \Gamma, \mathcal{D})$ is a support set, i.e., is a set of models m satisfying $m \models$ $\ulcorner S \equiv \tau_i(\alpha_i) \wedge \rho_i(\alpha_i) \wedge \%\overline{\eta}(\tau_i, \rho_i, p_i, q_i)\urcorner \wedge \forall p', q'(\Gamma \models \ulcorner \%\overline{\eta}(\tau_i, \rho_i, p', q')\urcorner \supset p_i \geq p' \wedge q' \geq q_i)$. Suppose that $p_i < p$. Let Δ^* be a set of statistical statements whose existence is called for by Definition 10.14. We know, by Theorem 9.8, that $\ulcorner \%\overline{\eta}(\tau_i, \rho_i, p, q)\urcorner$ is not in Δ^*. Δ^* must therefore sharpen any set Δ that contains this statistical statement. Two cases are possible:

Case I: $q_i \geq q$. Then $\Delta - \{\ulcorner \%\overline{\eta}(\tau_i, \rho_i, p, q)\urcorner\}$ sharpens Δ by precision. But then $\Lambda_i \notin$ $\Theta_\mathcal{D}(S, \Gamma)$, for if it were, it would be eliminated by weakness, $W(\Lambda_i, \Theta_\mathcal{D}(S, \Gamma))$, contrary to Definition 10.13.

Case II: $q_i < q$. Every statistical statement in Δ^* must conflict with $\ulcorner \%\overline{\eta}(\tau_i, \rho_i, p, q)\urcorner$ in the sense of #. It follows that $\Delta^* \bigcup \{\ulcorner \%\overline{\eta}(\tau_i, \rho_i, p, q)\urcorner\}$ is sharpened by either specificity or richness.

Case IIa. If the set of supporting statistical statements is sharpened by specificity, there must be some j such that $\ulcorner \%\overline{\eta}(\tau_j, \rho_j, p_j, q_j)\urcorner \in \Delta(S, \Gamma)$, satisfying Definition 9.17. But then $\Lambda_i \notin \Theta_\mathcal{D}(S, \Gamma)$, for if it were, it would be eliminated by generality, $G(\Lambda_i, \Theta_\mathcal{D}(S, \Gamma))$, contrary to Definition 10.13.

Case IIb. If the set of supporting statistical statements is sharpened by richness, a similar argument shows that Λ_i cannot be in $\Theta_\mathcal{D}(S, \Gamma)$, for if it were, it would be eliminated as a constraint by marginality.

In short, there can be no Λ_i containing a model ratio falling outside the interval $[p, q]$, since the statistical component of any such Λ_i would be eliminated by one of the varieties of sharpening from any set Δ establishing the partial proof, and each of those varieties of sharpening can be seen as leading to the elimination of the constraint Λ_i through a similar argument.   ∎

10.6   Remarks

Let us give some more thought to Γ. What we have in mind is a set of sentences representing a body of knowledge or a database. Typically it will be large. Typically it will have both sentences of the form $\ulcorner \rho(\alpha)\urcorner$ attributing a property to an object or a relation to a vector of objects, and sentences of the form $\ulcorner \%\overline{\eta}(\tau, \rho, p, q)\urcorner$ constraining the proportion of objects having one property or relation, among those that have another property or relation. It may also have sentences of other forms such as universally quantified conditionals ("All penguins are birds"), but it is sentences of the first two forms, together with biconditionals, that mainly concern us. We should think of Γ as a finite set of sentences given explicitly; for example, as axioms, or as a database or knowledge base.

What is the relation between this approach to partial entailment, and that which computes the ratio of the number of models of Γ in which S is true to the total number of models of Γ? After all, for a finite domain the ratio of the number of models of Γ in which S is true to the total number of models of Γ is logically determinate for any sentence S, and in some sense might be thought to provide a guide to belief that is based on more of the evidence than is provided by Δ. We argue, however, that while it uses more of the evidence, it is a less sensible guide.

Four considerations argue for looking at the ratio of S-models among a set of models broader than in the set of models of Γ: (a) the ratio of S-models in models of Γ varies according to the cardinality of the domain, (b) this ratio can be very difficult to calculate for an interesting language, (c) purely logical constraints can give misleading precision,[6] and finally, (d) the numbers we obtain from logical constraints can be very counterintuitive.

Simple combinatorial arguments can give you information about the relative numbers of models that can overwhelm statistical knowledge. Carnap's $c^\dagger$ provides an example: Given a reasonably large satisfaction set for $B(x)$, the number of models of $B(a)$ in which $A(a)$ is also true is, *a priori*, very near to a half, since there are many more ways in which half the Bs can be As than there are in which all (or none of) the Bs can be As. Furthermore, this ratio stays close to a half, however large a sample of Bs we take as evidence. Can we take this observation as implying that "$\%x(A(x), B(x), 0.5 - \varepsilon, 0.5 + \varepsilon)$" is part of our knowledge? No, because the quoted statement imposes severe constraints on the satisfaction sets of $A(x)$ and $B(x)$, for which we have no empirical evidence. It is a misleading fact that for combinatorial reasons it is true in most models. It is certainly false in those models in which a small proportion of Bs are As, or in which a large proportion of Bs are As.

Although, given a domain $\mathcal{D}$ of given cardinality, we could in principle calculate the exact proportion of models of Γ in which a given S is true, this does not seem to be what we want. The requirement that there be a known empirical basis for this proportion is what prevents this (hard-to-calculate and counterintuitive) proportion from being the degree of support. Not only must the proportion of models in which S is true lie between p and q, but p and q must be determined by knowledge in Γ to be the minimum and maximum of the relevant model proportions.

By focusing on empirical knowledge, characterized in terms of reference formulas and target formulas, we are able to impose constraints on the proportions of models that make S true in a set of support sets of models. By combining the information in these models in the way suggested, we can apply these general constraints to a particular instance without ending up with the worthless interval [0.0, 1.0]. We have here an intuition that is (at least vaguely) like Aristotle's intuition that rational belief about the particular must derive from rational belief about the general.

The object of primary interest to us, the object whose properties will provide the best indication of the epistemic stance we should take toward S, consists not of a single set of models, but of a set of sets of models. We want to consider the evidence represented by each support set of models, but neither their union nor their intersection will do, for in each of these, the frequency of the satisfaction of S depends on unknown factors. Surely the import of the evidence we have cannot be made to depend on evidence we don't have! The influence of our knowledge of the frequency of edibility among red berries and among soft berries cannot depend on the (unknown!) facts concerning berries that are both red and soft. We take it as a powerful desideratum that the formal validity of our arguments should not depend on facts about the world that are unknown to us. This is certainly true for deductive arguments; surely we may ask the same of inductive or uncertain arguments.

---

[6]Karl Popper [Popper, 1959] was among those who found this last aspect of "logical measure" approaches to uncertainty particularly disturbing.

Nevertheless, the true test of our semantics requires seeing whether it yields intuitively plausible results in simple cases, to begin with, and reasonable results in more complex cases in which our intuitions are not so strong.

10.7 Summary

A semantics allows us to specify truth conditions for the sentences of our formal language. It does not in general tell us what sentences are true, but what the conditions are that make a sentence true. An interpretation of a first order language consists of a domain and an interpretation function. Proper names are interpreted as members of the domain. Predicates are interpreted as subsets of the domain. Relations of n places are interpreted as n-tuples of members of the domain. An n-place function is interpreted as a function from n-tuples of the domain to the domain.

Under this scheme, the axioms of first order logic are true under any interpretation in any domain. Valid arguments preserve truth: If the premises of an argument are true in a domain, under an interpretation, then the conclusion of the argument will also be true under that interpretation. In particular, every theorem of the logic is true in every nonempty domain. This is called *soundness*. Furthermore, it turns out that every statement that is true in every nonempty domain is provable from the axioms for first order logic. This is called *completeness*.

The languages we consider are two-sorted: that is, they have two styles of variable, one of which takes as values real numbers, and the other of which ranges over the empirical objects of our domain. Similarly there are predicates and functions of each class. Because we are not interested in the interpretation of the mathematical part of the language, we say merely that this part of the language is to be given its "standard" interpretation. We focus on the empirical part of the language.

The empirical part of the language makes use of a primitive sentence form that we have informally taken to represent statistical knowledge: sentences of the form $\ulcorner\%\overline{\eta}(\tau, \rho, p, q)\urcorner$. In a given model, these sentences may be interpreted in terms of the cardinalities of the satisfaction sets of their formulas. Such a formula will be true, loosely speaking, if the ratio of the cardinality of the satisfaction set of $\ulcorner\tau \wedge \rho\urcorner$ to the cardinality of the satisfaction set of ρ lies in the interval denoted by "$[p, q]$."

Whereas argument validity in first order logic requires that the conclusion be true in every model in which the premises are true, we must obviously settle for less. We will require that the conclusion be true in a fraction of the set of relevant models that is bounded by an interval.

What is the set of relevant models? In simple cases this is determined simply by the appropriate reference set. In more complicated cases, what we must look at is a *set* of sets of models—one picked out by each possible appropriate reference set. The upper and lower bounds of the proportion of models in each of these sets that make the conclusion true are taken to be an appropriate measure of the cogency of the inference.

With this way of construing partial validity, we can establish the following relations, corresponding to soundness and completeness, between partial proof and partial entailment. If the sentence S is partially provable to the *degree* $[p, q]$ from the set of sentences Γ, then the minimum frequency or measure in any set of models that is picked out as relevant by the contents of Γ is p, and the maximum frequency or measure in any set of models that is picked out as relevant by the contents of Γ is q. This corresponds to soundness for our syntactical characterization of partial proof.

Suppose that in fact considerations of relevance given Γ pick out a set of sets of models $\{M_1, \ldots, M_k\}$ and that the maximum and minimum of the relative frequency with which S is true in these models are p and q. Then there is a set Δ of statistical statements that are in Γ, and there exist formulas τ and ρ, such that $\ulcorner \rho(\alpha) \urcorner$ and $\ulcorner S \equiv \tau(\alpha) \urcorner$ are in Γ, and for some t, $\mathrm{Support}(\Delta, S, \Gamma, \alpha, \tau, \rho) = [t, q]$, and there is no Δ' such that Δ' sharpens Δ in the context S, Γ. In other words, syntactically, the crucial set Δ contains q as an upper bound for some reference formula for S.

Similarly, p is a lower bound in the syntactically characterized set Δ.

These facts correspond to the completeness of our syntactical characterization of partial proof.

10.8 Bibliographical Notes

The semantic underpinnings of epistemological probability only began to appear in [Kyburg, 1994]. Nevertheless, the general idea was the driving force behind the publications cited in the last chapter: What is probable, or acceptable, is what holds "for the most part" or "almost always". Aristotle identified the probable with what holds for the most part, and the basic frequentist idea was supported by Reichenbach [Reichenbach, 1949], von Mises [von Mises, 1957], and Cramér [Cramér, 1951], of course; but only Reichenbach seems to have understood the importance of settling on a reference class in the case of conflict.

10.9 Exercises

(1) Discuss the testability of assertions of probability on the view presented here.

(2) Compare the semantics appropriate to subjective interpretations of probability with the semantics presented here.

(3) Compare the semantics appropriate to some empirical view of probability with the semantics presented here.

(4) What is problematic about the phrase "relative frequency of truth in an apropriate set of models"? What problems are raised, and how have we attempted to deal with them?

(5) Discuss the semantic plausibility of the principle of precision.

(6) Discuss the semantic plausibility of the principle of specificity.

(7) What is the relation between the principle of richness and the Bayesian approach to probability and statistics?

Bibliography

[Adams, 1966] Ernest Adams. Probability and the logic of conditionals. In Jaakko Hintikka and Patrick Suppes, editors, *Aspects of Inductive Logic*, pp. 265–316. North Holland, Amsterdam, 1966.

[Cramér, 1951] Harald Cramér. *Mathematical Methods of Statistics*. Princeton University Press, Princeton, 1951.

[Genesereth & Nilsson, 1987] Michael R. Genesereth and Nils J. Nilsson. *Logical Foundations of Artificial Intelligence*. Morgan Kaufmann, Los Altos, CA, 1987.

[Israel, 1983] David Israel. Some remarks on the place of logic in knowledge representation. In Nick Cercone and Gordan McCalla, editors, *The Knowledge Frontier*, pp. 80–91. Springer–Verlag, New York, 1983.

[Kyburg, 1994] Henry E. Kyburg, Jr. Believing on the basis of evidence. *Computational Intelligence*, 10:3–20, 1994.

[Mendelson, 1979] Elliott Mendelson. *Introduction to Mathematical Logic*. Van Nostrand, New York, 1979.

[Nilsson, 1986] Nils Nilsson. Probabilistic logic. *Artificial Intelligence*, 28:71–88, 1986.

[Popper, 1959] K. R. Popper. *The Logic of Scientific Discovery*. Hutchinson, London, 1959.

[Reichenbach, 1949] Hans Reichenbach. *The Theory of Probability*. University of California Press, Berkeley and Los Angeles, 1949.

[Suppes, 1966] Patrick Suppes. Probabilistic inference and the concept of total evidence. In Jaakko Hintikka and Patrick Suppes, editors, *Aspects of Inductive Logic*, pp. 49–65. North Holland, Amsterdam, 1966.

[von Mises, 1957] Richard von Mises. *Probability Statistics and Truth*. George Allen and Unwin, London, 1957.

11

Applications

11.1 Introduction

We are now in a position to reap the benefits of the formal work of the preceding two chapters. The key to uncertain inference lies, as we have suspected all along, in probability. In Chapter 9, we examined a certain formal interpretation of probability, dubbed evidential probability, as embodying a notion of partial proof. Probability, on this view, is an interval-valued function. Its domain is a combination of elementary evidence and general background knowledge paired with a statement of our language whose probability concerns us, and its range is of [0, 1]. It is *objective*. What this means is that if two agents share the same evidence and the same background knowledge, they will assign the same (interval) probabilities to the statements of their language. If they share an acceptance level $1 - \alpha$ for practical certainty, they will accept the same practical certainties.

It may be that no two people share the same background knowledge and the same evidence. But in many situations we come close. As scientists, we tend to share each other's data. Cooked data is sufficient to cause expulsion from the ranks of scientists. (This is not the same as data containing mistakes; one of the virtues of the system developed here is that no data need be regarded as sacrosanct.) With regard to background knowledge, if we disagree, we can examine the evidence at a higher level: is the item in question highly probable, given that evidence and our common background knowledge at that level?

There are a number of epistemological questions raised by this approach, and some of them will be dealt with in Chapter 12. For the moment and for our present purposes, it suffices to cast doubts on the superficial doctrine of subjectivity. It is *not* all in the eye of the beholder.

Chapter 9 provided some tools for finding the reference formulas from which the intervals of evidential probability derive. (The calculations provided there involved provability, and thus were not decidable, but in some (many?) practical cases we are in a "solvable" domain [Ackermann, 1962].) Chapter 10 provided some justification for the syntactical relation characterized in Chapter 9. We saw that the probability interval captured every relevant objective long run frequency of truth in a class of models *picked out* by our total body of evidence and background knowledge. More precisely, the syntactical rules pick out a *set* of sets of models, given background knowledge Γ and a sentence S, and in each of these sets of models the relative frequency of truth of the sentence S is included in the probability interval. This provides a tight connection between the syntactical notion of "partial proof" and the semantical notion of objective frequency of truth.

In this chapter, we will explore some of the elementary (and highly idealized) results we are led to by this approach, and discuss briefly the significance of the results. The most important general result concerns the justifiability of statistical inference; it is on the objective acceptability of general statistical statements that the viability of the whole approach depends. Statistical generalization, alone among inductive inferences, *need* not depend on background knowledge. This is not to say that in doing statistical inference we are free to ignore our background knowledge. Quite the contrary. We must take account of any knowledge we have. But it is possible to show that even without general background knowledge, we can assign high probabilities to statistical generalizations on the basis of purely enumerative evidence.

Along with this, a result of importance both in philosophy and computer science concerns the conflict between the popular Bayesian account of inference in the presence of uncertainty, and the objective long run frequencies that we see as essential to sound inference. On the usual Bayesian view, one only gets posterior probabilities by beginning with prior probabilities and conditioning on the evidence. Our view is that statistical testing and confidence interval estimation need not require general background knowledge. But in the process of learning about the world, we do acquire background knowledge, which can give rise to the probabilities required by the Bayesian application of conditionalization. A plausible view of uncertain inference must account for this transition between seemingly incompatible approaches to uncertain inference.

11.2 Elementary Results

The theorems that follow in this section mainly concern small bodies of knowledge. This is the framework within which inductive logic has almost always been considered. For example, we are given some empirical evidence, and asked to consider the "justification" of an inductive hypothesis. In the case of interpretations of probability as a logical measure, or as a subjective degree of belief, we are given a few sentences in a very impoverished language, a regimented prior probability, and invited to compute a conditional probability. In computational learning theory, we are given a sequence of 0s and 1s and asked to guess the next digit, or the general pattern. In the philosophy of science, we are given a hypothesis, some background knowledge, a deduced prediction, and asked to consider the effect of the success of that prediction on the hypothesis. In elementary statistics, we are given a sample space, and asked to construct a statistical test with "nice" properties. In this section we will look only at analogues of the Carnapian logical examples.

We are in good company in considering these elementary theorems, and we shall find that they provide useful insights into our logical constructions. Our goal, nevertheless, is to achieve an understanding of *real* inductive inference: Inference from rich bodies of evidence in the presence of real and defensible (and of course defeasible!) background knowledge.

The import of those theorems can sometimes be extended to large and realistic databases provided we can show that certain kinds of expansion of Γ will not undermine the semantic relations we have developed. Thus, in the absence of any knowledge connecting mortality and middle initial, the knowledge of an individual's middle initial will not, *and should not*, change our assessment of the chance that he will survive for an additional year. The fact that ducks are smaller than geese should not, absent knowledge that size is relevant, undermine analogical arguments based on their properties as water birds.

Suppose that Γ consists of only the two sentences, "$\%x(A(x), B(x), p, q)$" and "$B(j)$." For example, our knowledge base Γ might be:

{"47-year-old-male(John)"

"%(survive-12-months(x), 47-year-old-male(x), 0.987, 0.991)"}.

Then the exact number of models in which "$\%x(A(x), B(x), p, q)$" is true can be calculated for $\mathcal{D}$ of given size; it is merely a combinatorial problem to calculate the number of ways in which $|\mathcal{D}|$ objects can be assigned to four boxes, AB, $A\neg B$, $\neg AB$, and $\neg A\neg B$, in such a way that $|AB|/|B|$ lies in $[p, q]$.

There are 4^N ways of putting N objects into the four boxes. The number of ways of doing this that make "$\%x(A(x), B(x), p, q)$" true can be calculated as follows. Fix on a size for the interpretation of the reference set $|B|$—say n. There are $\binom{N}{n}$ ways of choosing an interpretation for B of this size. We must choose between pn and qn of the B's to be A's—$\sum_{k\geq pn}^{k\leq qn} \binom{n}{k}$—and then we are free to choose anywhere from 0 to $N - n$ non-B objects to be in the interpretation of A, which may be done in 2^{N-n} ways, for a total of $\binom{N}{n} \sum_{k\geq pn}^{k\leq qn} \binom{n}{k} 2^{N-n}$ ways. To get the total number of models that make "$\%x(A(x), B(x), p, q)$" true, we need merely sum over $0 < n \leq N = |\mathcal{D}|$. For $N = 30$, $p = 0.2$, $q = 0.4$, we have 351,325,760.

In that set of models satisfying the statistical statement, we pick out the subset in which "$A(j)$" and "$B(j)$" are both true, and thus for fixed $\mathcal{D}$ compute the exact proportion of models making the premises Γ true that also make "$A(j)$" true. But without even knowing the number of models satisfying the premises, we can be quite sure that between p and q of them also satisfy "$A(j)$", since the term "j" may denote any of the objects in the boxes AB and $\neg AB$, and between p and q of the choices must result in the truth of "$A(j)$". Note that "j" is a proper name here; it does not denote an "arbitrary" individual (whatever that may be) but our old friend John, about whom we only know his name, age, and sex. Nevertheless, in counting models we may assign any individual in the interpretation of B to the constant j. That is what leads us to the result that the statement "survive-12-months (j)" is partially proved by the premises to the degree [0.987, 0.991]. This is form of inference has been called "direct inference"; it is inference from a known frequency or measure to a statement concerning an individual. The example may be generalized as Theorem 11.1:

Theorem 11.1. {"$\%x(A(x), B(x), p, q)$", "$B(j)$"} $\models_{[p,q]}$ "$A(j)$".

Proof: Γ is {"$\%x(A(x), B(x), p, q)$", "$B(j)$"}. Take Δ to be {"$\%x(A(x), B(x), p, q)$"}. The only other possibilities would be statistical statements like "$\%x(A(x), x = j, 0, 1)$", which would be sharpened away by precision. Consider the set of models that satisfy $\Delta \cup \{"B(j)"\}$. The minimum of the model ratios $\{r_m(A(x), B(x)): m \in \mathcal{M}(\Delta)\}$ is p, and similarly the maximum is q. The only candidate for a support set Λ

is thus $\{$"$\%x(A(x), B(x), p, q) \wedge B(j)$"$\}$. (The set of models is not altered by adding the conjunct "$A(j) \equiv A(j)$".) Thus, $\Theta_\mathcal{D}$("$A(j)$", Γ) = Λ, and $\Gamma \models_{[p,q]}$ "$A(j)$".   ∎

Example 11.1. *Suppose we know that John is 47 years old, and that the chance that a 47-year-old male will survive for a year (L) lies between 0.987 and 0.991. In a given model of our premises, suppose that there are 1000 individuals in the satisfaction set of* "47-year-old-male(x)" *(M). In order for the statistical statement* "$\%x(L(x), M(x), 0.987,$ 0.991)" *to be satisfied, between 987 and 991 of those individuals must also be in the satisfaction set of* "survive-12-months(x)". *The interpretation of* "John" *must be one of the individuals in the satisfaction set of* "47-year-old-male(x)", *but he may be any one of them: we have a thousand choices. In all but from 9 to 13 of those choices, the interpretation of* "John" *is also a member of the satisfaction set of* "survive-12-months(x)". *Thus, the proportion of models of the premises in which* "survives-24-months(John)" *is true lies between 0.987 and 0.991. Note particularly that this is true regardless of the size of the satisfaction set of* "non-47-year-old-male(x)". *In particular, note that we do not have to fix on the size of the domain of our interpretation, or on the cardinality of the satisfaction set of* "47-year-old-male(x)".

Of course, in a richer context this result has consequences for statements of other forms as well. Thus, suppose that we also know that John's wife, Mary, has a lover whom she will marry if and only if John dies during the current year. That she will not marry her lover should therefore have the same probability as John's survival for a year. The appropriate reference class for Mary's marriage to her lover is that to which we are led by considering the chance that John will survive for a year. The principle of specificity concerns Mary and her lover as much as it concerns the liabilities of the insurance company: given the truth of the biconditional, to determine Mary's likelihood of not marrying her lover is exactly the *same problem* as that of determining John's likelihood of survival for a year: the set of models in which Mary marries her lover is exactly the set of models in which John fails to survive, and the proportion of those models is determined by all the constraints to which they are subject.

At this point you may feel your intuitions flagging: should our knowledge that Mary has a lover bear on John's chances of survival? Any reader of detective fiction may feel at least a slight temptation to answer in the affirmative. But recall how stark our background knowledge is: We have not supposed ourselves to have any knowledge of a relation between mortality and having a wife who has a lover. There may *be* such a connection, but a statement recording it is not among our premises. In the same way, there may be, of course, a metastasized cancer riddling John's bone marrow, unknown to him as well as to us. In some sense of "chance" this clearly affects his chances of surviving the year; but, since by hypothesis it is unknown, it does not affect the degree to which what we do know partially proves the proposition that he will survive the year, and, equally, the proposition that his wife will not marry her lover.

We have already mentioned that a single set of sentences Δ may support a given sentence S to various degrees. The following theorem says what happens in one very simple case in which the statistical information we have about two potential reference classes *differs*, in the sense that neither of the two intervals is included in the other. This is a particular case of "combining evidence."

Theorem 11.2. *If* $[p, q] \# [r, s]$ *then*

$$\{\text{"%}(A(x), B(x), p, q)\text{"},$$
$$\text{"}B(a)\text{"},$$
$$\text{"%}(A(x), C(x), r, s)\text{"},$$
$$\text{"}C(a)\text{"}\}$$
$$\models_{[t,u]} \text{"}A(a)\text{"},$$

where $t = \min(p, r)$ *and* $u = \max(s, q)$.

Proof: Take Γ to be the whole premise set, and Δ to be the two statistical statements mentioned. No subset of Δ sharpens Δ, since we know of no relation between "B" and "C". We have two support sets: $\Lambda_1 = \{\text{"%}(A(x), B(x), p, q) \wedge B(a)\text{"}\}$ and $\Lambda_2 = \{\text{"%}(A(x), C(x), r, s) \wedge C(a)\text{"}\}$. The set of support sets, $\Theta_{\mathcal{D}}(\text{"}A(a)\text{"}, \Gamma)$ has each of these sets as members. Now given our canonical domain $\mathcal{D}$, consider the set of nonempty equivalence classes $[r_{\mathcal{M}_{\mathcal{D}}(\Delta)}(A(x), B(x)) = z]$, z real. Similarly consider nonempty equivalence classes $[r_{\mathcal{M}_{\mathcal{D}}(\Delta)}(\text{"}A(x)\text{"}, \text{"}C(x)\text{"}) = y]$, y real. Note that we have both

$$\text{Support}(\Delta, \text{"}A(a)\text{"}, \Gamma, a, \text{"}A(x)\text{"}, \text{"}B(x)\text{"}) = [p, q]$$

and

$$\text{Support}(\Delta, \text{"}A(a)\text{"}, \Gamma, a, \text{"}A(x)\text{"}, \text{"}C(x)\text{"}) = [r, s]$$

The sets $[r_{\mathcal{M}_{\mathcal{D}}(\Delta)}(A(x), B(x)) = z]$ and $[r_{\mathcal{M}_{\mathcal{D}}(\Delta)}(A(x), C(x)) = y]$ constitute partitions of their respective sets of support models. In each set of each partition the proportion of models that are models of "$A(a)$" lies between t and u, and so we have $\Gamma \models_{[t, u]} \text{"}A(a)\text{"}.$ ∎

Example 11.2. *You know that the Potluck Casino is crooked, but you also know that its crookedness is constrained by rules. For example, their dice are either fair, or biased toward 2, or biased toward 5. Furthermore, a die that is biased toward 2 lands 2-up 50% of the time and 10% of the time on each of the other five sides, and a die that is biased toward 5 lands 5-up 50% of the time and 10% of the time on each of the other sides. You know nothing about the relative numbers of each of these three kinds of dice.*

Let t be the next throw of a die at the Potluck Casino. Consider the statement "t lands 4-up". There are two relevant statistical statements, one for normal dice, and one for biased dice. For normal dice, the frequency of 4's is $\frac{1}{6}$. For biased dice, of either sort, the frequency of 4's is $\frac{1}{10}$. The degree to which your evidence supports "t lands 4-up" is $[\frac{1}{10}, \frac{1}{6}]$. There are two support sets in $\Theta_{\mathcal{D}}(2\text{-up}(t), \Gamma)$.

Consider the statement "t lands 2-up". Now there are three relevant support sets, corresponding to the normal die, the die biased in favor of 2-up, and the die biased in favor of 5-up. The first case leads to a measure of $\frac{1}{6}$, the second to a measure of $\frac{1}{2}$, and the third to a measure of $\frac{1}{10}$. The degree to which your evidence supports "t lands 2-up" is $[\frac{1}{10}, \frac{1}{2}]$.

A natural constraint on uncertain or probabilistic inference, well known since Reichenbach,[1] is that given a choice between two reference classes for a probability,

---

[1] Hans Reichenbach offered the principle that posits (his version of propositional probabilities) should be based on the narrowest class about which we have "adequate statistics" [Reichenbach, 1949].

if we know one is a subclass of the other, it is (in some sense) a better guide to belief. In the theory of evidential probability [Kyburg, 1991b]), as in the theory of Pollock [Pollock, 1990], the included class is to be preferred in the case of conflict. In default reasoning about hierarchies, "specificity" is also deemed appropriate: the fact that Charles is a penguin is a better guide to his flying ability than the fact that he is a bird [Touretzky, 1986]. Again we restrict the application of the principle to potential measures that *differ* in the sense of # expressed by the antecedent.

Theorem 11.3. *If* $[p, q] \# [r, s]$ *then*

$$\{\text{``}\%(A(x), B(x), p, q)\text{''},$$
$$\text{``}\%(A(x), C(x), r, s)\text{''},$$
$$\text{``}B(a)\text{''},$$
$$\text{``}C(a)\text{''},$$
$$\text{``}\forall x(C(x) \supset B(x))\text{''}\}$$
$$\models_{[r,s]} \text{``}A(a).\text{''}$$

Proof: Clearly Γ supports "$A(a)$" both to the degree $[p, q]$ and to the degree $[r, s]$. We show that $\Delta_1 = \{\text{``}\%(A(x), C(x), r, s)\text{''}\}$ sharpens $\Delta_2 = \{\text{``}\%(A(x), B(x), r, s)\text{''},$ "$\%(A(x), C(x), r, s)$"$\}$. Definition 9.17 of "sharpens-by-specificity" requires that there exist a reference term ρ'' such that we may accept $\ulcorner \forall x(\rho'' \supset \rho') \urcorner$. The formula "$C(x)$" fills the bill, and we do indeed have Support("$A(a)$", Δ_1, a, A, C)$\# [p, q]$. The only further requirement of Definition 9.17 is that there be no more specific formula than "$C(x)$" that conflicts with $[r, s]$. That condition is clearly satisfied in this case with the Γ we have explicitly listed.

It follows that $\Gamma \models_{[r,s]} A(a)$. ∎

Example 11.3. *Suppose we know of Mark that he is a white collar worker, and a 47 year old male. We know that the proportion of 47 year old males who survive for 12 months lies between 0.987 and 0.991, and that the proportion of white collar workers among the 47 year old males who survive for a like period is between 0.990 and 0.994. This is the extent of our knowledge.*

We have in our knowledge base the statement "white-collar(John) $\wedge$ 47-year-old-male(John)" *and the statements* "$\%x$(survive-12-months(x), 47-year-old-male(x) $\wedge$ white-collar(x), 0.990, 0.994)" *and* "$\%x$(survive-12-months(x), 47-year-old-male(x), 0.990, 0.994)." *Then the appropriate measure of the degree to which our premises prove the sentence* "survive-12-months(John)" *is* [0.990, 0.994] *rather than* [0.987, 0.991]. *But if it were the case that we merely knew* less *about white collar workers who are 47 year old males than we know about 47 year old males—for example, that their survival rate lies between 0.800 and 1.000—then we would* not *want to use the more specific class to determine our beliefs. In the extreme case, when we consider that* "λy(Mark = y)Mark" *and* "$\%x$(survive-12-months(x), λy(Mark = y)(x), 0, 1)" *are in* any *corpus of knowledge, we realize that specificity can be pushed too far.*

This example brings out again the tension between precision and specificity already remarked in Chapter 9. The broader the reference class, the larger the sample on which our statistical knowledge may be based, and the more precise that knowledge—that is, the narrower the interval in which a frequency is known to fall. On the other hand, have that applies to it and the broader the interval into which the relevant frequency is known to fall. In practice, of course, we can pretend not to know as much as we do know about the object. But pretending not to know something we do know seems, except in delicate social situations, irrational and perhaps dishonest. The three kinds of sharpening are designed precisely to avoid that kind of pretense: much of what we know, in concrete situations, may *rationally* be ignored. It may be pruned away by considerations of precision.

Notice that in the example the two general evidential statements about males and about males who are white collar workers differ in that the truth frequency of "survive-12-months(Mark)" in some equivalence class of model ratios in $\{m : m \models \{$"47-year-old-male(Mark)", "%x(survive-12-months(x), 47-year-old-male(x), 0.988, 0.992)\}\}$ is incompatible with the truth frequency of "survive-12-months(Mark)" in any equivalence class of model ratios in

$$\{m : m \models \{\text{"47-year-old-male(Mark)"},$$
$$\text{"%x(survive-12-months(x), 47-year-old-male(x)}$$
$$\wedge \text{ white-collar(x), 0.990, 0.994)"}\}\}$$

and vice versa.

This situation is to be distinguished from the situation in which the statistical statements mention intervals one of which is included in the other—in our first example concerning Mrs. Jones, Example 9.5, the two intervals are [0.930, 0.975] and [0.945, 0.965]. In this case the truth frequency of $S =$ "survive-24-months(Mrs. Jones)" in every equivalence class of model ratios in $\Lambda_i(S, \Delta, \mathcal{D})$, where α_i is "Mrs. Jones", τ_i is "survive-24-months(x)", and ρ_i is "50-year-old-female(x) $\wedge$ US (x)"), is the same as the truth frequency of S in some equivalence class of models in the set of models $\Lambda_j(S, \Delta, \mathcal{D})$, where $\alpha_i = \alpha_j$, $\tau_i = \tau_j$, and $\rho_j =$ "50-year-old-female(x) $\wedge$ US(x) $\wedge$ Rochester(x)". There is no *difference* to be taken account of, only vagueness to be eschewed.

To show that the antecedent $[p, q] \# [r, s]$ of Theorem 11.3 is required, we note that the following theorem, which embodies a contrary assumption and reflects the analysis of the first example involving Mrs. Jones, leads to a different result. This result is at variance with the familiar (but unanalyzed) maximal specificity criterion.

Theorem 11.4. *If* $[p, q] \subset [r, s]$ *and* $[r, s] \neq [p, q]$, *then*

$$\{\text{"%(A(x), B(x), p, q)"},$$
$$\text{"%(A(x), C(x), r, s)"},$$
$$\text{"B(a)"}$$
$$\text{"C(a)"}$$
$$\text{"}\forall x(C(x) \supset B(x))\text{"}\}$$
$$\models_{[p,q]} \text{"A(a)"}.$$

Proof: In this case let the value of Δ_1 in the definition of partial proof be $\{$"%x(A(x), B(x), p, q)"$\}$, a weaker set of statements than Δ_2, containing both statistical statements. We need to show that that Δ_1 sharpens Δ_2. It does so by Definition 9.15, since $[p, q] = \text{Support}(\Delta_1, S, \Gamma, a, A, B) \subset [r, s] = \text{Support}(\Delta_2, S, \Gamma, a, A, C)$. The fact that we know "C(a)" and "$\forall x(C(x) \supset B(x))$" cuts no ice: it doesn't show us that there

Note that we always have *some* statistical knowledge—even, in the limit, of the form "%($A(x)$, $C(x)$, 0.0, 1.0)," which is true so long as something is C and C is in $\mathcal{R}$ and A is in $\mathcal{T}$. Such knowledge does not undermine the probability we obtain from Δ, since it only yields looser constraints on the partial entailment of "$A(a)$" by Γ.

11.3   Inference from Samples

We have been supposing that our background knowledge Γ contains statements of the form $\ulcorner\%(\tau, \rho, p, q)\urcorner$. An important part of our motivation is that we wish to include statements in our body of knowledge on the basis of their "practical certainty" or "high probability". We seek objectivity for these inferences. In the present section we will limit our concerns to *statistical* statements such as those we have imagined in our examples to be available to the insurance company. We now want to show that a body of knowledge that contains only data about a sample can confer a high degree of entailment on a general statistical statement, and thus warrant its inclusion in another body of knowledge on the basis of high probability. We leave to one side for the time being the question of whether a high degree of proof is sufficient for acceptance into a body of knowledge, as well as the question of the relation between the set of evidential statements containing the sample description and the set of practical certainties to which the general statistical statement is added.

Suppose that our background knowledge includes exactly the statements "$Q(a_1) \wedge P(a_1)$", "$Q(a_2) \wedge P(a_2)$", . . . ,"$Q(a_j) \wedge \neg P(a_j)$", where there are i unnegated and $j - i$ negated P-sentences. "Q" is understood to be a reference formula in $\mathcal{R}$, and "P" a target formula in $\mathcal{T}$.

Because our models are finite, in each model there will be a real number p that will make "%($P(x)$, $Q(x)$, p, p)" true. (Note that this position is available to real variables as well as constants in canonical form.)

We know that by and large subsets are like the populations of which they are subsets; the converse is also true: populations are like the subsets they contain. It is tempting to put it this way: samples are like the populations from which they are drawn; populations are like the samples they spawn. The first way of putting the matter is correct; the second is misleading, since "sample" is a somewhat loaded word.

In statistics, "sample" is often taken to mean a *random sample*, that is, a sample from the population obtained by a method that would select each possible sample (each subset) with equal frequency in the long run. For finite, present populations, randomness in this demanding statistical sense can be achieved with the help of tables of random numbers or the use of an auxiliary device whose behavior is well understood (but not too well understood). In most applications of sampling, this procedure is out of the question. One cannot sample a set of events this way: the events come into our purview when they do—events in the future, or in the distant past, cannot be part of our sample. The populations we are interested in, even if they are not events, are often distributed in the past and the future, as well as in the present.

Furthermore, even if the "long run" properties of the sampling method are such as to ensure that each possible subset of the population would be selected with equal frequency in the long run, that does not ensure that the sample serves its epistemic purpose as evidence. We can sample the population of a city to estimate the proportion of citizens of Irish descent by the most perfect random method imaginable, and come up with a sample of people whose names all begin with "O." Indeed, a method that comes up with each possible subset of the population with equal frequency *must* come up with such a sample with its appropriate (small) frequency.

It is our contention that what "random sampling" is intended to achieve is accomplished by our logical constraints on degrees of entailment. For example, the sample of Irishmen that we come up with by accident is, we know, not a sample from which we can infer with high degree of entailment a statement about the proportion of Irishmen in the city. Why? Because it is a member of that subset of samples that we know are less likely to be representative of the proportion of Irishmen than are other samples with a mixture of alphabetic characteristics. On the other hand, in testing a die for bias, a sequence of temporally contiguous tosses, though it cannot be considered a random member of the set of all tosses in the statistical sense, may serve perfectly well as the basis of an inference about bias.

From the knowledge base just described, containing no more than the statements stipulated, we may with some degree of confidence infer that the proportion of Q's that are P is close to i/j. This reflects the fact that the "sample" $\langle a_1, \ldots, a_j \rangle$ is, so far as we are concerned, a sound basis on which to infer the proportion of Q's that are P's.

Let us give mathematical precision to this statement in the following theorem:

Theorem 11.5. *For positive ε and δ, we can find a natural number j such that*

$$S_j \vdash_{[1-\gamma, 1]} \text{``}\%x\left(P(x), Q(x), \frac{i}{j} - \varepsilon, \frac{i}{j} + \varepsilon\right)\text{''}$$

where γ is less than δ, "$Q(x)$" is a formula in $\mathcal{R}$, "$P(x)$" is a formula in $\mathcal{T}$, and S_j is a set of j statements of the form "$Q(a_k) \wedge \pm P(a_k)$" in which "$\pm$" is replaced by nothing in i instances and by "$\neg$" in $j - i$ instances and $a_1, \ldots, a_j$ are distinct terms.

We approach the proof of this theorem through a sequence of easy lemmas. We establish that a certain statistical statement is a theorem of our first order language; we show that our statistical data support a certain biconditional; we show that the data provide support for the statement in question. The fourth lemma, the only messy one, shows that any attempt to take account of more of the data will merely weaken the conclusion, and that therefore our partial proof cannot be sharpened.

Lemma 11.1. "$\%y(|f_j(y) - F(P, Q)| \le k/2\sqrt{j}, Q^j(y), 1 - 1/k^2, 1)$" *is true in every model m of our language, where "P" and "Q" are target and reference formulas containing only x free, y is a j-sequence of variables of the language, $f_j(y)$ is the relative frequency of P's among these Q's, and $F(P, Q)$ is the ratio of the cardinality of the satisfaction set of "$P \wedge Q$" to the cardinality of the satisfaction set of "Q" in the model m:*

$$\frac{|SS_m(P(x) \wedge Q(x))|}{|SS_m(Q(x))|} = F(P, Q).$$

Proof: We show that

$$\%y\left(|f_j(y) - F(P, Q)| \le \frac{k}{2\sqrt{j}}, Q^j(y), 1 - \frac{1}{k^2}, 1\right). \tag{11.1}$$

Consider a model m in which the value of $F(P, Q)$ is p. For every value of j, the expected value of "$f_j(y)$" is p, and the variance of "$f_j(y)$" is $p(1-p)/\sqrt{j}$.

Chebychef's inequality tells us that a quantity is k standard deviations removed from its mean no more than $1/k^2$ of the time. This comes to the claim, in this case, that

$$\%y\left(|f_j(y) - F(P, Q)| \le k\sqrt{\frac{p(1-p)}{j}}, Q^j(y), 1 - \frac{1}{k^2}, 1\right)$$

Because $p(1 - p) \le \frac{1}{4}$, equation (11.1) is true in this model m.

It is also true, for the same reasons, in every model in the class of models in which $F(P, Q)$ has the value p. Now observe that, although $F(P, Q)$ is not a rigid designator, and has different values in different models, so that equation (11.1) *means* something different in models characterized by a different model ratio of "P" to "Q," the argument supporting (11.1), which does not depend on p, goes through just as well. Thus, equation (11.1) is true in every model of our language, regardless of the value of the mathematical term $F(P, Q)$ in that model. ∎

We may thus, without risk, add equation (11.1) to the data S_j. Next we need to establish that S_j, playing the role of Γ, entails a certain biconditional connecting the observed frequency in the sample to the statistical statement we are trying to infer.

Lemma 11.2. $S_j \vdash |f_j(\langle a_1, \ldots, a_j \rangle) - F(P, Q)| \le k/2\sqrt{j}$
$\equiv \%x(P(x), Q(x), i/j - k/2\sqrt{j}, i/j + k/2\sqrt{j}).$

Proof: Recall that $\langle a_1, \ldots, a_j \rangle$ is our sample. Let m be a model of S_j. By the characterization of S_j, $f_j(\langle a_1, \ldots, a_j \rangle)$ has the value i/j in this model. Recall that $F(P, Q)$ is the ratio of the cardinality of the satisfaction set of $P \wedge Q$ to the cardinality of the satisfaction set of Q. To say that this lies within a distance $k/2\sqrt{j}$ of i/j is exactly to say that the right hand side of the biconditional is true. ∎

The third lemma shows that S_j, together with equation (11.1) provides support for the statistical assertion concerning the measure of Ps among Qs. Its proof has a number of parts, corresponding to the clauses of Definition 9.13.

Lemma 11.3. Support(Δ_L, "$\%x(P(x), Q(x), i/j - k/2\sqrt{j}, i/j + k/2\sqrt{j})$", S_j, $\langle a_1, \ldots, a_j \rangle$, $|f_j(y) - F(P, Q)| \le k/2\sqrt{j}, Q^j(y)) = [1 - \gamma, 1]$, *where* $1 - \gamma \ge 1 - k/2\sqrt{j}$, *and* Δ_L *contains only logically true statistical statements.*

Proof: In Definition 9.13, instantiate Δ to "$\%y(|f_j(y) - F(P, Q)| \le k/2\sqrt{j}, Q^j, 1 - 1/k^2, 1)$", S to "$\%x(P(x), Q(x), i/j - k/2\sqrt{j}, i/j + k/2\sqrt{j})$", Γ to $\{S_j, \%y(|f_j(y) - F(P, Q)| \le k/2\sqrt{j}, Q^j, 1 - 1/k^2, 1)\}$, α to "$\langle a_1, \ldots, a_j \rangle$", τ to "$|f_j(y) - F(P, Q)| \le k/2\sqrt{j}$", and ρ to "$Q^j(y)$". Note that "$f_j(\langle a_1, a_2, \ldots, a_j \rangle) = i/j$" is included in the consequences of $\Gamma = S_j$. Because "$\%y(|f_j(y) - F(P, Q)| \le k/2\sqrt{j}, Q^j(y), 1 - 1/k^2, 1)$" is true in every model, it is among the consequences of $\emptyset$ and so of Γ. Let the strongest statistical statement with this subject matter true in every model with domain of cardinality $|\mathcal{D}|$ be "$\%y(|f_j(y) - F(P, Q)| \le k/2\sqrt{j}, Q^j(y), 1 - \gamma, 1)$". It has an upper bound of 1, because there are models—those in which the model ratio of P to Q, $r_m(P(x), Q(x))$, is 1 or 0—in which all the samples are precisely representative. Its lower bound, $1 - \gamma$, is the proportion of samples that are $k/2\sqrt{j}$-representative in the worst case—i.e., when the model ratio of $P(x)$ in $Q(x)$ is about a half. In any event, Chebychef's theorem assures us that $1 - \gamma$ is greater than $1 - 1/k^2$. For any p' and q' such that $\%y(|f_j(y) - F(P, Q)| \le$

is among the consequences of S_j. The necessary biconditional is a consequence of S_j by Lemma 11.2. ∎

What is most difficult to see, and what has escaped the notice of many writers, is that the support just mentioned *sharpens* any body of knowledge that identifies some particular members of the sample as Ps. Let us call the members of Q^j that are constrained to contain j' specific Qs, of which i' are specified to be Ps constrained samples, and take Δ' to contain only a statistical statement concerning constrained samples whose target formula is equivalent to that of the single statistical statement in Δ. The details of the equivalence will be given in the proof of the following lemma:

Lemma 11.4. *Let* $\phi(y)$ *represent a constraint on* y *that is satisfied by our sample* $\langle a_1, \ldots, a_j \rangle$. *Then* $\Delta = $ *"%y($|f_j(y) - F(P, Q)| \leq k/2\sqrt{j}, Q^j, 1 - \gamma, 1$)" sharpens-by-precision* $\Delta' = \Delta \cup \{$ *"%y($|f_j(y) - F(P, Q)| \leq k/2\sqrt{j}, Q^j(y) \wedge \phi(y), 1 - \delta, 1$)"$\}$ *in the context* "%y($|f_j(\langle a_1, \ldots, a_j \rangle)) - F(P, Q)| \leq k/2\sqrt{j}, \Gamma$)", *where* δ *is determined, as we shall see, by the constraint* $\phi(y)$.

Proof: Observation: In any model m of our language almost all large samples exhibit a relative frequency that is close to the model ratio of P in Q. If we look at the subset of Q^j in which it is true that a sample exhibits a relative frequency of i/j, this is no longer true. In a model in which the model ratio is far from i/j, *almost none* of the samples exhibit relative frequencies close to the model ratio, since they all exhibit relative frequencies of i/j. Thus, among the models of S_j, the proportion of samples that are representative may lie anywhere from 0 to 1. We show something stronger: that the inclusion of *any* empirical constraints at all entailed by S_j—beyond the membership of the sample in Q^j—in the reference formula of a statistical statement in Δ' allows $\Delta' \cup \Delta$ to be sharpened-by-precision by Δ, where $\Delta = \{$"%y($|f_j(y) - F(P, Q)| \leq k/2\sqrt{j}, Q^j, 1 - \gamma, 1$)"$\}$.

We must show that any more complete utilization of our knowledge will merely lead to looser constraints on the degree to which the observed sample supports the conclusion—i.e., we must show that for any Δ', if $S_j \vdash \Delta'$ and Δ' is not equivalent to Δ, then Δ sharpens-by-precision Δ' in the context "%x($P(x), Q(x), i/j - \varepsilon, i/j + \varepsilon$)," S_j.

It is clear that $Q^j(\langle a_1, \ldots, a_j \rangle)$ entails no conjunct "$Q(a_k) \pm P(a_k)$" in S_j, since any particular conjunct of this form can be made false in a model of our language. Conjuncts of this form are our first candidates for ϕ. (Weaker sentences will be dealt with later.) Suppose that in fact $\phi(y)$ consists of j' such sentences, in which i' of the "P"s are unnegated. Because the interpretations of the a_i are arbitrary, the import of the additional information is simply that i' specific sampled items out of j' specific sampled items are P. Without loss of generality, take the objects mentioned in the additional sentences to be $a_1, \ldots, a_{j'}$. S_j entails the equivalence of

$$|f_j(\langle a_1, \ldots, a_j \rangle) - F(P, Q)| \leq \frac{k}{2\sqrt{j}}$$

and

$$\frac{j}{j - j'} \left(F(P, Q) - \frac{i'}{j} - \frac{k}{2\sqrt{j}} \right) \leq f_{j-j'}(\langle a_{j'+1}, \ldots, a_j \rangle)$$

$$\leq \frac{j}{} \left(F(P, Q) - \frac{i'}{} + \frac{k}{\phantom{2\sqrt{j}}} \right)$$

since

$$i = jf_j(\langle a_1, \ldots, a_j \rangle) = j'f_{j'}(\langle a_1, \ldots, a_{j'} \rangle) + (j - j')f_{j-j'}(\langle a_{j'+1}, \ldots, a_j \rangle)$$
$$= i' + (j - j')f_{j-j'}(\langle a_{j'+1}, \ldots, a_j \rangle).$$

The real-valued quantity $f_{j-j'}(\langle y_{j'+1}, \ldots, y_j \rangle)$ is defined on $Q^{j-j'}$ (and the assertion that it has a particular value is equivalent to a formula in T). Let $[p]$ be the set of models in which the model ratio of P in Q is p. In any of the models in $[p]$ the quantity $f_{j-j'}(\langle y_{j'+1}, \ldots, y_j \rangle)$ has a mean of $F(P, Q)$ and a variance of $p(1 - p)/j - j'$. Let $F+$ be the quantity

$$\left(\frac{j}{j - j'} \left[F(P, Q) - \frac{i'}{j} - \frac{k}{2\sqrt{j}} \right] \right),$$

and let $F-$ be the quantity

$$\left(\frac{j}{j - j'} \left[F(P, Q) - \frac{i'}{j} + \frac{k}{2\sqrt{j}} \right] \right).$$

What we need to show is that

$$\min\{r_m(F- \leq f_{j-j'}(y) \leq F+, Q^{j-j'}):$$
$$m \in \mathcal{M}(Q^j(\langle a_1, \ldots, a_j \rangle) \wedge \phi(\langle a_1, \ldots, a_j \rangle))\} = 1 - \delta < 1 - \gamma \quad (11.2)$$

and

$$\max\{r_m(F- \leq f_{j-j'}(y) \leq F+, Q^{j-j'}):$$
$$m \in \mathcal{M}(Q^j(\langle a_1, \ldots, a_j \rangle) \wedge \phi(\langle a_1, \ldots, a_j \rangle))\} = 1. \quad (11.3)$$

It is only at this point in the proof that we must think carefully about statistical issues. Note that $1/k^2$ is not the measure of nonrepresentative samples; it is an upper bound for that measure. What we require in order to show that Δ sharpens-by-precision Δ' is that the bounds generated by Δ' are broader.

To show that equality (11.3) is true is easy: the same argument we used before will work.

To show inequality (11.2), note that $f_{j-j'}(\langle y_{j'+1}, \ldots, y_j \rangle)$ is distributed about $F(P, Q) = p$ in every model in $[p]$, with variance $p(1 - p)/(j - j') > p(1 - p)/j$; the quantity $f_{j-j'}(\langle y_1, \ldots, y_{j-j'} \rangle)$ lies in $[F-, F+]$ if and only if it lies in the interval

$$I = \left[\frac{j}{j - j'} \left(p - \left(\frac{k}{2\sqrt{j}} \right) - \frac{i'}{j} \right), \frac{j}{j - j'} \left(p + \left(\frac{k}{2\sqrt{j}} \right) - \frac{i'}{j} \right) \right].$$

But since we know the distribution of $f_{j-j'}$ in the equivalence class of models $[p]$, we can clearly find a value of p in which $f_{j-j'}$ is less likely than $1 - \gamma$ to have a value in the interval I. We look for a value of p that is far from i'/j but near 1 or 0 in order to have a small variance. The frequency with which $f_{j-j'}(\langle y_{j'+1}, \ldots, y_j \rangle)$ is close to p will then be high, and the frequency with which it will be close to $p - i'/j'$ will be low: certainly less than $1 - \gamma$. Thus, there are models of Δ' with frequencies below $1 - \gamma$ as well as with frequencies equal to 1. Let the minimum such value be p'. Thus, (perhaps surprisingly) Δ sharpens-by-precision Δ' in the context "%$x(P(x), Q(x), i/j - \varepsilon, i/j + \varepsilon)$", S_j.

Now suppose that $\phi(y)$ is a disjunction of conjunctions of the form previously conjunctions; since the minimum frequency with which "$\%x(P(x), Q(x), i/j - k/2\sqrt{j}, i/j + k/2\sqrt{j})$" will be less than $1 - \gamma$ in each of the models of the conjunctions, the minimum frequency will also be less than $1 - \gamma$ in their union: Δ will also sharpen any Δ' that takes account of a condition $\phi(y)$ that is not satisfied in every model of Q^j. ∎

Given these four lemmas, Theorem 11.5 is immediate.

11.4 Example

Let us consider an extended example. Suppose that the cardinality of the interpretation of "Q" in a model is n, and the cardinality of the satisfaction set of "$Q(x) \wedge R(x)$" in that model is m. We can calculate the proportion of subsets of Qs of size 20 (let the corresponding open sentence be "subsets-of-20-$Q(y)$") whose proportion of Rs lies within a distance δ of m/n. Let us call such a subset y δ-*representative* of the proportion of Rs among Qs, or δ-rep (R, Q, y) for short. We can calculate this proportion for every possible value of m/n. The proportion will vary, according to the value of m/n (it will be 1 for $m = n$, for example, and a minimum for m approximately equal to $n/2$), but it will fall in an interval that depends only slightly on m and n for reasonably large values of m and n.

To illustrate this for our example, let us suppose that n is large enough so that there is no great loss in using the binomial approximation to the hypergeometric distribution, and no great loss in using the normal approximation to the binomial for the parameter $\frac{1}{2}$. The smallest value of the ratio of the cardinality of the satisfaction set of "δ-rep(R, Q, y)" to the cardinality of the satisfaction set of "subsets-of-20-$Q(y)$" is given by the sum over all k in $[10 - 20\delta, 10 + 20\delta]$ of

$$\binom{20}{k} \left(\frac{1}{2}\right)^k \left(\frac{1}{2}\right)^{20-k}$$

since, as just observed, the smallest value corresponds to $m/n = \frac{1}{2}$. If we take δ to be 0.10, this smallest value is 0.737; the largest is of course 1.0.[2]

Thus, we know a priori that "$\%y(\delta$-rep(R, Q, y), subsets-of-20(y), 0.737, 1.0)" is true in every model of our language, where y is a vector of length 20, and thus, of course, it is true in every model in which our premises are true. Let our premises Γ contain the information that we have chosen 20 Qs and that of our sample of 20, four specified individuals satisfy the predicate R. Now note that $a = \langle a_1, \ldots, a_{20} \rangle$ is in the satisfaction set of "δ-rep(R, Q, y)" if and only if "$\%x(R(x), Q(x), 0.10, 0.30)$" holds in the corresponding model. This biconditional, as well as the assertion that $\langle a_1, \ldots, a_{20} \rangle$ is a vector of Qs, is entailed by Γ, as is, of course, the logical truth "$\%y(\delta$-rep(R, Q, y), subsets-of-20(y), 0.737, 01.0)$."

The proportion of models of Γ in which δ-rep(R, Q, a) holds may vary from 0 to 1, because the proportion of members of a that satisfy R is specified. If the interpretation of Q is of large cardinality, as it will be in most models whose domain $\mathcal{D}$ is large, in most models the proportion of the interpretations of Q that are in the interpretation of R will be close to $\frac{1}{2}$. In most models, then, a is *not* representative, since only 20% of a is in R. If the our model domain is small, however, for example 20 or close thereto, a will be representative in almost every model.

---

[2] Actually, the smallest value is $\frac{823004}{\text{...}} = 0.7848777771$. For simplicity, we will stick to the normal

This is a misleading line of thought, however. The very fact that the proportion of representative models may vary from 0 to 1, given our sample ratio, renders the set of all models in which a sample of 20 is drawn of which four are R too weak to be of interest. We can do better by disregarding some of the information available.

If we consider the less specific reference class corresponding to the formula subsets-of-20(y), the proportion of *these* models in which "δ-rep(R, Q, a)" is satisfied varies only from 0.737 to 1.0 in virtue of the general truth of {"%$y(\delta$-rep(R, Q, y), subsets-of-20(y), 0.737, 1.0)"}. Thus, in a proportion between 0.737 and 1.00 of the relevant models that make our premises true, "%$x(R(x), Q(x), 0.10, 0.30)$" will be true. It follows that our observations lead us to assign to the uncertain conclusion "%$x(R(x), Q(x), 0.10, 0.30)$" a degree of certainty of at least 0.737.

We are supposing that $a = \langle a_1, \ldots, a_{20} \rangle$ is a vector of proper names (so that our sample is being drawn "without replacement"); the proportion of assignments of objects to these names that makes "δ-rep(R, Q, a)" true is between 0.737 and 1.0. Note that there is no question here of "random sampling"—of attempting to see to it that the procedure by which we obtain our sample of 20 is such that each sample of 20 will be drawn equally often in the long run.[3] We are simply counting models, though we are disregarding certain facts about the sample. Now in real life it is quite true that we know a lot about the subset of Q that constitutes our sample. We know which entities it consists of, we know who selected them, we know when they were selected, etc. We *might* know or discover that we obtained the sample of 20 by a method that yields biased samples. That possibility is real, but it comes down to having additional evidence, in the light of which the inference we are discussing would no longer go through. (For an example, see Section 11.7.)

We are taking as our evidence Γ, but we are deriving the statistical support for our partial proof from a tautology. (The corresponding Λ does contain content from Γ.) Suppose we thought to strengthen Δ with the information that two of four specified Qs sampled were Rs. This would *undermine* the strength of our inference, since the frequency of δ-rep(R, Q, y) where y is subject to the condition that two of its components are Rs is less precisely known than the frequency of δ-rep(R, Q, y) when y is not subject to that condition. This reflects the contents of Lemma 11.4.

Now let us add to our database ten more observations: $Q(a_{21}) \wedge R(a_{21})$, $Q(a_{22}) \wedge R(a_{22}), \ldots, Q(a_{30}) \wedge R(a_{30})$. Suppose that (surprisingly) they comprise six unnegated and four negated Rs. Now let us again see in what proportion of models consistent with our premises "%$x(R(x), Q(x), 0.10, 0.30)$" is true. (Of course, this is not the statement we would be interested in, once we had the new information; but it is of interest to see what its status is.) Given our statistical data (it is available in Γ), "%$x(R(x), Q(x), 0.10, 0.30)$" will be true if and only if $p + 0.03 \leq 0.33 \leq p + 0.23$, since $0.20 - 0.1 \leq p \leq 0.20 + 0.1$ if and only if $0.33 - 0.23 \leq p \leq 0.33 - 0.03$. The proportion of relevant models satisfying our expanded data set in which "%$x(R(x), Q(x), 0.10, 0.30)$" is true can thus be calculated (as a function of p) to be

$$\sum_{k=30(p-0.03)}^{30(p+0.23)} \binom{30}{k} p^k (1 - p)^{30-k}.$$

---

[3] It has been argued [Kyburg, 1974] that random sampling in this strong sense is neither sufficient nor necessary for the cogency of statistical inference. This is not to say that sampling may be treated casually in real science; it is merely to say that the statistical conception of randomness is inappropriate here.

EXAMPLE   261

Recall that p may have any value in our models. Thus, this sum has a minimum value of 0.0 for $p = 0$, and a maximum of 0.447 for $p = 0.77$: the proportion of models, relative to these new premises, in which "$\%x(R(x), Q(x), 0.10, 0.30)$" is true is in [0.0, 0.447].

The statement "$\%x(R(x), Q(x), 0.26, 0.40)$" is one we would more likely be interested in; the proportion of relevant models in which this is true ranges from 0.737 to 1.0.

11.5 Statistical Induction

Given a sample from a population, it is not difficult to see how the present approach, if successful, could generate probabilities for statistical hypotheses. As already illustrated by the preceding example, though, it will generate probabilities, both high and low, for many contrary hypotheses. Thus, there will be many pairs of constants (δ_1, δ_2) such that for given observations, the support for the statistical hypothesis "$\%x(R(x), Q(x), r - \delta_1, r + \delta_2)$" is at least $1 - \varepsilon$.

A natural way of approaching induction is to suggest that we accept whatever is highly supported—i.e., whatever is true in almost all the models of our premises. We have already noted one way in which we may wish to modify this notion of support: we may want to regard as highly supported what is true in almost all models of each of the model sets constituting $\{\Lambda_1, \Lambda_2, \ldots \Lambda_n\}$ rather than in almost all models of our complete body of evidence Γ, or even in almost all models of the sharpest statistical evidence Δ.

But now we have a reason to restrict the scope of our inductive inference. If we accept every statistical generalization that is well supported, we will be accepting too much: given a body of evidence, there are many interval hypotheses, for the same subject matter, that are highly supported. To accept the conjunction of all these hypotheses would be to accept, if not a contradiction, a statement whose degree of support, relative to our evidence, is approximately 0. It is natural to select one of these as being the most useful, or most interesting, to accept. This function is performed in classical statistical testing by focusing on *shortest* confidence intervals. Thus, a natural formulation of the inductive principle at the level of acceptance $1 - \varepsilon$ is: Accept "$\%x(R(x), Q(x), p, q)$" when the support your evidence gives to it is at least $1 - \varepsilon$, *and* among such statements concerning R and Q, $q - p$ is a minimum.

This assumes, as does nonmonotonic logic, that the result of uncertain inference is the acceptance of the conclusion, rather than the assignment of some number to it indicating its relative certainty. The present approach thus falls in the category of logical or qualitative approaches to uncertainty, as opposed to quantitative approaches. The assumption of (tentative) acceptance on the basis of a high degree of entailment is thus in conflict with the approach of many probabilistic or Bayesian approaches to induction, according to which the role of evidence (which, curiously, is itself "accepted" even though at a later time it may be undermined) is to confer some measure of certainty on a conclusion, and not to yield the categorical, if temporary, acceptance of that conclusion. The conclusion of an induction is not of the form "Believe H to the degree p (or $[p, q]$)", as Bayesians would have it, but rather "Believe the conclusion H itself, supported by the evidence and the high degree of entailment that our knowledge, including the evidence, confers on H".

11.6 Bayesian Induction

A few further remarks on the relation between our approach to induction and the Bayesian view will be to the point. The uncertain inference from a sample of objects to an approximate statistical generalization about the class from which the objects were drawn made no use of any prior assumptions or degrees of belief. It may thus be a matter of some interest to explore the relation between our approach and the approach that begins with prior distributions.

The classical Bayesian approach would begin (in the artificially sparse example of Section 4.5) with a distribution of beliefs over the possible worlds corresponding to our language.[4] (See, for example [Carnap, 1950; Carnap, 1952; Carnap, 1971; Carnap, 1980; Bacchus, 1992; Bacchus et al., 1992].) The simplest view, corresponding to Carnap's $c^\dagger$ ([Carnap, 1950]; this also corresponds to the random worlds of [Bacchus et al., 1992]), simply counts each model once: each possible world gets equal weight. As Carnap discovered, on the basis of this prior probability distribution, induction in the sense of learning from experience is impossible. Regardless of how large a sample of Q's you have drawn, and what proportion of R's you observed in it, the probability that the *next* Q you draw is an R is $\frac{1}{2}$—just what it was before you did any sampling. This led Carnap to introduce c^* (also in [Carnap, 1950]), in which, first, each structure description (collection of statistically similar models) receives equal weight, and, second, each possible world in a given structure description receives an equal share of that weight. This does allow learning from experience, in a certain sense. There are a variety of modifications of this approach: [Hintikka, 1966; Halpern, 1990a; Grove et al., 1992].

The crucial question from our point of view is how to explain the apparent conflict between the measures assigned by our method to statistical generalizations and the measures assigned by what I will call generically the Carnapian approach.

First, it should be noted that our inference procedure does not proceed by conditionalization; it employs direct inference—inference from a *known* frequency or proportion to a degree of certainty. Because we have not assigned measures to our models, there is no updating of those measures going on. No a priori or prior probability functions are needed or introduced.

Second, because we have not assigned measures to the models of our language, the only uncertainty it makes sense to assign to "$R(a)$" on the Γ whose only element is "$Q(a)$" is [0, 1]. Now let us add knowledge of a large sample of Qs to our knowledge base. No matter what the observed frequency of Rs in this observed sample may be, the only uncertainty we can assign to "$R(a)$" is still [0, 1]: the proportion of models in which "$R(a)$" is true may be anywhere from 0 to 1. "Observed frequencies" lead to inductive probabilities only indirectly, by leading to the acceptance of statistical generalizations. On the other hand our system is designed to allow for acceptance: if we accept as practically certain a statistical hypothesis about the frequency of Rs among Qs, for example, that it lies between 0.32 and 0.37, then that statistical hypothesis may determine the uncertainty of the next Q being an R, relative to that body of knowledge or a database in which that statistical hypothesis has been accepted as part of the evidence.

---

[4]It is not clear that such a distribution of beliefs makes any sense at all in the absence of empirical data.

Third, whereas it may possible to find a set of distributions on the structure classes of models that will emulate our results (in the sense that the conditional probability of the statistical statement at issue will match our degree of certainty), to say, because the numbers match, that it is "really" or "implicitly" assuming those measures is highly misleading. If confidence interval methods lead to a 0.95 confidence interval of $[p, q]$ for a certain parameter, *of course* there will exist a distribution of prior probabilities which is such that the posterior probability that that parameter falls in that interval is 0.95. That can hardly be taken as warrant for claiming that the confidence interval method "requires" that distribution of prior probability.

Fourth, there are circumstances under which the logical or subjective prior probability of a class of models must be irrelevant to direct inference. For example, in a large domain the number of models in which the proportion of Rs among Qs is close to a half is vastly larger than the number of models in which the proportion is close to 0 or to 1. That makes no difference to the direct inference concerning representativeness, since the direct inference is based on the statistical statement (11.1) in the proof of Lemma 11.1, which is true in *every* model. In fact, whatever be the prior distribution of the proportion of Rs among Qs in the class of models, the frequency of representative samples will be approximately the same. In particular, the conditional subjective probability of the hypothesis "%$x(R(x), Q(x), 0.55, 0.65)$", relative to Γ containing a report of a large sample of Qs *all* of which are Rs, may be extremely high (as when each world is counted once), while the degree of proof of that hypothesis, relative to the same Γ, may be extremely low.

Fifth, it does make sense to apply Bayes' theorem when there exists in our knowledge base a joint distribution from which sequential selections are being made. For example, if we choose an urn from a population of urns, in each of which the relative frequency of black balls is known, and then draw a sample from that urn with a view to estimating that frequency (i.e., estimating which urn we have), then conditionalization is surely the way to go. But in this case the "prior" probability has a sound frequency interpretation, and we can cash that out in a long run frequency of being correct in similar situations. It is sharpening-by-richness. Bayesian inference makes perfectly good sense when we have the statistical background knowledge it calls for, as we have already pointed out; this fact underlies the relation sharpens-by-richness. If we allow approximate and practically certain statistical hypotheses to count as knowledge, we often have the kind of knowledge we need for Bayesian inference. But then it boils down to just another case of direct inference, as was pointed out in the early statistical literature [Neyman & Pearson, 1932]. This is particularly the case in situations in which robust Bayesian inference can be applied—inference in which very loose approximations will serve to generate, by conditioning on empirical evidence, relatively precise posterior distributions.

Finally, and most importantly, there are cases in which the Bayesian and the non-Bayesian methodologies appear to conflict. Suppose we have an approximate prior distribution for the parameter p of a binomial distribution. (For example, we know the approximate distribution of the proportion of black balls over a set of urns, and we are concerned to guess the proportion of black balls in an urn we are allowed to sample from.) We can also approach the problem from a confidence interval point of view. Suppose the confidence interval approach yields an interval [0.34, 0.45], and the conditionalization approach yields an interval [0.23, 0.65]. This reflects the precision of the particular confidence analysis and the vagueness of the prior probability. In this case, the interval to adopt as a measure of degree of entailment is [0.34, 0.45] for reasons of precision. On the other hand, if the intervals are interchanged, so that by confidence we get [0.23, 0.65] and by conditionalization we get [0.34, 0.45], then the interval to adopt as a measure of degree of entailment becomes the interval arising from conditionalization. Finally, if we have conflict (for example, [0.23, 0.45] as a confidence interval, and [0.34, 0.65] from conditionalization), it is conditionalization that will rule, by richness. These cases are sorted out by the rules we already have for the choice of a reference class.

It is generally the case that when both confidence methods and Bayesian conditional methods of inference are applicable and conflict—that is, when we have a sufficiently sharp prior distribution concerning the parameter at issue—then the Bayesian method is the appropriate one according to our semantic framework. To cash these considerations out in terms of the machinery we developed earlier requires that we show that a Δ embodying these constraints of prior distributions sharpens any Δ' not embodying them. This is simply a matter of showing that Δ sharpens-by-richness in the context S, Γ.

Example 11.4. *We sell pizzas, and we are considering starting a pizza club whose members will receive a discount. We have surveyed 100 customers and discovered that 32 of them are willing to become members of our pizza club. If this represents all the relevant information we have, we can perform a confidence interval analysis and assign a probability of 0.95 to the hypothesis that in general between 20% and 50% of our customers will want to join the club. If 0.95 counts as practical certainty for us in this circumstance, we will simply accept that between 20% and 50% of our customers will join the club.*

We now consult a friend in the survey sampling business. He points out that promotions of this sort are quite common and that quite a bit is known about them. For example, an approximate distribution is known for the long run response rates. We can take this approximate prior distribution into account and apply a Bayesian analysis that yields a very much smaller probability than [0.95, 1.0] for the hypothesis that between 20% and 50% of the customers will want to join the club.

More precisely, long experience may have established that the response rate is between 0% and 20% about 70% of the time, between 20% and 50% about 20% of the time, and between 50% and 100% only about 10% of the time. Furthermore, we will get the survey rate of 32 positive responses out of 100 in each of these cases respectively, 10% of the time, 60% of the time, and 30% of the time. These rates are, of course, approximate, but the result of using them would yield, approximately, an evidential probability of 0.54 ± 0.25 for the hypothesis that between 20% and 50% of our customers will join the pizza club.

This conflicts with the interval [0.95, 1.0] we took to characterize that hypothesis on the basis of confidence methods. Richness dictates that the probability based on prior experience sharpen the statistical background knowledge that includes the statistics on which the confidence analysis is based.

11.7 Sequences of Draws

Here is another case in which we may have conflicting candidates for inference. This is an important and historically influential example, with a very simple structure. Nagel [Nagel, 1939] pointed out long ago that it is one thing to know the proportion of black balls *in* an urn, and quite another to know the proportion of black balls *drawn* from an urn. We argued earlier that knowledge that the proportion of black balls in an urn U is p, and knowledge that a is a ball in the urn (perhaps singled out by the definite description "the next ball to be drawn"), supports the sentence "a is black" to the degree p. For example, if what we know is that 60% of the balls are black, the sentence "the next ball to be drawn is black" (note that it *is* a ball in the urn) is supported to degree 0.60. Does this conclusion fly in the face of Nagel's warning?

No, because we have supposed ourselves to know nothing about the frequency of black balls drawn. There is no conflict until we do know this. Suppose we wanted to add such a fact to our knowledge base; how would we express it? A natural procedure is to assign an ordinal number to each ball drawn from the urn: first ball, second ball . . . . This allows us to draw a ball, observe its color, and return it to the urn. Formally, the set of draws, DR, is now a function from an initial segment of N, the set of natural numbers, to the set of balls in the urn. (We restrict the domain of the function to an initial segment of N to avoid the irrelevant problems of infinite reference classes, and because we have required that all our models be finite.) A specific draw, say the seventh, is now represented by the pair $\langle 7, b \rangle$, where b is a specific ball in the urn. Because draws are represented by pairs, they can hardly be black; but we can define a predicate BP that applies to a pair just in case its second member is black. More simply, using our syntactical machinery, we can talk about the frequency of black balls among the pairs consisting of a draw and a ball.

We can now add to our knowledge base the knowledge that between 40% and 45% of the draws result in a black ball and that $\langle 7, b \rangle$ is the draw of a ball:

$$\%xy(B(x), DR(y, x), 0.40, 0.45)$$
$$DR(7, b).$$

It will follow, on the same principles we have already employed, that the proportion of support models in which "$B(b)$" holds is between 0.40 and 0.45. Now suppose we also have in our knowledge base the knowledge that 60% of the balls in the urn are black:

$$\%x(B(x), U(x), 0.60, 0.60).$$

Because we also know that b is a ball in the urn, we also have our initial argument to the effect that 60% of the models are ones in which b is black. Of course, we have more premises (a larger database) than we had then, but how does that help us?

Theorem 11.1, which gave us the initial result, allowed us to ignore all aspects of the models of our language that were not involved in the premises. Let us introduce draws of balls as a consideration even in that simple case. In that case, we don't know anything about the proportion of draws that result in black balls: what we have concerning the seventh ball drawn from the urn is exactly

$$\%x, y(B(x), DR(x, y), 0.0, 1.0)$$
$$DR(7, b).$$
$$\%x(B(x), U(x), 0.60, 0.60)$$
$$U(b).$$

Because the first two elements of our database do not, by their truth, in any way restrict the number of models in which "$B(b)$" holds, the last statistical statement sharpens the set of the two statistical statements, and our earlier conclusion is perfectly valid.

Now let us restrict the initial set of models to those in which the statement "%x, y(B(x), DR(x, y), 0.40, 0.45)" is true. Now how many relevant models are there in which "$B(b)$" is true? What we want to show is that the first statistical statement sharpens the full set. Referring to the last disjunct of the definition of sharpening, we see that this requires that we find a predicate of pairs, and a target property that applies only if the second term of the pair is B, and that leads to disagreement with [0.60, 0.60], while being compatible with (or identical to, in this case) the interval [0.40, 0.45]. The predicate of pairs is just DR itself, and the second member of any pair in DR is a member of U; the first statistical statement sharpens the set of two by richness.

Thus, if we have both information about the contents of the urn and information about the sequence of draws, *and these two items of information conflict*, the probability that a drawn ball is black will be given by the latter, rather than the former, in conformity with Nagel's intuitions.

But observe that the original assessment of evidential probability is unaffected. *Before* we know anything about the frequency of draws of black balls, the proportion of black balls in the urn gives rise to a perfectly reasonable probability.

There is one feature of this analysis that is worth noting. Our rational epistemic attitudes, as embodied in evidential probability, exhibit discontinuities. Suppose you are making draws from an urn that contains 60% black balls. Suppose that in fact 42.5% of the draws yield black balls. As you draw balls, you make confidence interval estimates of the proportion of draws that are black, at, say, the 0.95 level. The probability that the next ball drawn will be black changes thus: [0.6, 0.6], [0.6, 0.6], ... , [0.6, 0.6], [0.25, 0.59]. There is a sudden change from a precise probability determined by our knowledge of the proportion of black balls in the urn to a less precise probability, determined by our knowledge of the relative frequency of black draws. Whereas some may find this discontinuity disturbing, we think it is natural and something we must learn to live with.

11.8 Summary

In each of the theorems we have discussed, we have taken the listed members of Γ to be the only members of Γ. None of these theorems thus reflect more than ideally impoverished bodies of knowledge that cannot be taken very seriously. In any realistic situation, we must take account of far richer background knowledge.

The problems that this entails are the problems of developing lemmas that allow us to rule large parts of our knowledge base irrelevant, and thus allow us to approximate the idealized knowledge bases we have been considering. We have assumed that the individuals in our domain are known by their proper names. This is extremely unrealistic. We may know some objects by their proper names, but in general we know objects by definite descriptions. To identify an individual by a definite description, however, involves knowing that it satisfies a unique constellation of many properties and relations. It is thus bound to be false that we know of an individual "only that it is a ball in the urn."

In any realistic case, our database must be finitely axiomatizable. Consider a sentence S of $\mathcal{L}$. Although the database may imply a denumerable set of well-formed support statements of the form "%$(\phi(x), \psi(x), p, q) \wedge \psi(\alpha) \wedge S \equiv \phi(\alpha)$", we may expect them to be finitely axiomatizable and to mention only a finite number of intervals $[p, q]$. We need, therefore, only consider a finite number of sentences of that form, corresponding to a finite number of support sets. Because the cases we have considered include the most basic relations, it is reasonable to conjecture that the problem of calculating the proportion of models in which S is true, among those in which the sentences of Λ_i hold, is soluble. The project of devising algorithms for doing this—that is, developing the "proof" theory for uncertain inference—is one that is being undertaken by a number of researchers, for example [Loui, 1986; Murtezaoglu, 1998; Kyburg, 1991b; Kyburg, 1991a].

From the point of view of the present inquiry, what is central is the plausibility of using the relation of partial proof as a basis for uncertain inference. We know our conclusion cannot be true in *all* the models of our premises $\Gamma(=$ background knowledge), because then it would be formally provable, due to the completeness of first order logic. We cannot demand, literally, that the conclusion be true in *almost all* models of our premises, because what we know about any instance makes it unique, and in its unique class it either does or does not satisfy the target formula in question. The proportion of the models of the premises in which the conclusion is true is thus either 1 or 0.

Although it is not feasible to demand that an uncertain conclusion be true even in almost every model of the database Γ constituting the set of premises of the inference, we can achieve something like that effect by broadening the set of models with which we are concerned, first, by disregarding all but precise, rich, and specific models (the models of Δ), and second, by focusing on the *set of sets* of models $\Theta_D(S, \Gamma)$ that correspond to the remaining possible ways to take the evidence.

We have tried to motivate each move both by examples and by abstract argument. What we have arrived at seems a reasonable first cut at producing a set of semantic constraints that can be taken to embody principles characterizing acceptable first order uncertain inference.

We have attempted to show, in severe detail, how it is that a relatively virginal body of knowledge, supplemented by observations constituting a sample, could rationally come to assign a high evidential probability to a general statistical hypothesis. It is noteworthy that the analysis shows that for the argument to go through, we must ignore some of the information in the sample. The problem is parallel to that of attempting to take too much of our knowledge into account in assessing the probability of survival of a candidate for insurance.

It must be remembered (and some of the examples of this chapter have called this to our attention) that one of the important functions of high probability is to warrant acceptance. We are interested in highly probable statistical hypotheses because they may form part of our background knowledge for further inquiries. Chapter 12, in which we examine scientific inference more generally, will treat the question of acceptance on the basis of high probability in greater detail.

11.9   Bibliographical Notes

Classical philosophical works concerning the relation between probability and induction are to be found in [Nagel, 1939] and [Reichenbach, 1949]; more recent work, still in a philosophical tradition, is to be found in [Carnap, 1950], [Carnap, 1952], [Hintikka, 1966], [Carnap, 1971], and [Carnap, 1980].

Coming from the computer science side of things, the short tradition includes [Pearl, 1988], [Halpern, 1990b], [Grove et al., 1992], [Geffner, 1992], and [Touretzky, 1986].

Work that is more closely related to what is discussed here than that found in either traditional computer science or traditional philosophy is to be found in [Kyburg, 1974], [Kyburg, 1990], [Kyburg, 1997], [Pollock, 1990], [Loui, 1986], and [Murtezaoglu, 1998].

11.10   Exercises

(1)   In Theorem 11.5 suppose that j, the size of the sample, is 10,000. Suppose i is 3,450. What is the largest interval within which we may suppose the parameter in question lies, if we take 0.95 as practical certainty?

(2)   Follow through with the calculation of equation (11.3) in this case, on the assumption that we have incorporated the knowledge that of 1000 specified Qs, 500 have also been Ps.

(3)   Give an example of the use of some nonstatistical uncertainty (e.g., analogy), and discuss the question of whether it can or can not be reduced to statistics in (a) the object language and (b) the metalanguage.

(4)   Show that in fact 0.54 is (approximately) correct as the Bayesian conclusion in the example of the pizza club.

Bibliography

[Ackermann, 1962] W. Ackermann. *Solvable Cases of the Decision Problem*. North Holland, Amsterdam, 1962.

[Bacchus, 1992] Fahiem Bacchus. *Representing and Reasoning with Probabilistic Knowledge*. The MIT Press, 1992.

[Bacchus et al., 1992] F. Bacchus, A. J. Grove, J. Y. Halpern, and D. Koller. From statistics to degrees of belief. In *AAAI-92*, pp. 602–608, 1992.

[Carnap, 1950] Rudolf Carnap. *The Logical Foundations of Probability*. University of Chicago Press, Chicago, 1950.

[Carnap, 1952] Rudolf Carnap. *The Continuum of Inductive Methods*. University of Chicago Press, Chicago, 1952.

[Carnap, 1971] Rudolf Carnap. A basic system of inductive logic, part I. In Richard C. Jeffrey and Rudolf Carnap, editors, *Studies in Inductive Logic and Probability I*, pp. 33–165. University of California Press, Berkeley, 1971.

[Carnap, 1980] Rudolf Carnap. A basic system for inductive logic: Part II. In Richard C. Jeffrey, editor, *Studies in Inductive Logic and Probability*, pp. 7–155. University of California Press, Berkeley, 1980.

[Geffner, 1992] Hugo Geffner. High probabilities, model-preference and default arguments. *Minds and Machines*, 2:51–70, 1992.

[Grove et al., 1992] Adam J. Grove, Joseph Y. Halpern, and Daphne Koller. Random worlds and maximum entropy. In *7th IEEE Symposium on Logic in Computer Science*, pp. 22–33, 1992.

[Halpern, 1990a] Joseph Y. Halpern. An analysis of first-order logics of probability. *Artificial Intelligence*, 46:311–350, 1990.

[Halpern, 1990b] Joseph Y. Halpern. An analysis of first-order logics of probability. *Artificial Intelligence*, 46:311–350, 1990.

[Hintikka, 1966] Jaakko Hintikka. A two-dimensional continuum of inductive logic. In Jaakko Hintikka and Patrick Suppes, editors, *Aspects of Inductive Logic*, pp. 113–132. North Holland, Amsterdam, 1966.

[Kyburg, 1974] Henry E. Kyburg, Jr. *The Logical Foundations of Statistical Inference*. Reidel, Dordrecht, 1974.

[Kyburg, 1990] Henry E. Kyburg, Jr. Probabilistic inference and non-monotonic inference. In Ross Shachter and Todd Levitt, editors, *Uncertainty in AI 4*, pp. 319–326. Kluwer, Dordrecht, 1990.

[Kyburg, 1991a] Henry E. Kyburg, Jr. Beyond specificity. In B. Bouchon-Meunier, R. R. Yager, and L. A. Zadeh, editors, *Uncertainty in Knowledge Bases. Lecture Notes in Computer Science*, pp. 204–212. Springer–Verlag, 1991.

[Kyburg, 1991b] Henry E. Kyburg, Jr. Evidential probability. In *Proceedings of the Twelfth International Joint Conference on Artificial Intelligence (IJCAI-91)*, pp. 1196–1202. Morgan Kaufman, Los Altos, CA, 1991.

[Kyburg, 1997] Henry E. Kyburg, Jr. Combinatorial semantics. *Computational Intelligence*, 13:215–257, 1997.

[Loui, 1986] Ronald P. Loui. Computing reference classes. In *Proceedings of the 1986 Workshop on Uncertainty in Artificial Intelligence*, pp. 183–188, 1986.

[Murtezaoglu, 1998] Bulent Murtezaoglu. Uncertain Inference and Learning with Reference Classes, PhD thesis, The University of Rochester, 1998.

[Nagel, 1939] Ernest Nagel. *Principles of the Theory of Probability*, University of Chicago Press, Chicago, 1939, 1949.

[Neyman & Pearson, 1932] Jerzy Neyman and E. S. Pearson. The testing of statistical hypotheses in relation to probabilities a priori. *Proceedings of the Cambridge Philosophical Society*, 29:492–510, 1932.

[Pearl, 1988] Judea Pearl. *Probabilistic Reasoning in Intelligent Systems*. Morgan Kaufmann, San Mateo, CA, 1988.

[Pollock, 1990] John L. Pollock. *Nomic Probability and the Foundations of Induction*. Oxford University Press, New York, 1990.

[Reichenbach, 1949] Hans Reichenbach. *The Theory of Probability*. University of California Press, Berkeley and Los Angeles, 1949.

[Touretzky, 1986] D. S. Touretzky. *The Mathematics of Inheritance Systems*. Morgan Kaufman, Los Altos, CA, 1986.

12

Scientific Inference

12.1 Introduction

We have abandoned many of the goals of the early writers on induction. Probability has told us nothing about how to find interesting generalizations and theories, and, although Carnap and others had hoped otherwise, it has told us nothing about how to measure the support for generalizations other than approximate statistical hypotheses. Much of uncertain inference has yet to be characterized in the terms we have used for statistical inference. Let us take a look at where we have arrived so far.

12.1.1 Objectivity

Our overriding concern has been with objectivity. We have looked on logic as a standard of rational argument: Given evidence (premises), the validity (degree of entailment) of a conclusion should be determined on logical grounds alone. *Given* that the Hawks will win or the Tigers will win, and that the Tigers will not win, it *follows that* the Hawks will win. *Given* that 10% of a large sample of trout from Lake Seneca have shown traces of mercury, and that we have no grounds for impugning the fairness of the sample, *it follows with a high degree of validity* that between 8% and 12% of the trout in the lake contain traces of mercury.

The parallel is stretched only at the point where we include among the premises "no grounds for impugning ...." It is this that is unpacked into a claim about our whole body of knowledge, and embodied in the constraints discussed in the last three chapters under the heading of "sharpening." We have no grounds for impugning the fairness of the sample exactly when there is no alternative inference structure that would sharpen the inference structure based on the representativeness of most equinumerous samples of trout. For *uncertain* inference, we need to take as premises all we know, not just one or two special premises.

There are two kinds of objectivity involved here. First, there is the objectivity of the relation of *partial proof*—given a corpus of statements taken as evidence, whether or not a given statement is supported to a certain (interval) degree is being construed as a matter of (inductive) logic, and thus, according to our reconstruction, an objective matter in the same sense as the validity of deductive arguments is an objective matter.

Second, uncertain inferences all make use of what we have called "statistical state-ments" among their premises. These statements are construed as embodying objective empirical facts about the world. This is a second kind of objectivity underlying our uncertain inferences—what one might call *empirical* objectivity.

Thus, given a set of premises Γ and a conclusion S, uncertain inference is objective both in the logical sense that it reflects a relation between premises (evidence) and conclusion that is independent of what anyone thinks or believes or hopes, and in the empirical sense that its force or strength depends on premises that represent empirical facts about the world.

To arrive at agreement about what beliefs are warranted, we also have to agree about the evidence. The evidence need not be regarded as incorrigible (we reject C. I. Lewis's claim that "there can be no probability without certainty" [Lewis, 1940]). But the evidence must be *acceptable*. We have been unpacking this in terms of high probability. How can we avoid infinite regress on the one hand and subjectivity on the other?

The answer is that *at some level* we have the same evidence, or that we are willing to *share* evidence. With regard to the former, we are all born into the same physical world, and have been subject to very similar streams of experience. This is a matter of foundational and philosophical interest, but not particularly germane to the justifi-cation of particular inductive inferences. The latter consideration, the willingness to share each other's evidence, on the other hand, is central to the scientific enterprise. It is this that makes matters of integrity and objectivity so crucial in the social fabric of science. When you tell me that your studies of lake trout have indicated that, at the 0.99 level, no more than 10% have detectable amounts of mercury, other things being equal I accept your data as mine and dispute your claim only if I have *further* data that undermines your inference. And I expect you to accept my additional data. Thus, we both expect that in the final analysis we will achieve agreement concerning the levels of toxicity in lake trout.

At any point along the way, data may be questioned. I may question your sampling techniques, the measurement techniques employed, even the error functions you have employed. The questioning is not flat rejection, but a calling for evidence, or the pro-viding of additional evidence. The point is that disagreements about conclusions here, just as in any logical matter, can be traced to computational blunders or disagreements about premises. Disagreements about premises, in turn, can be resolved by retreating to a level at which all relevant evidence can be shared. That there is such a level is what objectivity in science comes to.

Note that we do not achieve such a level by retreating to phenomenalism and subjectivity. You and I cannot share our sensations.[1] In general, agreement is reached at or before the ordinary language description of things and their properties. People still make mistakes; no kind of sentence is immune from retraction; but an important and frequent form of uncertain inference allows us to accept observation sentences even though they can be in error. I accept "Bird b is blue" on the basis of experience (not on the basis of evidence) even while acknowledging that further experience might lead me to withdraw my acceptance.

---

[1] In fact, of course, we would not even know about sensations were it not for the kind of shared information that makes psychological science possible. A "sensation" is a useful construction devised to help account for the way in which *people* perceive *things*.

12.1.2   Evidential and Practical Certainty

What counts as evidence for an uncertain inference? In the schematic cases of statistical generalization, this isn't problematic. We assume that we can tell a crow when we have one, and that we can agree on its color. So "a_1 is a crow and a_1 is black," "a_2 is a crow and a_2 is black," and their ilk constitute the evidence for "At least 90% of crows are black."

As soon as we leave the schematic and imaginary, things get more complicated.

First, there is the issue of "grounds for impugning...." Whether or not we have grounds for impugning the fairness of the sample depends on everything else we know. It is hard to imagine any fact that is ruled out a priori as relevant. The color of my coat? Suppose it is black, and that I know that nonblack crows are frightened by black objects. These facts could sharpen the canonical inference: A set of crows constituting a sample collected by a black object (me), is less likely than sets of crows in general to be *representative* of the distribution of color among crows. But of course we know no such thing about crows. Or if we do, we also know of precautions that can be taken to circumvent that bias.

Second, the observations themselves may be no more than *probable*—that is, they themselves may be the results of uncertain inference. Consider the previous example concerning mercury levels in lake trout (a rather more realistic example than our canned examples about crows). You and I have just pulled a trout from our net. Does it have a detectable level of mercury in it? You and I can't agree on that; we must hand the trout over to the chemist who will measure its mercury level. The chemist gives us an answer: Yes, this trout has a detectable level of mercury. You and I believe him. We *accept* his pronouncement. But we know his testimony could be wrong. Accepting his word is, on our part, a matter of uncertain inference. Even authoritative testimony is uncertain.

Even for the chemist himself, the result is uncertain, as is the result of any measurement. Any measurement is subject to error. The procedure by which it is determined that the fish contains less than m ppm of mercury is a procedure that is characterized by a certain distribution of error: It returns a certain frequency of false positives, and a certain frequency of false negatives. This is to say, that among fish containing more than m ppm of mercury, f_p, approximately, will be classed as containing less than m ppm of mercury, and among those containing less than m ppm of mercury, approximately f_n will be classed as containing m ppm of mercury or more.

These parameters f_p and f_n are of course not known precisely. In Section 12.3, we will discuss how such numbers can be based on empirical data. What is to the point here is that these numbers themselves are known not only *approximately*, but also *uncertainly*.

We have already introduced the idea of distinguishing between a body of *evidential certainties* and a body of *practical certainties*. Statements enter the latter body by having a probability of at least $1 - \epsilon$, say, relative to the evidence in the evidential corpus. The contents of the evidential corpus is not up for dispute: it is what the participants in the discussion of the practical corpus take for granted. But objectivity is preserved in the following way: If some item of evidence is disputed, we can examine the evidence for it, in turn. What this comes to is treating the evidential corpus as a body of practical certainties of level $1 - \epsilon/2$, and considering the body of evidence that supports items in that corpus of level $1 - \epsilon/2$.

Why level $1 - \epsilon/2$? The intuitive reason is that, in standard probability theory, if the probability of S is at least $1 - \epsilon/2$ and the probability of T is at lest $1 - \epsilon/2$, then the probability of $S \wedge T$ is at least $1 - p$. The conjunction of a couple of evidential certainties is still a practical certainty.[2]

12.1.3 Statistical Inference

In the last chapter we looked in painful detail at a very crude example of statistical inference. In the interest of simplicity (and of getting definite results) we considered a case in which the contents of the evidential corpus was limited to defensibly "observational" sentences: "The first object is Q and it is also a P".

Of course, one may well ask how we learned our Ps and Qs, as well as how reliably we can judge whether a certain Q is a P. In the model we are developing we don't have to be incorrigibly certain about our data; we merely have to be "evidentially certain" of them. But now we have opened the door to the use of sentences of all kinds as evidential: statistical generalizations as well as observations. We may be evidentially certain that the quantity M has an approximately normal distribution (with mean approximately m and standard deviation approximately s) in the population P. (Because we have rejected the reality of infinite domains, nothing can have an *exactly* normal distribution.) We may well be evidentially certain that the population we are concerned with is one of a family of populations, in which the quantity Q has one of a certain class of distributions, and we may in addition be evidentially certain that this class of distributions itself has a certain distribution. We may, in short, be in a position to apply Bayes' theorem on the basis of objective empirical knowledge. This is no more anomalous, no more a departure from objectivity, than the fact that many classical statistical techniques can only be applied in the presence of general objective knowledge concerning the form of the underlying distribution. Although such facts are sometimes called "assumptions," it is better to avoid that terminology, since if they are worth employing, if we are going to depend on them, they must be justified, that is, based on empirical evidence.

Here is a final example of statistical inference in the presence of background knowledge: Suppose that we know that the weights of eggs of a certain species of bird are approximately normally distributed, with a mean and variance that is characteristic of their nutritional environment. We wish to estimate the mean of the distribution in the environment E.

The quantity $t = \sqrt{n-1}[(\overline{X} - m_X)/s]$, where X has a normal distribution with unknown mean m_X and unknown variance σ_X, $\overline{X}$ is the mean in a sample of n, and s is the sample standard deviation, has a Student's t-distribution. Student's t is tabulated; $\tau(t_1, t_2)$ is the measure of the set of points such that $\sqrt{n-1}(\overline{X} - m_X)/s$ lies between t_1 and t_2. Given that we have observed $\overline{X}$ and s, this is equivalent to the assertion that m_X lies between $\overline{X} - t_1(s/\sqrt{n-1})$ and $\overline{X} + t_2(s/\sqrt{n-1})$: $\Gamma \vdash t_1 \le \sqrt{n-1}(\overline{X} - m_X)/s \le t_2 \equiv \overline{X} - t_1(s/\sqrt{n-1}) \le m_X \le \overline{X} + t_2(s/\sqrt{n-1})$.

Thus, if we sample 20 eggs, and discover a mean weight of 12.45 grams, with a sample standard deviation of 2.33 grams, we may be 95% confident that the long run mean weight will be $12.45 \pm 2.093 \times 2.33 = 12.45 \pm 4.98$ grams.

---

[2] To show that something like this holds for evidential probability is nontrivial, since the probabilities may be referred to different sets of models. It was established for an earlier version of the system in [Kyburg. 1983]

Note that the same thing is true here as was true of our simplified binomial example. The proportion of models of our full database in which the proposition in question is true may be 0 or may be 1, according to the actual mean value of the weight of eggs. This is because what we know about the sample renders it unique: the sample was collected by Robin Smith at 10:00 A.M., March 12, 1999. The mean weight in that sample either is or is not within 5 grams of the mean weight of eggs from environment E in general. In some models the model ratio corresponding to this claim is 0; in other models it is 1. But in *all* the models the quantity t has a Student's distribution, and falls between t_1 and t_2 with a frequency of approximately 0.95.

The bridge between this class of models and the claim concerning limits for the mean weight of eggs is provided by a biconditional supported by our full body of knowledge: that $\sqrt{n-1}(\overline{X} - m_X)/s$ lies between t_1 and t_2 if and only if m_X lies between $\overline{X} - t_1(s/\sqrt{n-1})$ and $\overline{X} + t_2(s/\sqrt{n-1})$.

12.2   Demonstrative Induction

Demonstrative induction is induction in which the conclusion follows deductively from the premises. A simple example is that in which we determine the reproductive characteristics of a newly discovered species S. From the fact that one of the females is ovovivaparous, we can feel quite free to infer that this property is characteristic of the species as a whole. We can do this because we know—we are evidentially certain that—each species reproduces in a single characteristic way.

Formally, we would have an inference that could be represented in first order logic as follows:

(1) $\forall x$(species $(x) \wedge \exists y$(female.instance$(y, x) \wedge$ ovoviviparous$(y)) \supset$
     $\forall y$(female.instance $(y, x) \supset$ ovoviviparous$(y))$),
(2) species$(S) \wedge$ female.instance$(s_1, S) \wedge$ ovoviviparous(s_1),

from which we conclude, deductively,

(3) $\forall y$(female.instance$(y, S) \supset$ ovoviviparous$(y))$.

Similarly, we may obtain the melting point of a new organic salt from a single observation. If one sample of O melts at $t°$C under standard conditions of pressure, then all samples of X share this characteristic.

(1) $\forall x, k$(Orgsalt$(x) \wedge \exists y$(specimen$(y, x) \wedge$ MP$(y) = k°$C$) \supset \forall y$(specimen$(y, x) \supset$
     MP$(y) = k°$C$))$,
(2) Orgsalt$(O) \wedge$ specimen$(s_2, O) \wedge$ MP$(s_2) = 32.4°$ C,

from which we conclude

(3) $\forall y$(specimen$(y, O) \supset$ MP$(y) = 32.4°$ C$)$.

The determination of many physical constants follows this same pattern. To determine the density, under standard conditions, of a new alloy, we need merely (in principle) examine a single instance of that alloy.

In real life, things are not this simple. We do not, in fact, use one sample of a new alloy to determine its conductivity or melting point. There are several reasons for this, which are apparent if we reflect on the formal reconstruction of these inferences suggested above.

The broad generalization, concerning organic salts, species, or alloys, strikes us as very dependable, perhaps even "true by definition." It is part of what *defines* an organic salt that it should have a fixed melting point. We shall in due course argue that such generalizations are indeed much like definitions. The questionable statement is the second in each case—the statement that *appears* to involve only what is immediately before our eyes.

It is particularly clear in the second example that more than one measurement on more than one sample is required to establish the singular empirical statement. On closer examination, we realize that however extensive our data we do not conclude that the melting point is *exactly* 32.4°C, but only something like: the melting point is almost certainly 32.4 ± 0.1°C.

It is curious: The *exact* sciences are precisely the sciences based on measurement, and measurement is inevitably probabilistic and uncertain. The most we have reason to believe, based on a procedure of measurement, is that a quantity falls in a certain interval; and the goodness of the reason is a function of the size of the interval (for given data)—it never becomes a deductive reason. Furthermore, the inference itself is uncertain. From the data, we can be only practically certain that the interval captures the true value.

That the less exact sciences allow a higher degree of confidence is an illusion. The *terms* involved in the nonquantitative sciences admit of a certain fuzziness or imprecision that leads to the same general conclusion: an ineliminable element of uncertainty infects the less general premise of a demonstrative induction.

What the exact sciences do offer often is the possibility of indefinite refinement: Of reducing the error to an arbitrarily small amount, for a given degree of certainty. We can replicate and refine our measurements to yield the conclusion that the melting point of O is almost certainly within an interval of length 0.0002°C. In quantum mechanics, as we all know, there are bounds on this reduction of error; but our generalization, nevertheless, holds within practical limits in most disciplines. Making ever more accurate measurements is a question largely of engineering within an accepted body of knowledge.

12.3 Direct Measurement

There is much to be learned from the study of measurement as an example of the complex interaction of accepted statements, convention, observation, and probabilistic inference. To be able to examine the logical transitions that lead from what is clearly empirical observation to what is clearly a theoretical structure, it is helpful to begin with an imaginary very primitive state of scientific culture.

Suppose we (you and I, our brothers and sisters, and friends) have learned to make judgments of relative length: we can judge that A is longer than B, $L(A, B)$, and that C is not longer than D, $\neg L(C, D)$. Sometimes we just cannot tell one way or the other. Most of the time we agree; perhaps sometimes we don't agree. All of this can be thought of as learning the usage of the English "longer than."

There are some statistical generalizations at which we could arrive. We could infer almost certainly that almost all the time, if A is judged longer than B, B will not be judged longer than A.[3] That is, we may find the formula "$\%x, y(\neg L(x, y), L(y, x),$ 0.90, 1.0)" proved to a high degree by our observational evidence. There are infrequent exceptions. At this point, that is what they remain: they need not be "explained away."

There are some more interesting statistical generalizations at which we may arrive: We can be practically certain that if A is judged to be longer than B, and B is judged to be longer than C, then if A and C are compared, A will be judged to be longer than C. Symbolically, we may be practically certain of the statistical generalization "$\%x, y, z(L(x, z), L(x, y) \wedge L(y, z), 0.95, 1.00)$". By and large, the relation of being longer than is transitive.

So far, but for the talk of "practical certainty" and "approximation," we have been emulating the classical treatment of "longer than" as giving rise to equivalence classes, which we then use to generate numerical lengths. What is lacking and what spikes that attempt is that there are significantly frequent failures of the transitivity of *not* being longer than. We have evidence that allows us to be practically certain that "not longer than" frequently fails to be transitive: "$\%(x, y, z(L(x, z), \neg L(x, y) \wedge \neg L(y, z), 0.1, 0.2)$". A home example can be found in carpentry: In constructing a set of bookshelves, we are sure that shelf 1 is not longer than shelf 2, that shelf 2 is not longer than shelf 3, . . . , and that shelf 6 is not longer than shelf 7. But when we put the bookcase together we find that shelf 1 is definitely longer than shelf 7.

We want $\neg L$ to be transitive, reflexive, and asymmetric, and it is not. What can we do? We can take L to have the properties we want *by stipulation*—we can stipulate that the relation of "being at least as long as" generates an equivalence relation— but there is a cost to this. The discrepancy between what we observe and what we *stipulate* to be true of L must be laid down to *error*; L is no longer quite an observational term. We can no longer incorrigibly attribute length relations to the objects of our experience. This is a cost; it is a cost in certainty. For this cost to be worthwhile, there must be a correlative gain. Merely having universal generalizations such as "$\forall x, y, z(L(x, y) \wedge L(y, z) \supset L(x, z))$" in place of the corresponding statistical generalizations "$\%x, y, z(L(x, z), L(x, y) \wedge L(y, z), 0.95, 1.00)$" is not reward enough. The statistical uncertainty of observation just replaces the statistical uncertainty embodied in our knowledge of the properties of L.[4]

Let us proceed. Length is a particularly nice quantity (and particularly useful) because there is a way of *combining* lengthy objects: collinear juxtaposition, or the operation of placing objects end to end in a straight line. (It would be a mistake to call this a method of combining *lengths*; lengths will turn out to be abstract objects.)

Now that we have *stipulated* that L generates an equivalence relation ($\neg L(x, y) \wedge \neg L(y, x)$), we can observe (infer with practical certainty) that the collinear juxtaposition of any k elements from an equivalence class belongs to the same equivalence class as the collinear juxtaposition of any other k elements. Because in any instance of this principle we must be uncertain that the k elements really *are* from the same equivalence class—i.e., that they stand in the equivalence relation—it is natural to raise this generalization also to the level of a stipulation. The way is now cleared for the quantitative treatment of length, again at the cost of the introduction of uncertainty.

---

[3] We assume here that there is no problem with the reidentification of objects in the world. At the cost of some additional complications, we could take that into account, but doing so would shed no new light on measurement.
[4] This claim is made more precise in [Kyburg, 1984].

Choose an equivalence class: Say the class of objects equal in length to a certain platinum object at a certain time. (Today lengths are determined by a wavelength; to follow this through would make this long story even longer.) We take account of all the things we know that cause lengths to vary to ensure that that object remains in its equivalence class. If the platinum object were to warm up, that would not cause all the other objects in the world to shrink: It would merely mean that that particular object is no longer a member of the equivalence class we singled out in the beginning.

We can now *define* the abstract quantity *length-of*. The length of an object at a time is m/n units if and only if the collinear juxtaposition of n objects from its equivalence class is equal to the collinear juxtaposition of m objects from the unit equivalence class.

Fine. We have defined the quantity, but we are a long way from having understood its measurement.

Again, we indulge in some oversimplification. Let us suppose that a given object—say, the table before us—stays in the equivalence class determined by its length at time t. We suppose that the meterstick we have is in the equivalence class of the standard unit. Both of these assumptions *may* be false; but in both cases we may also have good reason to believe them. The table may change length in response to change in temperature and humidity, but (so far as we can tell!) the temperature and humidity have remained unchanged. The meterstick has been designed to be a member of the unit-meter class, and its gradations have been designed to come from an equivalence class, the juxtaposition of a thousand members of which is in the unit-meter class.

When we measure the table we find that it is 1.350 meters in length. But when we measure it again, we find that it is 1.355 meters in length. When we measure it a third time, we find that it is 1.348 meters in length. We assumed, however, that its length remained unchanged; So we have arrived at a contradiction.

The way out of this contradiction is quite parallel to the way in which we resolved the difficulty of the nontransitivity of $\neg L(x, y)$. We suppose that what we observe embodies errors: errors of observation. This time, however, the error is quantitative rather than qualitative. We suppose that the table has a "true length" and that our *measurements* of the table each embody error.

There is this difference: where, in confronting failures of transitivity of $\neg L(x, y)$, we need to suppose only that rare judgments were in error, in the case of measurement we must suppose that *all* our measurements are in error. It is worthwhile reflecting on this. The closer we approach to the quantitative ideal of objective science, the more unreliable do our observations get. In the end we can be essentially certain that our observations (measurements) are wrong! What kind of way is this to base knowledge on experience? What kind of empiricists are we?

One reason that this is worth thinking about is that it is right here, in the analysis of measurement, that we can most clearly see the common sense resolution of this problem. Sure, all our measurements are wrong, but some are further off than others, and by and large and in general the errors are fairly small. That's common sense; what does it come to in foundational terms?

Let us suppose that we have a large collection of measurements of the same object. Let us say we obtain the sequence of measurements ⟨1.34932, 1.35351, 1.34991, 1.35202, 1.35182, 1.34965, 1.34981, 1.33893, 1.35244, 1.3493, 1.35043, 1.34855, 1.35187, 1.34943, 1.35128⟩. These measurements, if grouped by intervals, will exhibit

a pattern like that of the famous bell shaped curve. Most of them will be fairly close
to each other and to the average. They are essentially *all* wrong, though; why doesn't
this cause us to throw up our hands in despair and abandon the whole measurement
project?

There is a natural empiricist principle that saves us: We should attribute no more
error to our observations than is required to make sense of them. Under water the fish
looks as long as my arm, but when I catch it it is only a foot long. I attribute error to
my original estimate, rather than supposing that fish shrink dramatically.

In the case of the failures of transitivity of $\neg L(x, y)$, we just took note of those
failures and supposed that they represented errors. Even there, we could not always
tell which of our comparative judgments was wrong: If we have observational warrant
for each of $\neg L(a_i, a_{i+1})$ and for $L(a_0, a_{100})$, we know we have to regard at least one of
those statements as false, but we may have no clear ground for regarding one as false
rather than another. (The bookcase provides an illustration: We judge the first board
to be the same length as the second, the second to be the same length as the third ...
but we compare the first and the last, and are surprised.) We *could* have rejected
all the judgments.[5] In the case of measurement, we have a quantitative measure of
falseness. Each of our measurements is in error, but the amount of error depends on
the true length of the table.

Suppose the true length of the table is l_t. Then the error of the ith measurement of
the table T, $m(T, i)$, is $|m(T, i) - l_t|$. Adopting the principle of minimizing the error
we must attribute to our observations in order to make sense of them amounts in this
case to minimizing the error we must suppose to be embodied in our measurements
in order to make sense of the assumption that they are all measurements of the same
thing. It is often actually easier to minimize the *square* of the error. In this example,
the total squared error is

$$\sum_{i < n} (a_i - l_t)^2.$$

Differentiating and setting the result equal to zero, we have

$$\sum_{i < n} a_i = nl_t.$$

The value of the true length that minimizes the long run squared error is the average
of the measurements.

Given a value for l_t, what our measurements have given us is a sample of the
errors generated by our measuring process. We may apply our standard techniques of
statistical inference to make inferences about this distribution. In particular, we can
become *practically* certain that the distribution of the errors of measurement generated
by the procedure we are applying to the object we are measuring is *approximately* a
normal distribution with a mean *close to*[6] 0.0 and a variance of *about* 0.003.

We may observe that this same distribution of errors (so far as we can tell) characterizes the process of measurement, whatever object (within limits) we are measuring.
It is characteristic of the particular process of measurement.

---

[5]We need reject only one, but it seems arbitrary to reject only one. An alternative is to accept them all, but this
may require a change of logic or a change of epistemology. See [Kyburg, 1997b].

[6]Why not *exactly* 0.0? Because we must take account of more than the consistency of measurements of a fixed
length; we want to be sure that the square of the true length of an hypotenuse of a right triangle is the sum of the

Phrasing the matter thus suggests that we are generalizing on our experience. That is not the case. We are not arguing that so and so has always worked out in the past, and therefore we may assume it will work out in the future. This is the classic trap of seeking inductive justification where it is hard to find. Rather, we are arguing that we have found no way of discriminating subclasses of the general run of the mill cases. We have found no *reason*, where we could have found reason, to attribute a different distribution of error to a certain class of measurements. Or rather, we do have such reasons in the case of earthworms, mountains, and microbes, at the level of direct measurement, but we do not have any such reasons when it comes to measurements of our tables, our rake handles, and our doorways.

This general process is characteristic of the process of developing techniques of measurement. Note that *quantities* are being treated as theoretical, unobservable kinds of things. To say this, of course, is useful for drawing attention to the question of observability rather than for making a deep philosophical point. We cannot observe the length of the table directly, although we can perfectly well make measurements and use the results of those measurements to generate objectively acceptable hypotheses concerning its length, hypotheses of which we can all be practically certain.

It would be conventional now to ask "What assumptions are these conclusions based on?" But because we have adopted a point of view from which only *warranted* assumptions may be made, we can better phrase the question: What are the acceptable grounds on which these conclusions are based? Because measurement is the prototype of theoretical scientific inference, it behooves us to be very cautious in answering this question. Many of the same kinds of answers will be appropriate in the case of other, more difficult instances of scientific inference.

On what grounds do we take the pattern of measurements of a particular object over a particular interval of time to be representative of the overall pattern of measurements of that object?

We don't know that the pattern is representative, but we do know that it is *very probably* representative. This is so, because almost all large subsets of the set of all potential measurements of the object are statistically representative. The formal reconstruction of this inference follows the general pattern presented in Chapter 11. In almost all the models in which a sample is taken, the distribution of a quantity in the sample is close to the distribution of that quantity in the population. We assume we have no positive grounds (precision, specificity, or richness) for looking at a smaller set of models. Therefore, we suppose that *this* sample is probably representative in the specified sense. Because this sample is approximately normal, with mean and variance of so and so, so also (within the sense of "representative") is the population of measurements.

Why doesn't specificity undermine the argument? The set of measurements has a mean of 1.340, and we do not know the frequency with which *such* sets of measurements are representative. The frequency may, within the class of models satisfying our body of knowledge, have any value from 0 to 1. But within the class of models that remain after pruning by precision, specificity, and richness, the frequency of representativeness will have a determinate (though approximate) value. Refer again to the arguments of Chapter 11: it is only conflicting knowledge that can undermine an argument. Mere possibilities alone cannot.

Don't we have to assume that the selection of the subset is *random*? It has been argued that "random selection," though often stipulated, is, in fact, neither necessary

Surely we have to assume that the object we are measuring remains the same—that it doesn't change length?

There are two answers to this question. The first is that if the object were expanding and contracting like a live earthworm, we would not regard it as a rigid body and thus as a fit subject for measurement. It is, after all, only *rigid* bodies that have lengths. If we found the object changing length (to our surprise), we would not use it for calibrating our measuring procedures. No one would suggest adopting an earthworm, for instance, as the standard unit of a system of lengths.

The other answer is more interesting. Suppose that the object were expanding and contracting very slowly, or by very small amounts, so that we continued to *judge* it rigid and of constant length. How might we discover that it was nevertheless changing? On the basis of other general knowledge—that trees grow, that heated things expand, and cooled things contract, etc. It is possible, in this case, that the very same measurement data from which we were inferring the distribution of error could also be used, together with other data, to arrive at some of these conclusions. Of course, this depends on our having the right data available and the right language in which to express it—always a problematic question, and one to which we shall return shortly.

Nevertheless, if we do not have that data, then we have no *grounds* for rejecting the inferred distribution of error. Like anything else we infer from data, it *might* be wrong; but given that it is supported by the data we have, given that its evidential probability, relative to our body of evidential certainties is high, it is something we *ought* to believe.

What right have we to suppose that the distribution of error obtained from a group of measurements of one object or a set of objects should apply to measurements made on a completely new object?

Again, this is an inappropriate question. The relevant question is: What right have we to suppose that this distribution does *not* apply to the new object? Is there some reference set of measurements whose error distribution would be more appropriate than the one we have? The answer to that question, in this hypothetical case, is no. What we need *grounds* for is not applying the error distribution in a class to which we know the instance belongs, but for applying some *different* distribution. We can do this only if (a) we know that distribution, and (b) one of the three structures, precision, specificity, and richness, rules out the original distribution by sharpening.

Of course, there are such circumstances. It would be inappropriate to suppose that our fine analytical balance that we use for weighing fractions of a milligram should weigh a boot with equal accuracy. But the inappropriateness depends on the fact that we *know*—have rational grounds for believing—that the weight of the boot lies beyond the range in which the balance is designed to operate. This represents positive knowledge, not negative speculation. In the presence of positive knowledge, we have grounds for a different probability rather than no probability.

12.4   Indirect Measurement

A quantity is measured indirectly when the measurement makes use of some lawlike quantitative connection between that quantity and another that is easier to measure directly. Thus, we typically measure temperature by measuring the length of a column of mercury; we measure the area of a rectangle by measuring its length and width.

We measure voltage by measuring torque, which in turn we measure by the angular displacement of a pointer. Many measurements today are reduced to observing the distribution of charges in LEDs.

Indirect measurement is treated extensively in [Kyburg, 1984]. Our interest in it here is that it provides a fairly clear bridge from the theoretical aspects of measurement, which already enter into the treatment of direct measurement, to the general considerations that underly the choice between scientific theories. The measurement of temperature provides an enlightening example.

It has been thought that temperature only admits of indirect measurement. Although that can be challenged [Kyburg, 1984], it is clearly the case that there is no good way of measuring temperature that is not indirect. Most commonly, we measure temperature by means of the length of a column of mercury (or alcohol). This procedure takes advantage of the fact that these fluids, over some useful range of temperature, satisfy a law of linear thermal expansion: The change in linear dimension of a quantity of mercury is proportional to the change in its temperature. This change in turn is magnified by confining the mass of mercury in a bulb whose coefficient of thermal expansion is small, and allowing the mercury to expand only through a very fine tube.

Simple enough. But we have several questions to take account of. First, this "law" of thermal expansion is hardly one we can establish before we have a way to measure temperature. And in general, we can only measure temperature accurately through the aid of some device that relates temperature to some other, more readily measurable quantity, such as length. Nevertheless, we can, within a certain range of temperatures, directly judge one body to be hotter than another. This is enough to get us started.

Second, the "law" of thermal expansion that underlies our common method of measuring temperature is false: Whereas most substances, in most states, expand when heated, there are exceptions (near changes of state, as ice to water); and no substance (that I know of) satisfies a law of linear thermal expansion exactly.

That would be a funny thing to say if we *defined* temperature in terms of linear expansion (change of temperature = change in length of mercury column). But what is the relation if not one of definition? There are two observations we can make. There is a process of refinement that takes place even in the absence of general theoretical input. As one can imagine, the distribution of error in judgments of temperature as determined by direct measurement (comparisons with standard temperatures) would embody a much greater variance than our thermometer measurements do. On the other hand, if we took the linear expansion of mercury as *defining* temperature, the only source of error in measuring temperature would be the error in measuring the length of the mercury column, and that doesn't seem right, either. But if we take into account the thermal behavior of various substances, under various conditions, we can make adjustments that yield a body of knowledge—a body of practical certainties—about thermal behavior that is *maximally predictive.*

A different and more important set of considerations arise from the role that temperature plays in more abstract parts of physics. The discovery, as recently as the nineteenth century, of the mechanical equivalence between heat and work laid the foundations for thermodynamics, and tied "degrees" to "feet," "pounds," and "seconds." At the level of thermodynamics the possibilities for the indirect measurement of temperature become very much greater; the network of laws and relations into which temperature enters, directly or indirectly, becomes correspondingly enlarged. It is this whole fabric that we seek to simplify and clarify by our decisions with regard to the measurement of temperature.

What is important from the point of view of uncertain inference or inductive logic is that despite all these ramifications, despite the fact that many statements involving temperature become "true by definition" (different ones, of course, in different epochs), the results of the measurement of temperature remain, as they have ever been, approximate and uncertain. The same is true of our predictions involving temperature. We know that the boiling point of water is a function of ambient pressure, but we can measure neither temperature nor pressure without error. Nevertheless, we can discover empirically the relation between pressure and boiling point.

Within the falsificationist approach to the philosophy of science, for example that of Popper [Popper, 1959], it is an anomaly pointed out by Duhem [Duhem, 1954] and Quine [Quine & Ullian, 1970], among others, that when a hypothesis is put to the test, it never stands alone, but always has a lot of company in the form of auxiliary hypotheses, statements concerning boundary conditions, bits of theory bearing on the operation of instruments of observation, etc. This has led some philosophers to argue that science confronts experience only as a whole, and that, just as no piece of empirical evidence can, by itself, establish a law or a theory, so there is no piece of empirical evidence that can by itself refute a law or theory.

Generalized to the context of this volume, we may read "establish" and "refute" in terms that allow for uncertainty. This does change the game significantly. Whereas we cannot be *certain* of the boundary conditions and auxiliary hypotheses, they can be as certain as our observations or our measurements of quantities.

We have seen one case where this kind of analysis works to establish a generalization: The case of determining the general melting point of an organic compound. In this case, as in any other, we require premises that are stronger than the conclusion we reach, insofar as the argument has a deductive structure. With self-evident symbolism, we have:

$$\forall x, t(\text{OrgComp}(x) \wedge \exists y(\text{Specimen}(x, y) \wedge \text{Melts}(y, t))) \supset$$
$$\forall z(\text{Specimen}(x, z) \supset \text{Melts}(z, t))$$

$\text{OrgComp}(C)$

$\text{Specimen}(C, s)$

$\text{Melts}(s, t)$

$\forall z(\text{Specimen}(C, z) \supset \text{Melts}(z, t)).$

Let us suppose, as we suggested, that the major premise is to be construed as "analytic"—as constituting part of the technical meaning of organic compound." There is certainly nothing "analytic" or "a priori" about the conclusion: it is a universal generalization that characterizes arbitrarily many specimens of compound C; but we have tested only one. Each of the other three premises clearly has empirical content.

"Melts(s, t)" is a relatively straightforward measurement statement. To be sure, we should unpack it as "Melts$(s, t \pm \delta)$", because no measurement procedure is perfectly accurate. And we must recognize that even thus hedged about with δ's, the claim is uncertain. The interval mentioned is a confidence interval; it is characterized by a certain level of confidence, which our reconstruction interprets as a probability high enough for acceptance. It is in conformity with standard laboratory procedures that we infer the melting point to be $t \pm \delta$.

This does contribute to the empirical content of our conclusion, but it is hard to see that, without the other premises, it warrants the belief that the next specimen of C will melt in the same range.

The premise "Specimen(C, s)" is one that is easy to imagine undermined. In fact, one can easily imagine arguments pro and con concerning whether or not s is in fact a specimen of C. (In real life, we would demand that s be a "pure" specimen of C—something that is quite clearly a matter of degree and of uncertainty.) What we demand of the argument is at least that this premise be acceptable at the level of evidential certainty.

Again, however, because this premise concerns only the specific specimen s, it is clear that the generality of the conclusion cannot be laid to this premise alone, nor this premise in conjunction with the measurement premise.

It is in virtue of the second premise, "OrgComp(C)" that the conclusion gets its power. But what have we here? This seems to be just the same *sort* of uncertain premise as the one claiming that s is a specimen of C. It differs, however, in terms of generality. There are a number of highly dependable grounds we can have for the claim that s is a specimen of C: the provenance of the sample, the procedures by which it was purified, the label on the bottle, the testimony of the lab supervisor. The evidence appropriate for the support of "OrgComp(C)" is somewhat different, precisely because this sentence says so much about specimens of C. On the one hand, it is arguably part of the *meaning* of "OrgComp" that anything that is one has a determinate melting point (boiling point, etc.) under standard conditions. But the reason that the term "OrgComp" is useful is that we *can* often identify things as organic compounds (otherwise than by testing their uniformity of melting point extensively), and in doing so we make errors relatively rarely.

This rather simple example can be raised to another level of abstraction. Consider celestial mechanics, for example. Here we may say that Newton's laws provide all the generality we need.[7] To apply these laws, however, we have to provide more than data: A misleading way to put it is to say that we must make assumptions that Newton's laws themselves tell us are false—for example, that the earth and the sun exert gravitational attraction only on each other and that each can be regarded as a point mass. A more accurate way to put it is to say that we are introducing some *approximations* that will introduce relatively small errors in our predictions of potential measurements.

It is in the predictive content of a theory, or a subtheory, that we find a measure of its worthiness as a scientific structure. We do not, as Carnap once hoped, have grounds for assigning a probability to a general universal hypothesis. But we *can* have good reasons to suppose that the predictive empirical content of one structure is greater than that of another.

As an example, consider the transition from Newtonian mechanics to special relativity. As is well known, for most ordinary calculations and predictions, which concern relatively low velocities, the difference between the two theories is negligible.

---

[7]We leave to one side the successors to Newton's laws. At this point, we are merely trying to illustrate part of the internal structure of a simple system.

This has led some people to say that the earlier theory is merely a "special case" of the latter theory, so that there is no absolute conflict between them. This is a mistake: Even at low velocities the two theories flatly contradict each other. There is no way to regard Newtonian theory as a "special case" of the special theory of relativity. What is true, and what these people have in mind, is that the quantitative difference in predictions between the two theories is too small to be measured at ordinary earthly velocities.

If this were true for all phenomena visible from the earth, then there would be no reason to prefer one theory to the other. But there are some anomalies in the story given by Newton and his successors. One concerns the precession of the perihelion of Mercury. Within the Newtonian framework for solar mechanics, it cannot be accounted for. There has been a rich history of efforts, which have involved such imaginative ploys as the postulation of a dust cloud, or an invisible planet (Vulcan), but none worked. By "none worked," we mean that no acceptable hypothesis led to predictions that were satisfactorily close to what was observed. This is a quantitative matter: "Close" is a matter of degree. As we have noted, the result of any measurement is logically compatible with the quantity being measured having any value.

Although we *can* account for anything with measurement error, we don't. Large errors are incredible. Except in rare and special circumstances, we rationally reject their occurrence. Large and systematic errors call for explanation, but lacking an explanation, we may have no choice but to admit to puzzlement while we continue to cast about for an account. What this amounts to in our framework is this: In the vast majority of instances of measurement (of a given sort) we have every reason to believe that the distribution of error is approximately normal, with mean of about 0 and variance characteristic of that sort of measurement. But measurements made in connection with observations of Mercury belong to a special subclass: One characterized by non-zero mean and much larger variance.

The way this works is that the Newtonian theory tells us what those values *ought* to be or *really are*, because we are taking that theory as an a priori framework for celestial knowledge. The observations we make tell us about the distribution of error. We find that we can isolate one group of measurements that is anomalous with regard to its errors, namely those concerned with orbit of Mercury.

If we adopt the special theory as a framework, however, we can reconcile our observations with the theoretical values. We don't need to suppose that the errors characteristic of a certain subset of the class of measurements are different. We can more nearly treat measurement uniformly. Our treatment is still imperfectly uniform, because special relativity doesn't account for the full quantitative discrepancy in predictions of the perihelion of Mercury, but it is certainly better.

Note that this characterization of the value of a theory in terms of its predictions is tied closely to the quantitative treatment of error that must go with the theory. The theory itself is being taken as a priori, as empty of empirical content; but what we think of as the empirical content of a theory does not disappear: It reappears in the distribution of errors of measurements, both historical and prospective—measurements that *might have been* predicted by the theory, and measurements that predict how things will be in the future.

Because this perspective on theories and on theory change is quite new, relatively little attention has been paid to treating the predictive empirical content of a theory in a quantitative or quasiquantitative way.[8] It is an problem that might well reward more detailed consideration.

12.5   Theory, Language, and Error

It is clear that although it does make sense to regard most general theories in science (one might say *all* exceptionless generalizations) as acceptable on the basis of stipulated meanings, rather than on the basis of supporting evidence, that does not solve or even simplify the problem of arriving at desirable stipulations. *Why* should we take mass to be the ratio of acceleration to force?

Total holism, to many a seriously threatening consequence of any sort of conventionalism, is avoided here, because there are few scientific principles that are so pervasive that they impinge on every domain. That people act out of momentarily perceived self-interest, and that a body free of external forces tends to continue in its state of rest or rectilinear motion, have very little in common other than their stipulative character. The desirability of adopting a language in which they are features depends, respectively, on the usefulness of the body of economic or mechanical doctrine they ultimately contribute to. Evaluating the usefulness of a theoretical doctrine is a multistep and multilevel process: can we find a way of reducing the errors we must attribute to our observations?

What can we say about content? We are always free to save a generalization by making it analytic or a priori, but the cost is that the reliability of measurement or observation may be disrupted or undermined. "All swans are white" could perfectly well be taken as true by stipulation, but there are consequences to doing so: We must suppose that we make errors in identifying colors or in identifying swans. It would be awkward to suppose we have difficulty in identifying whiteness only when we are looking at swans. The difficulty of knowing a swan when we see one is not at all as bad: We know that there are many species that only experts can tell apart. But the indirect consequences of assuming that the ornithologists cannot tell a swan without observing its color would be very bad. The taxonomy of bird species is based on morphological traits that are systematically related,[9] and the criteria of species identity have to do, only a little vaguely, with cross-fertility. The disruption of this body of knowledge that would be entailed by taking "All swans are white" as a priori would be major. We would suffer a net loss in predictive empirical content.

On the other hand, taking being-the-same-length-as to be a priori transitive makes an enormous contribution to our ability to predict, all the way from predicting when a refrigerator will fit through a doorway, to predicting the consequences of having a large mass of uranium-235 in a confined volume. Without this assumption, we would have no measurement of length, and because many other measurements are parasitic on the measurement of length, our ability to measure in general would be undermined.

---

[8] Some preliminary discussion will be found in [Kyburg, 1990a].

[9] How these traits are related and how important this relationship is are questions open to dispute among taxonomists.

A major factor in the usefulness of supposing that being-the-same-length-as is transitive is our ability to control our errors of observation: It turns out to be possible for us to develop a quantitative statistical theory of errors of measurement, as we have seen. Given such a theory, it also becomes possible for us to control and reduce errors of measurement (for example by replication of measurements).

12.6   Summary

All interesting inference is uncertain. It is uncertain in one or both of two ways. Either some of the propositions on which the conclusion is based are uncertain, in which case that uncertainty also infects the conclusion, or (more often) both some of the premises and the principle of inference (e.g., accept what is probable enough) lead to uncertainty.

Is there any way we can think ourselves into a state in which we have access to incorrigible data? It does not seem so. The more we learn about human perception, the more of a fragile, although generally dependable, baroque process it appears to be. This fact has both positive and negative features. On the negative side, we know that perception is profoundly influenced by many factors, from digestion to state of mind. On the positive side, we also know that training is an important factor in many forms of perception: The histologist can see the health of the cell, the South Sea islander can see the shape of the waves.

One common response to these uncertainties has been to retain the structure of classical deductive logic, importing into the premise set whatever is needed. If what is imported is plainly questionable, it is called an "assumption," the idea being that no one can be held responsible for making an assumption so long as it is plainly labeled as such. But an unwarranted assumption is, in our view, no less unwarranted for being admitted to.

Another common response is to insist that uncertain inference be replaced by the manipulation of probabilities in accord with the classical probability calculus. We have suggested some drawbacks to this approach in earlier chapters. The most serious, from our point of view, is the fact that it depends on probabilities that reflect no more than subjective assessment or stipulated arbitrary measures on a language. Such probabilities undermine the objectivity, both logical and empirical, we take to underlie justified inference.

Each of these approaches has its place. Assumption-based reasoning is just what one needs for hypothetical reasoning, and we have many occasions to consider hypothetical arguments. But even in these cases, it might well be desirable to be able to admit more than classical deductive procedures in making inferences from "assumptions."

Similarly, one of the concerns of many who have written in this area has been what one might call the psychology of evidence. One may be concerned with the explanation of a choice in the face of some new evidence, or with the marshaling of evidence with regard to its rhetorical effect, or simply with articulating in a coherent way the role that evidence plays in altering beliefs. Subjective Bayesian methods fit such circumstances very well. Furthermore, there is no requirement that the "degrees of belief" that are represented by subjective probabilities be unfounded or based on whimsy.[10] Nevertheless, even in this context the added richness provided by interval values of our objectively based notion might well prove useful.

Our own approach is based on the principle that the justification for uncertain inference lies in the frequency with which the conclusion is made true among the set of models picked out by the premises. Probability, correctly defined, allows us to do this. Although the approach we have outlined in the last few chapters doesn't settle all of the practical questions of uncertain inference, we believe it sets the groundwork for a research program that shows a great deal of promise. More importantly, it provides a framework within which questions of empirical validity or partial validity can be constructively discussed without invoking question-begging "assumptions." In this respect, it is our intent that it function in practice in a way similar to that in which first order logic functions in practice: As a rarely used, but often in some degree approached, final standard for generating explicit argument.

12.7 Bibliographical Notes

12.7.1 Measurement

The literature on measurement includes Norman Campbell's classic [Campbell, 1928], as well as [Helmholtz, 1977]. The idea that quantities, as distinct from real numbers, should be eschewed is found in [Nagel, 1960]. This point of view is also described in [Hempel, 1952]. A treatment of measurement that explicitly ties it to the development of theory is to be found in [Kyburg, 1984]. A discussion of magnitudes will be found in [Kyburg, 1997a].

12.7.2 Theories

One of the most influential treatments of scientific inference as concerns theories is due to Karl Popper [Popper, 1959]. Illustrations of how theories can be supported with the help of some of their proper parts is to be found in [Glymour, 1980]. The thesis that theories and generalizations can be construed as a priori appears in [Kyburg, 1977] and [Kyburg, 1990b].

12.7.3 Datamining

At the other end of the spectrum of uncertain inference from Popper's approach, is the blind mining of bodies of data for useful associations. This mining is not, of course, completely "blind": the qualification "useful" tells us that. It is being developed for commercial use, and much of it is practically oriented. The procedures we have outlined—particularly in the discussion of statistical inference—are applied to a body of data. As we have described statistical inference, one must have an idea of what one wants to test to start with. One cannot examine each possible empirical hypothesis

---

[10]The frequent requirement that subjective probabilities be real valued does preclude identifying them with our empirically founded probabilities in most cases, because we can hardly claim to know empirical frequencies or measures with perfect accuracy.

that would be appropriate to a large body of data. There are too many possibilities. It would clearly be desirable to be able to automatically limit the number of possibilities one needs to consider, but little work has yet been done along those lines.

A recent book on the new field of datamining is [Weiss & Indurkhya, 1999]. Although relatively practical in orientation, this book gives a good general overview and contains an extensive bibliography.

12.8   Exercises

(1)   Show that in probability theory, if $P(S) > 1 - p/2$ and $P(T) > 1 - p/2$ then $P(S \wedge T) > 1 - p$.

(2)   Imagine a pair of competent and responsible scientists who trust one another's data discussing the issue of whether remedy R is efficacious in the cure of the common cold. The two scientists share a large general body of knowledge, but each also has his own empirical data. You are to make up the data and the relevant parts of the general body of knowledge, and show how an initial disagreement can become resolved.

(3)   Is the Newtonian law $F = ma$ to be taken as a priori? Why, or why not? Support your claim.

(4)   Suppose that a traveler newly returned from the depths of the Amazon reports having found a species of bird that has three sexes—that is, it takes three-way conjugation for the female to produce fertile eggs. What would you say to this story, and why?

(5)   How many experiments will it take to determine the electrical conductivity of the newly discovered metal albinium? Why will it take so many? Why will it take so few?

(6)   Discuss the many ways in which scientific knowledge is involved in establishing an astronomical conclusion concerning a chemical process in stars. Where in such arguments does uncertainty enter explicitly, and where does it enter implicitly. What would a "Bayesian" argument for such a conclusion be like?

(7)   Discuss the issues involved in adopting an inductive or nonmonotonic rule of acceptance. What are the strongest arguments you can think of in favor of doing this? What are the strongest arguments you can think of against doing this? Where do you think the truth lies?

Bibliography

[Campbell, 1928] Norman R. Campbell. *An Account of the Principles of Measurement and Mensuration.* Longmans, New York, 1928.

[Duhem, 1954] Pierre Duhem. *The Aim and Structure of Physical Theory.* Princeton University Press, Princeton, 1954.

[Glymour, 1980] Clark Glymour. *Theory and Evidence.* Princeton University Press, Princeton, 1980.

[Helmholtz, 1977] H. von Helmholtz. Numbering and measuring from an epistemological point of view. In Cohen and Elkana, editors, *Helmholtz: Epistemological Writings.* Reidel, Dordrecht, 1977.

[Hempel, 1952] Carl G. Hempel. *Fundamentals of Concept Formation in Empirical Science.* University of Chicago Press, Chicago, 1952.

[Kyburg, 1964] Henry E. Kyburg, Jr. Logical and fiducial probability. In *Proceedings on the 34th*

Session of The International Statistical Institute, Volume 40, pp. 884–901. Toronto University Press, Toronto.

[Kyburg, 1977] Henry E. Kyburg, Jr. All acceptable generalizations are analytic. *American Philosophical Quarterly*, 14:201–210, 1977.

[Kyburg, 1983] Henry E. Kyburg, Jr. The reference class. *Philosophy of Science*, 50:374–397, 1983.

[Kyburg, 1984] Henry E. Kyburg, Jr. *Theory and Measurment.* Cambridge University Press, Cambridge, 1984.

[Kyburg, 1990a] Henry E. Kyburg, Jr. Theories as mere conventions. In Wade Savage, editor, *Scientific Theories*, Volume XIV, pp. 158–174. University of Minnesota Press, Minneapolis, 1990.

[Kyburg, 1990b] Henry E. Kyburg, Jr. Theories as mere conventions. In Wade Savage, editor, *Scientific Theories*, pp. 158–174. University of Minnesota Press, Minneapolis, 1990.

[Kyburg, 1997a] Henry E. Kyburg, Jr. Numbers, quantities, and magnitudes. *Philosophy of Science*, 64:377–410, 1997.

[Kyburg, 1997b] Henry E. Kyburg, Jr. The rule of adjunction and reasonable inference. *The Journal of Philosophy*, 94, pp. 109–125, 1997.

[Lewis, 1940] Clarence Irving Lewis. *An Analysis of Knowledge and Valuation.* The Free Press, Glencoe, IL, 1940.

[Nagel, 1960] Ernest Nagel. Measurement. In Arthur Danto and Sidney Morgenbesser, editors, *Philosophy of Science*, pp. 121–140. Meridian Books, New York, 1960.

[Popper, 1959] K. R. Popper. *The Logic of Scientific Discovery.* Hutchinson, London, 1959.

[Quine & Ullian, 1970] W. V. O. Quine and J. S. Ullian. *The Web of Belief.* Random House, New York, 1970.

[Weiss & Indurkhya, 1999] Sholom M. Weiss and Nitya Indurkhya. *Predictive Data Mining.* Morgan Kaufman, San Francisco, 1999.

Names Index

Index

Printed in the United States
By Bookmasters